THERMODYNAMICS

AN ADVANCED TREATMENT FOR CHEMISTS AND PHYSICISTS

THERMODYNAMICS

AN ADVANCED TREATMENT
FOR CHEMISTS AND PHYSICISTS

BY

E. A. GUGGENHEIM, M.A., Sc.D., F.R.S., F.R.I.C.

PROFESSOR OF CHEMISTRY
IN THE UNIVERSITY OF READING
1946–1966

Fifth, revised edition

1967

NORTH-HOLLAND PUBLISHING COMPANY
AMSTERDAM

© NORTH-HOLLAND PUBLISHING COMPANY, AMSTERDAM, 1967

No part of this book may be reproduced in any form by print, photoprint, microfilm or any other means without written permission from the publisher

First edition   1949
Second edition 1950
Third edition   1957
Fourth edition 1959
Fifth edition   1967

PUBLISHERS:

NORTH-HOLLAND PUBLISHING CO. – AMSTERDAM

SOLE DISTRIBUTORS FOR U.S.A. AND CANADA:
INTERSCIENCE PUBLISHERS, A DIVISION OF
JOHN WILEY & SONS, INC. – NEW YORK

Library of Congress Catalog Card Number 67-20003

PRINTED IN THE NETHERLANDS

PREFACE

In view of the large number of books on thermodynamics it may seem surprising that there should be any need for yet another. A cursory survey of all the existing books will however show that only very few are at all comparable. The total number is considerably reduced if we reject the ones which G. N. Lewis so aptly described as containing 'cyclical processes limping about eccentric and not quite completed cycles' and consider only those which present thermodynamics as an exact science. Many of these, including some of the best, are out of date. No book written before 1929 even attempts an account of any of the following matters: the modern definition of heat given by Born in 1921; the quantal theory of the entropy of gases and its experimental verification; Debye's formulae for the activity coefficients of electrolytes; the use of electrochemical potentials of ions; the application of thermodynamics to dielectrics and to paramagnetic substances. The first textbook on thermodynamics to include any of these matters is that of Schottky published in 1929. The number of textbooks on thermodynamics written since then is in single figures and of these fewer than half a dozen are in English. The only two available bearing any appreciable resemblance to this book are Zemansky's 'Heat and Thermodynamics' and Macdougall's 'Thermodynamics and Chemistry'. I have a great admiration for both these books, but they are quite different from each other and from this book. Zemansky's book is supremely good on the fundamentals of thermodynamics and should be equally useful to physicists, chemists and engineers. It includes especially thorough discussions on the meaning of heat, on calorimetry, on thermometry, on steam engines and on refrigerators. On the other hand there are important applications to physical chemistry, such as solutions, interfaces, electrochemistry, the third principle, entropy constants which are dealt with sketchily or not at all. Macdougall's book on the other hand is, as its title indicates, devoted mainly to applications of thermodynamics to chemistry. Less attention has been paid to a logical formulation of the fundamental principles and there is no application to dielectrics or to paramagnetics.

The present book is addressed equally to physicists and to chemists, but not to engineers. It is thus in a sense intermediate between the other two books mentioned. It is written for graduates, but much of it should be useful to undergraduates intending to specialize in physical chemistry or chemical physics.

There are several novel or unusual features in the treatment, notably the following. The third principle of thermodynamics is introduced near the beginning and is then used throughout the book. As the third principle can be properly understood only through statistical mechanics an early chapter is devoted to a digression on this subject. Considerable use is made of a function λ, called the absolute activity, related to the chemical potential μ by $\lambda = \exp(\mu/RT)$ or $\mu = RT \log \lambda$. This function plays an important part in statistical mechanics, more especially in Bose-Einstein and in Fermi-Dirac statistics, but its close relation to μ has not always been appreciated. At the same time λ is often more convenient than μ for formulating equilibrium conditions, especially those of chemical reactions. Physico-chemical systems are classified in chapters according to the number and nature of the components, not according to the number or nature of the phases. Interfaces are treated as thin phases after the manner of van der Waals and Bakker, not as fictitious geometrical surfaces after the manner of Gibbs. There is no separate chapter on interfaces, but they are dealt with according to the number of components. The thermodynamics of an interface in a two component system is much more complicated than that of a single component system and conveniently comes at a later stage. The treatment of electrolyte solutions is split into two chapters; in the first of these electric potential need not be mentioned, while the second, entitled electrochemistry, is by contrast devoted to electrochemical cells. The treatment of systems in electric and in magnetic fields, especially the latter, is more detailed than usual.

Choice of notation always leads to difficulties. No notation is perfect, but some are better than others. I have tried to be guided by the principle that the symbols used should be as simple as possible provided they are unambiguous. I will mention two examples. The symbol V_i for the partial molar volume of the species i is better than $\overline{V}_i$ because the bar serves no useful purpose and in fact does harm by suggesting an average. Again if the superscript ^{0} is used to refer to a component in the pure state, then to denote the value at a standard pressure a different superscript, such as †, should be used. It is unfortunate that there is not yet uniformity in the use of symbols for Gibbs' four thermodynamic potentials $\varepsilon, \chi, \varphi, \zeta$. I have used those recommended both by the International Union of Chemistry (1947) and by the

International Union of Physics (1948), namely U, H, F, G. In my opinion there is no disrespect to Gibbs, nor to anyone else, in finding these symbols more convenient than their alternatives.

Experimental data and detailed calculations have been included here and there for illustrative purposes only. In all such examples care has been taken to use the most reliable modern data available.

Copious references have been given to modern literature, but references have usually not been given for theorems and formulae now become classical unless there seemed to be a special reason for so doing, as for example to emphasize a point of historical interest or to throw light on a controversial matter.

There is an author index of the references to the literature. The omission of a subject index is deliberate as it would have had to be either excessively long or incomplete. There is however a detailed table of contents which should almost always enable an experienced reader to find what he is looking for.

A few of the diagrams have by permission been copied from diagrams in other books or in journals. I am grateful for such permission to the Royal Society, the American Institute of Physics, the Cambridge University Press and Messrs. Taylor and Francis.

I want to thank Dr. G. S. Rushbrooke for reading, checking and correcting the last four chapters of the book. The rest of the book has been checked by Mr. B. Topley, to whom my debt of gratitude cannot be adequately expressed in words. He has been of inestimable help to me in eliminating not only misprints and errors of transcription but also poor English, bad grammar, false reasoning and obscurity. If, as I hope, there remain but few examples of these, the credit is his.

Reading University, August 1949 	E. A. G.

PREFACE TO SECOND EDITION

The text of this edition is essentially the same as that of the first edition, the only significant change being at the end of § 4.06. Several typographical errors in the first edition have been corrected. A subject index compiled by Mr. M. L. McGlashan has been added.

A distinguished American reviewer of the first edition has proposed that this book might have the sub-title 'Pride and Prejudice'. Each reader must decide for himself on the merit of this proposal.

Reading University, June 1950 	E. A. G.

PREFACE TO THIRD EDITION

Apart from numerous corrections and changes of detail, which I hope are improvements, there are only two major changes in the structure of this book.

The first change relates to the treatment of mixtures. In the previous editions systems of two components were treated in a separate chapter from systems of more than two. This necessitated an appreciable amount of repetition which some reviewers found tedious. The new edition contains instead a chapter on 'mixtures' in which all components are treated on a par with the emphasis on mole fractions and another chapter on 'solutions' in which one component is singled out as the solvent with the emphasis on mole ratios or molalities. With this new arrangement the chapter on 'extremely dilute solutions' becomes superfluous and is omitted.

The other significant change is the inclusion of a short chapter on Onsager's relations. The discussion has been confined to isothermal systems so as not to increase unduly the length and so the price of the book. I hope that reviewers who deplore the omission of thermal diffusion will advise me what I ought to have omitted to make room for it. On the other hand I offer no apology for ignoring the application of Onsager's relations to coupled chemical reactions under conditions close to equilibrium. This application is in my opinion a fruitless one. I still adhere to the established view that any study of chemical kinetics which ignores the mechanism is profitless.

It gives me pleasure to conclude by thanking my friend Dr. M. L. McGlashan for his indefatigable help in preparing the new edition and his ever constructive criticism, and to congratulate the publishers on what I consider to be beautiful printing.

Reading University, June 1956   E. A. G.

PREFACE TO FOURTH EDITION

Typographical errors have been corrected on pages 43, 63, 175, 178, 225, 286, 451, and 452. I am grateful to my friend N. K. Adam and others for drawing my attention to these errors. There are no other changes.

Reading University, 1959   E. A. G.

PREFACE TO FIFTH EDITION

The whole text has been thoroughly revised. The resulting changes are neither striking nor numerous. The most important are the following. The elementary discussion of partial differentiation in chapter 3 has been omitted and the rest of chapter 3 has been put into chapter 1. In the chapter on mixtures the definitions of excess functions have been clarified since the previous text has sometimes been misinterpreted. Chapter 13 now includes the Onsager relations for thermoelectricity and for isothermal diffusion. There is still no reference in the text to Onsager relations for 'coupled' chemical reactions since I believe such relations to have no bearing on real chemical reactions. A short chapter has been added on systems in motion treated according to the special theory of relativity. Throughout the book almost all formulae are written in a single line. Any loss in appearance is compensated by saving in printing costs.

Internationally recommended notation is used throughout with three exceptions. The Helmholtz function is denoted by $\mathcal{F}$ leaving A free for area and F free for the Faraday constant. The fugacity of i is denoted by p_i while the partial pressure of i is denoted by $y_i P$, where y_i is the mole fraction of i in the gas phase and P is the total pressure; when the gas may be regarded as perfect then $p_i = y_i P$. The symbol °K for degree Kelvin is abbreviated without ambiguity to K.

I am extremely grateful to Professor M. L. McGlashan for reading the whole of the text and suggesting numerous improvements. I also thank Mr. N. F. Judd and Mr. I. C. McKinnon for reading and correcting proofs.

Reading, June 1966   E. A. G.

TABLE OF CONTENTS

Preface. V
Preface to second edition. VII
Preface to third edition. VIII
Preface to fourth edition VIII
Preface to fifth edition IX
Table of contents X
Important symbols XXI
General physical constants XXIV
Defined values and conversion factors XXIV

Introduction concerning notation and terminology 1

CHAPTER 1
Fundamental principles

SECTION

1.01	Scope of thermodynamics	5
1.02	Thermodynamic state. Phases.	6
1.03	Thermodynamic process	7
1.04	Infinitesimal process	7
1.05	Insulating walls. Adiabatic processes	7
1.06	Conducting walls. Thermal equilibrium	8
1.07	Zeroth law. Temperature	8
1.08	Thermostats and thermometers	8
1.09	Temperature scales	9
1.10	Energy and heat. First law	9
1.11	Conversion of work to heat	11
1.12	Natural and reversible processes	12
1.13	Reversible process and reversible change	13
1.14	Equilibrium and reversible changes	14
1.15	Closed systems and open systems	14
1.16	Entropy	14
1.17	Thermal equilibrium	15

1.18	Thermodynamic temperature	15
1.19	Entropy and heat	16
1.20	Second law	17
1.21	Units	18
1.22	Extensive properties	18
1.23	Intensive properties	19
1.24	Chemical content of phase	19
1.25	Chemically inert species	19
1.26	Partial and proper quantities	20
1.27	Chemical potentials	21
1.28	Characteristic functions. Fundamental equations	22
1.29	Mole fractions	24
1.30	Gibbs–Duhem relation	25
1.31	Multiphase systems	25
1.32	Adiabatic changes in closed system	26
1.33	Isothermal changes in closed systems	26
1.34	Equilibrium conditions. General form	27
1.35	Stability and metastability	28
1.36	Thermal internal stability	30
1.37	Hydrostatic equilibrium	31
1.38	Hydrostatic internal stability	31
1.39	Equilibrium distribution between phases	32
1.40	Phase stability	33
1.41	Gibbs' phase rule	33
1.42	Membrane equilibrium	34
1.43	Chemical reactions. Frozen equilibrium	34
1.44	Chemical equilibrium. Affinity	35
1.45	Choice of independent variables	38
1.46	Thermal expansivity and isothermal compressibility	38
1.47	Maxwell's relations	39
1.48	Dependence of thermodynamic functions on pressure	39
1.49	Gibbs–Helmholtz relation	40
1.50	Dependence of thermodynamic functions on T, V	41
1.51	Use of Jacobians	41
1.52	Reversible cycles	43
1.53	Surface phases	45
1.54	Interfacial tension of plane interface	46
1.55	Helmholtz function of surface layer	47
1.56	Integrated relation. Gibbs function of surface phase	47
1.57	Analogue of Gibbs–Duhem relation	48
1.58	Invariance of relations	49
1.59	Gibbs geometrical surface	49
1.60	Interfacial tension of curved interface	50

1.61	Pressure within a bubble	52
1.62	Determination of interfacial tension	53
1.63	Independence of interfacial tension of curvature	54
1.64	Mathematical analysis of curved interface	55
1.65	Basis of thermodynamic laws	58
1.66	Third law	60

CHAPTER 2

Digression on statistical thermodynamics

2.01	Microdescriptions and macrodescriptions of a system	61
2.02	System of given energy, volume, and composition	62
2.03	Characteristic of macroscopic system	63
2.04	System of given temperature, volume, and composition	65
2.05	Further characteristics of macroscopic system	68
2.06	System of given temperature, pressure, and composition	69
2.07	System of given temperature, pressure, and chemical potential	70
2.08	Recapitulation	72
2.09	Extension to several components. Absolute activities	73
2.10	Antisymmetric and symmetric eigenfunctions	73
2.11	Fermi–Dirac and Bose–Einstein statistics	74
2.12	Boltzmann statistics	75
2.13	Partition functions of units and thermodynamic functions	76
2.14	Separable degrees of freedom	77
2.15	Classical and unexcited degrees of freedom	77
2.16	Translational degrees of freedom	78
2.17	Third law of thermodynamics	80

CHAPTER 3

Systems of a single component

3.01	Single components and single phases	82
3.02	Dependence of entropy on temperature	82
3.03	Heat capacity at constant pressure	85
3.04	So-called mechanical equivalent of heat	86
3.05	Dependence of entropy on pressure	87
3.06	Heat capacity at constant volume	88
3.07	Relation between heat capacities	89
3.08	Adiabatic compressibility	89
3.09	Condensed phases and gases	90
3.10	Isothermal behaviour of a gas	90
3.11	Throttling	92
3.12	Measurement of thermodynamic temperature	93

3.13	The gas constant and the mole	94
3.14	Isothermal equation of a gas	95
3.15	Absolute activity	95
3.16	Thermodynamic functions of a gas	96
3.17	Fugacity	97
3.18	Gases at high temperatures	97
3.19	Slightly imperfect gases	98
3.20	Joule–Thomson coefficient	99
3.21	Temperature dependence of second virial coefficient	99
3.22	Boyle temperature and inversion temperature	100
3.23	Relation between heat capacities of slightly imperfect gas	101
3.24	Adiabatic change of a gas	101
3.25	Temperature dependence of $\mu^{\ominus}$ and $\lambda^{\ominus}$	102
3.26	Monatomic molecules	103
3.27	Numerical values in entropy constant	103
3.28	Linear molecules	104
3.29	Non-linear molecules	105
3.30	Pressure dependence for condensed phases	108
3.31	Temperature dependence for liquids	110
3.32	Crystals at very low temperatures	111
3.33	Crystals at intermediate temperatures. Debye's approximation	112
3.34	Corresponding temperatures of crystals	116
3.35	Comparison of Debye's functions with Einstein's functions	116
3.36	Equilibrium between two phases	118
3.37	Relation between temperature and pressure for two-phase equilibrium	118
3.38	Clapeyron's relation applied to two condensed phases	119
3.39	Clapeyron's relation applied to saturated vapour	120
3.40	Heat capacities of two phases in equilibrium	121
3.41	Heat capacities at saturation	122
3.42	Temperature dependence of enthalpies of evaporation and of fusion	123
3.43	Triple point	124
3.44	Critical points	126
3.45	Continuity of state	129
3.46	Two phases at different pressures	132
3.47	Fugacity of a condensed phase	134
3.48	Corresponding states of fluids	135
3.49	Corresponding states of solids	140
3.50	Two simple equations of state	141
3.51	Zero-temperature entropy in crystals	144
3.52	Two numerical examples	146
3.53	Statistical-mechanical interpretation	148
3.54	Simple typical exceptions	148

3.55	Isotopic mixing	150
3.56	The exceptional case of hydrogen	151
3.57	Third law of thermodynamics and the Nernst heat theorem	154
3.58	Thermal expansion at low temperatures	156
3.59	Unattainability of zero temperature	157
3.60	Interfacial layers	159
3.61	Temperature dependence of surface tension	160
3.62	Invariance of relations	161
3.63	Simplifying approximation	161
3.64	Vapour pressure of small drops	162
3.65	Empirical temperature dependence of surface tension	163
3.66	Corresponding states of surface tension	166
3.67	Sorption of a single gas	166
3.68	Temperature dependence of sorption	167

CHAPTER 4

Mixtures

4.01	Introduction	170
4.02	Composition of mixture	170
4.03	Partial and proper quantities	171
4.04	Relations between partial quantities	173
4.05	Partial quantities at high dilution	173
4.06	Perfect gaseous mixture	173
4.07	Slightly imperfect gaseous mixture	175
4.08	Fugacities of gases	177
4.09	Liquid mixtures	177
4.10	Liquid–vapour equilibrium	177
4.11	Azeotropy	178
4.12	Relative activities and fugacities in liquids	180
4.13	Pressure dependence	182
4.14	Osmotic equilibrium	182
4.15	Pressure on semi-permeable membrane	184
4.16	Duhem–Margules relation	185
4.17	Temperature coefficients	186
4.18	Ideal mixtures	186
4.19	Thermodynamic functions of ideal mixtures	187
4.20	Fugacities in ideal mixtures	188
4.21	Osmotic pressure of ideal solution	190
4.22	Non-ideal mixtures	190
4.23	Functions of mixing and excess functions	191
4.24	Volatility ratio	192
4.25	Internal stability with respect to composition	194

4.26	Critical mixing	195
4.27	Excess functions expressed as polynomials	196
4.28	Symmetrical mixtures	196
4.29	Simple mixtures	197
4.30	Critical mixing in simple mixtures	200
4.31	Critical mixing in symmetrical mixtures	203
4.32	Example of unsymmetrical mixture	203
4.33	Athermal mixtures of small and large molecules	205
4.34	Osmotic pressure in athermal mixtures	206
4.35	Interfacial layers	207
4.36	Liquid–vapour interface	207
4.37	Invariance of relations	208
4.38	Temperature coefficient of surface tension	209
4.39	Variations of composition	210
4.40	Example of water+ethanol	211
4.41	Interface between two binary liquids	213
4.42	Temperature dependence of interfacial tension	214
4.43	Accurate formulae	215
4.44	Solid mixtures	217
4.45	Stationary melting points	218
4.46	Solid ideal mixtures	218
4.47	Excess functions	219

CHAPTER 5

Solutions, especially dilute solutions

5.01	Introduction	220
5.02	Mole ratios and molalities	220
5.03	Partial and apparent quantities	221
5.04	Gibbs–Duhem relation	222
5.05	Partial quantities at high dilution	223
5.06	Ideal dilute solutions	223
5.07	Thermodynamic functions of ideal dilute solutions	224
5.08	Real solutions	226
5.09	Activity coefficients of solute species	226
5.10	Osmotic coefficient of solvent	227
5.11	Relation between activity coefficients and osmotic coefficient	227
5.12	Temperature dependence	227
5.13	Pressure dependence	228
5.14	Temperature dependence of fugacity of solvent	229
5.15	Temperature dependence of fugacity of solute	229
5.16	Osmotic pressure	229
5.17	Freezing point	230

5.18	Boiling point	232
5.19	Distribution between two solvents	233
5.20	Solubility of pure solid	233
5.21	Experimental determination of ϕ	234
5.22	Determination of γ from ϕ	235
5.23	Fugacity of saturated solution	236
5.24	Surface tension	237
5.25	Temperature dependence	238
5.26	Variations of composition	238
5.27	Interfacial tension between two solutions	238
5.28	Volume concentrations	239

CHAPTER 6

Systems of chemically reacting species

6.01	Notation and terminology	240
6.02	Enthalpy of reaction	241
6.03	Hess' law	242
6.04	Kirchhoff's relations	243
6.05	Prescription of standards	243
6.06	Construction of tables	244
6.07	Gaseous equilibria	245
6.08	Equilibria between gases and solids	246
6.09	Temperature dependence	247
6.10	Numerical example	248
6.11	Reactions between pure solids or liquids	248
6.12	Transition of sulphur	250
6.13	Homogeneous equilibrium in solution	251
6.14	Temperature dependence	252
6.15	Use of volume concentrations	252
6.16	Heterogeneous equilibria involving solutions	252
6.17	Transitions of second order	252
6.18	Cooperative systems	255
6.19	Alternative notation	257
6.20	Lambda point	257
6.21	Comparison with phase change and critical point	261
6.22	Dependence of lambda point on pressure	262
6.23	Transitions of higher order	264
6.24	Components and degrees of freedom	265

CHAPTER 7

Solutions of electrolytes

7.01	Characteristics of strong electrolytes	268
7.02	Ionic mole ratios and ionic molalities	268
7.03	Electrical neutrality	269
7.04	Ionic absolute activities	270
7.05	Ideal dilute and real solutions	272
7.06	Osmotic coefficient of the solvent	272
7.07	Freezing point and boiling point	272
7.08	Osmotic pressure	273
7.09	Ionic activity coefficient	273
7.10	Mean activity coefficient of electrolyte	273
7.11	Temperature dependence	275
7.12	Distribution of electrolyte between two solvents	275
7.13	Solubility	276
7.14	Chemical reactions	277
7.15	Gibbs–Duhem relation for electrolyte solutions	277
7.16	Limiting behaviour at high dilutions	280
7.17	Limiting law of Debye and Hückel	281
7.18	Aqueous solutions	283
7.19	Less dilute solutions	284
7.20	Formulae of Debye and Hückel	284
7.21	Specific interactions	286
7.22	Chemical reactions involving solvent	289
7.23	Acid–base equilibrium	290
7.24	Weak electrolytes	293
7.25	Surface phases	293

CHAPTER 8

Electrochemical systems

8.01	Electrically charged phases	298
8.02	Phases of identical composition	299
8.03	Electrochemical potentials	299
8.04	Absolute activities of ions	302
8.05	Dilute solutions in common solvent	302
8.06	Volta potentials	304
8.07	Membrane equilibrium (non-osmotic)	304
8.08	Osmotic membrane equilibrium	305
8.09	Contact equilibrium	307
8.10	Examples of galvanic cell	307
8.11	General treatment of electromotive force	311

8.12	Temperature dependence	313
8.13	Application of Nernst's heat theorem	313
8.14	Cells without transference	316
8.15	Standard electromotive force	317
8.16	Numerical example	318
8.17	Standard electromotive force of half-cell	320
8.18	Cells with transference with two similar electrodes	321
8.19	Cells containing single electrolytes	324
8.20	Cells with transference having two dissimilar electrodes	326

CHAPTER 9

Gravitational field

9.01	Nature of gravitational field	327
9.02	Phases in gravitational field	327
9.03	Thermodynamic functions in gravitational field	327
9.04	Equilibrium in gravitational field	328
9.05	Dependence of μ_i on T and P	328
9.06	Single component in gravitational field	329
9.07	Mixture in gravitational field	330
9.08	Mixture of gases	331
9.09	Ideal dilute solutions	331
9.10	Chemical reaction in gravitational field	332

CHAPTER 10

Electrostatic systems

10.01	Introduction	333
10.02	Parallel-plate capacitor in vacuo	333
10.03	Parallel-plate capacitor in fluid	333
10.04	Work of charging a capacitor	334
10.05	Characteristic functions	334
10.06	Analogues of Maxwell's relations	335
10.07	Constant permittivity. Dielectric constant	335
10.08	Integrated formulae	336
10.09	Application to perfect gas	336

CHAPTER 11

Magnetic systems

11.01	Introduction	338
11.02	Fundamental electromagnetic vectors	339
11.03	Permittivity and permeability in a vacuum	339

11.04	Simplest examples of field in a vacuum	341
11.05	Presence of matter	342
11.06	Electric and magnetic work	344
11.07	Formula for Helmholtz function	346
11.08	Other thermodynamic functions	347
11.09	Case of linear induction	348
11.10	Specimen in uniform external field	348
11.11	Other thermodynamic functions	350
11.12	Specimens of simple shape	350
11.13	Diamagnetic, paramagnetic, and ferromagnetic substances	351
11.14	Simple paramagnetic behaviour	351
11.15	Entropy of simple paramagnetic substances	355
11.16	Adiabatic demagnetization	355
11.17	Unattainability of zero temperature	356

CHAPTER 12

Radiation

12.01	General considerations	357
12.02	Energy and entropy in terms of g_i's	357
12.03	Thermodynamic functions	359
12.04	Evaluation of integrals	359
12.05	Stefan–Boltzmann law	360
12.06	Adiabatic changes	360

CHAPTER 13

Onsager's reciprocal relations

13.01	Introduction	362
13.02	Electric insulators and conductors	362
13.03	Onsager's reciprocal relations	365
13.04	Electrokinetic effects	366
13.05	Electric double layer	368
13.06	Electrochemical cells with transference	369
13.07	Thermoelectricity	371
13.08	Seebeck effect and thermoelectric power	372
13.09	Peltier effect	374
13.10	Kelvin's second relation	374
13.11	Thomson effect	375
13.12	Interdiffusion of two fluids	375
13.13	Interdiffusion of two solutes in dilute solution	376

CHAPTER 14

Systems in motion

14.01 Introduction 378
14.02 Mechanics and hydrodynamics 378
14.03 Entropy 379
14.04 Thermal equilibrium 379
14.05 Temperature 379
14.06 Fundamental equations 380

Author index 381
Subject index 383

IMPORTANT SYMBOLS

A	area
$\mathcal{F}$	Helmholtz function
B	second virial coefficient
B_n	nth virial coefficient
C	heat capacity
E	electromotive force
F	Faraday constant
$\mathcal{J}$	characteristic function for variables T, $\boldsymbol{B}_e$
G	Gibbs function
H	enthalpy
I	ionic strength; moment of inertia
J	Massieu function
K	equilibrium constant
L	Avogadro constant
L_{ik}	Onsager coefficients
M_i	proper mass of species i
N	number of molecules or photons
P	pressure
Q	partition function; electric charge
R	gas constant; electric resistance
S	entropy
T	thermodynamic temperature
U	energy
V	volume
X	extensive property
Y	Planck function
$\boldsymbol{A}$	magnetic potential
$\boldsymbol{B}$	magnetic force vector (induction)
$\boldsymbol{D}$	electric derived vector (displacement)

IMPORTANT SYMBOLS

$\boldsymbol{E}$	electric force vector (field)
$\boldsymbol{H}$	magnetic derived vector (field)
$\boldsymbol{J}$	electric current density
$\boldsymbol{M}$	magnetization
$\boldsymbol{P}$	electric polarization
a	cohesive energy density in equation of state; radiation density constant
a_i	relative activity of species i
b	excluded volume in equation of state
c	speed of light
c_i	volume concentration of species i
d	distance
e	elementary charge
g	gravitational acceleration
g_r	statistical weight of energy level ε_r
h	Planck constant
i	electric current
k	Boltzmann constant
l	distribution coefficient
m	mass; molality
n_i	amount of species i
p_i	fugacity of species i
q	heat absorbed by system
r	solute–solvent mole ratio
s	solubility; characteristic length of ionic interaction; symmetry number
t	transport number; time
w	work done on system
x	mole fraction
y	mole fraction in vapour
z	charge number of ion
Γ	surface concentration
Θ	characteristic temperature
Λ	molar conductance
Π	osmotic pressure
Φ	gravitational potential
Ω	partition function of isolated system

IMPORTANT SYMBOLS

α	thermal expansivity; volatility ratio; coefficient in Debye–Hückel formulae
$\beta_{R,X}$	ionic specific interaction coefficient
γ	surface tension
γ_i	activity coefficient of species i on mole ratio or molality scale
ε	permittivity; energy level of molecule; thermoelectric power
η	viscosity
θ	freezing-point depression; boiling-point elevation
κ	reciprocal of characteristic length in Debye–Hückel theory
κ_T	isothermal compressibility
κ_S	adiabatic compressibility
λ_i	absolute activity of species i
μ	permeability
μ_i	chemical potential of species i
ν	frequency
ν_i	stoichiometric number of species i
ξ	extent of reaction
o	zero-temperature degeneracy
π	Peltier coefficient
ϱ	density
σ	function in Debye–Hückel formula for osmotic coefficient; Thomson coefficient
τ	thickness of interfacial layer
ϕ	osmotic coefficient
ψ	electric potential

GENERAL PHYSICAL CONSTANTS*

Speed of light	c	2.997925×10^8 m s^{-1}
Permittivity of a vacuum	ε_0	8.85416×10^{-12} C m^{-1} V^{-1}
Planck constant	h	6.6256×10^{-34} J s
Proton charge	e	1.60210×10^{-19} C
Bohr magneton	μ_B	9.2732×10^{-24} A m^2
Water triple-point temperature	T_{tp}	273.16 K
Gas constant	R	8.3143 J K^{-1} mole^{-1}
Boltzmann constant	k	1.38054×10^{-23} J K^{-1}
Avogadro constant	L	6.02252×10^{23} mole^{-1}
Faraday constant	F	9.64870×10^4 C mole^{-1}
Second radiation constant	hc/k	1.43879×10^{-2} m K^{-1}
Stefan–Boltzmann constant	σ	5.6697×10^{-8} W m^{-2} K^{-4}

DEFINED VALUES AND CONVERSION FACTORS

Atmosphere	1.01325×10^5 J m^{-3}
Thermochemical calorie	4.184 J

* Pure and Applied Chemistry 1964 **9** 453.

INTRODUCTION CONCERNING NOTATION AND TERMINOLOGY

We consistently use symbols to denote physical quantities, not their measure in terms of particular units. For example, we may write of a pressure P

$$P = 1.2 \text{ atm}$$
$$= 912 \text{ mmHg}$$
$$= 1.216 \text{ bar}$$
$$= 1.216 \times 10^6 \text{ dyn cm}^{-2}$$
$$= 1.216 \times 10^5 \text{ J m}^{-3}$$

or alternatively $P/\text{atm} = 1.2$, etc., but under no circumstances shall we equate P to 1.2 or to any other number. This method of notation called 'quantity calculus' is far from new. It was recommended by Alfred Lodge*. It has been used by some of the greatest theoretical physicists, in particular Planck and Sommerfeld. The advantages of this notation have been emphasized by Henderson[†] who attributed it to Stroud.

This notation is especially clear and tidy for labelling the axes of a graph, e.g.

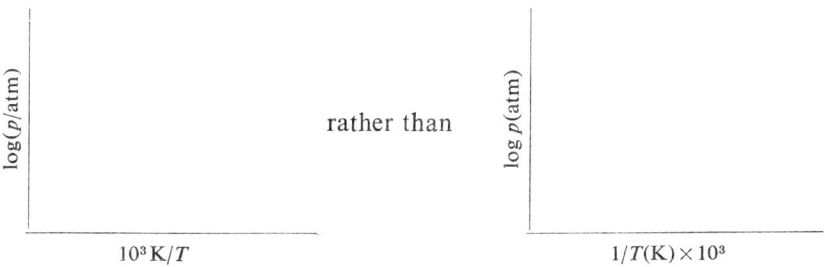

So far as possible we adhere in our notation to the recommendations of the Conférence Générale des Poids et Mesures[‡], the International Standards

---

* Lodge, Nature 1888 **38** 281.
[†] Henderson, Math. Gaz. 1924 **12** 99.
[‡] C.R. Conférence Générale des Poids et Mesures, 1948, 1954, 1960.

Organisation*, The International Union of Pure and Applied Chemistry[†], and the International Union of Pure and Applied Physics[‡], these being in almost complete mutual agreement. In particular we use italic (sloping) letters for physical quantities and roman (upright) letters for mathematical operators and units. The notation recommended by these bodies has been simplified in one respect. The symbol °K for the temperature unit in the Kelvin scale is abbreviated to K. This economy cannot lead to any ambiguity. The symbol °C for the unit in the Celsius scale is on the contrary retained. For example we write for the specific heat capacity C of water

$$C = 4.1793 \text{ J g}^{-1}\text{K}^{-1} \quad \text{at} \quad 25 \text{ °C}.$$

When discussing matters involving electricity or magnetism we use the internationally recommended rationalized system and units based on the ampere and the volt. In this system the electric potential difference ψ in a parallel plate capacitor is given by

$$\psi = \sigma d/\varepsilon$$

where d is the distance between the plates, $\pm \sigma$ is the charge density on the plates, and ε is the permittivity of the medium. If there is a vacuum between the plates

$$\varepsilon = \varepsilon_0 = 8.85416 \times 10^{-12} \text{ C m}^{-1} \text{ V}^{-1}.$$

The ratio $\varepsilon_r = \varepsilon/\varepsilon_0$ is called the *relative permittivity* or the *dielectric constant*. A reader who prefers the obsolescent unrationalized system[‡] can translate any formula into it by substituting $\varepsilon'/4\pi$ for ε and $\varepsilon_0'/4\pi$ for ε_0 while ε_r remains unchanged.

In 1961 the International Union of Pure and Applied Physics made the following recommendations:[§]

'In the field of *chemical and molecular physics*, in addition to the basic quantities defined above having been defined by the Conférence Générale des Poids et Mesures, *amount of substance* is also treated as a basic quantity. The recommended basic unit is the mole, symbol: mol. The mole is defined as the amount of substance, which contains the same number of molecules (or ions, or atoms, or electrons, as the case may be), as there are atoms in exactly 12 gramme of the pure carbon nuclide ^{12}C.'

---

* I.S.O. Recommendation R.31, 1956–.
† I.U.P.A.C. Manual of Physico-chemical symbols and Terminology, 1959.
‡ I.U.P.A.P. Symbols, Units and Nomenclature in Physics, 1965.
§ I.U.P.A.P. Symbols, Units and Nomenclature in Physics, 1961 p. 19.

In 1965 the International Union of Pure and Applied Chemistry adopted the almost identical recommendation:*
'A mole is an amount of a substance of specified chemical formula, containing the same number of formula units (atoms, molecules, ions, electrons, quanta, or other entities) as there are in 12 grams (exactly) of the pure nuclide ^{12}C.'

In this book the recommendation has been accepted that amount of substance is a distinct quantity different from mass. There remains an unresolved question of terminology. There is a need for a word to denote per amount of substance analogous to *specific* meaning per mass and *density* meaning per volume. In the absence of any recommendation we have used the word *proper*. We accordingly say that the 'proper energy' of a substance is so many joules per mole instead of saying that the 'molar energy' is so many joules per mole, which is as clumsy as if we spoke of 'gramme energy' instead of 'specific energy'.

One further change in terminology must be mentioned. In contrast to the previous editions of this book, but in conformity with usual American practice[†], molality denoted by m is here a dimensionless quantity defined by

$$m = r/r^\ominus$$

where r denotes the solute–solvent mole ratio and $r^\ominus$ is a standard value of r.

We choose for $r^\ominus$ the value

$$r^\ominus = M_1 \text{ mole kg}^{-1}$$

where M_1 denotes the proper mass of the solvent. This revised interpretation of molality leads to shorter formulae.

---

* I.U.P.A.C. Information Bulletin Number 24, 1965, p. 4.
† Lewis and Randall, Thermodynamics, revised by Pitzer and Brewer, McGraw-Hill 1961 p. 200.

CHAPTER 1

FUNDAMENTAL PRINCIPLES

§1.01 *Scope of thermodynamics*

The most important conception in thermodynamics is temperature. The essential properties of temperature will be described below. Anticipating this we may define thermodynamics as that part of physics concerned with the dependence on temperature of any equilibrium property. This definition may be illustrated by a simple example. Consider the distribution of two immiscible liquids such as mercury and water in a gravitational field. The equilibrium distribution is that in which the heavier liquid, mercury, occupies the part of accessible space where the gravitational potential is lowest and the lighter liquid, water, occupies the part of the remaining accessible space where the gravitational potential is lowest. This equilibrium distribution, if we neglect the effect of thermal expansion, is independent of temperature. Consequently the problem does not involve thermodynamics, but only hydrostatics. Now consider by contrast the distribution in a gravitational field of two completely miscible fluids such as bromine and carbon disulphide. The relative proportions of the two substances will vary from place to place, the proportion of the heavier liquid, bromine, being greatest at the lowest gravitational potential and conversely. The precise relation between the composition and the gravitational potential depends on the temperature, assumed uniform, of the mixture. Clearly this is a problem in thermodynamics, not merely hydrostatics.

We shall now mention a few other typical examples to show that thermodynamics has a bearing on most branches of physics, including elasticity, hydrodynamics, electrostatics, and electrodynamics. In the relation, known as Hooke's law, of proportionality between tension and extension the coefficient of proportionality will in general be temperature dependent. In so far as its variation with temperature is relevant thermodynamics is involved. To study the temperature dependence of the compressibility of a fluid, that of the permittivity of a dielectric, that of the permeability of a paramagnetic material, that of the electromotive force of a cell, and in fact the temperature dependence of any equilibrium property, thermodynamics is needed,

The name 'thermodynamics' is too firmly established to be changed, but a better name is 'thermophysics' containing as branches 'thermodynamics' or 'thermomechanics', 'thermoelasticity', 'thermoelectrostatics', 'thermomagnetics', 'thermochemics', and so on.

§1.02 *Thermodynamic state. Phases*

The simplest example of a system to which thermodynamics can be applied is a single homogeneous substance. In this simplest case a complete description of its thermodynamic state requires a specification of its content, i.e. amount of each chemical substance contained, and further a specification of two other quantities such as for example volume and viscosity, or density and pressure. If all the physical properties of the system in which we are interested were independent of whether the system is hot or cold, then in order to describe its state it would be sufficient to specify, apart from the amount of each chemical substance contained, only one quantity, such as volume. Usually some, if not all, of the properties of interest do depend on whether the body is hot or cold and the specification of one extra independent quantity fixes the degree of hotness or coldness. Thus this simple *thermo-hydrodynamic* system has one more degree of freedom than the corresponding *hydrodynamic* system.

If the system is not homogeneous, then in order to describe its thermodynamic state we have to consider it as composed of a number, small or large, of homogeneous parts called *phases* each of which is described by specifying its content and a sufficient number of other properties; the sufficient number for each *thermo-physical* phase is always one more than in the corresponding hypothetical *physical* system with all its properties of interest independent of whether it is hot or cold.

In some cases the complete description of the thermodynamic state of a system may require it to be regarded as composed of an infinite number of infinitesimal phases. If the physical properties vary continuously over macroscopic parts of the system, this procedure offers no difficulty. An example is a high column of gas in a gravitational field. If on the other hand there are infinitely many discontinuities over finite regions, it may be difficult if not impossible to give a complete description of the thermodynamic state. An example is a gas flowing turbulently through an orifice.

In considering the properties of interfaces, we shall have to include phases which are extremely thin in the direction normal to the interface.

To sum up, the complete description of the thermodynamic state of any system involves a description of the thermodynamic state of each of its homogeneous phases, which may be few or many or infinite in number. The description of the thermodynamic state of each phase requires the specification of one more property than the description of the physical state of an analogous hypothetical phase all of whose properties of interest are independent of whether it be hot or cold.

§1.03 *Thermodynamic process*

If on comparing the state of a thermodynamic system at two different times it is found that there is a difference in any macroscopic property of the system, then we say that between the two times of observation a *process* has taken place. If, for example, two equal quantities of gas are allowed to intermix, this will constitute a *process* from a thermodynamic point of view provided the two initially separate gases are distinguishable by any macroscopic property, even though their difference is very slight as would be the case for two isotopes. If, on the other hand, the two initially separate gases are not distinguishable by any macroscopic property, then from a thermodynamic point of view no process takes place although from a molecular point of view there is a never-ceasing intermixing.

§1.04 *Infinitesimal process*

A process taking place to such an extent that there is only an infinitesimal change in the macroscopic properties of a system is called an *infinitesimal process*.

§1.05 *Insulating walls. Adiabatic processes*

The boundary or wall separating two systems is said to be *insulating* if it has the following property. If any system in complete internal equilibrium is completely surrounded by an *insulating* wall then no change can be produced in the system by external agency except by

(a) movement of the containing wall or part of it, or

(b) long range forces, e.g. movement of electrically charged bodies.

When a system is surrounded by an insulating boundary the system is said to be *thermally insulated* and any process taking place in the system is called *adiabatic*. The name adiabatic appears to be due to Rankine*.

---

* Maxwell, Theory of Heat, Longmans 1871 ed. p. 129.

§1.06 Conducting walls. Thermal equilibrium

The boundary or wall separating two systems is said to be *thermally conducting* if it has the following property. If any two separate systems each in complete internal equilibrium are brought together so as to be in contact through a *thermally conducting* wall then in general the two systems will be found not to be in mutual equilibrium, but will gradually adjust themselves until eventually they do reach mutual equilibrium after which there will of course be no further change. The two systems are then said to have reached a state of *thermal equilibrium*. Systems separated by a conducting boundary are said to be in *thermal contact*.

§1.07 Zeroth law. Temperature

We are now ready to formulate one of the important principles of thermodynamics.

If two systems are both in thermal equilibrium with a third system then they are in thermal equilibrium with each other.

This will be referred to as the *zeroth law* of thermodynamics.

Consider now a reference system in a well-defined state. Then all other systems in thermal equilibrium with it have a property in common, namely the property of being in thermal equilibrium with one another. This property is called *temperature*. In other words systems in *thermal equilibrium* are said to have the *same temperature*. Systems not in thermal equilibrium are said to have different temperatures.

§1.08 Thermostats and thermometers

Consider now two systems in thermal contact, one very much smaller than the other, for example a short thin metallic wire immersed in a large quantity of water. If the quantity of water is large enough (or the wire small enough), then in the process of attaining thermal equilibrium the change in the physical state of the water will be negligible compared with that of the wire. This situation is described differently according as we are primarily interested in the small system or in the large one.

If we are primarily interested in the small system, the wire, then we regard the water as a means of controlling the temperature of the wire and we refer to the water as a *temperature bath* or *thermostat*.

If on the other hand we are primarily interested in the large system, the water, we regard the wire as an instrument for recording the temperature of the water and we refer to the wire as a *thermometer*. This recording of temperature can be rendered quantitative by measuring some property of the thermometer, such as its electrical resistance, which varies with temperature.

§1.09 *Temperature scales*

The choice of thermometers is very wide especially as there is a choice both of the substance constituting the thermometer and of the property measured. Consequently there is a wide, effectively infinite, choice of temperature scales. There is however one particular scale which has outstandingly simple characteristics which can be described in a manner independent of the properties of any particular substance or class of substances. This temperature is called *thermodynamic temperature* or *absolute temperature*. It was first defined by Kelvin* and is denoted by T. It is the only scale that we shall use. It will be defined by its properties, especially its relation to entropy. The question how T can best be measured must necessarily be postponed to chapter 3.

§1.10 *Energy and heat. First law*

Leaving temperature for the moment, we must now say something about energy. The conception of energy arose first in mechanics and was extended to electrostatics and electrodynamics. When these branches of physics are idealized so as to exclude friction, viscosity, hysteresis, temperature gradients, temperature dependence of properties, and related phenomena, the fundamental property of energy can be described in two alternative ways.

I. When several systems interact in any way with one another, the whole set of systems being isolated from the rest of the universe, the sum of the energies of the several systems remains constant.

II. When a single system interacts with the rest of the universe (its surroundings) the increase of the energy of this system is equal to the work done on the system by the rest of the universe (its surroundings).

Under the idealized conditions mentioned above these two descriptions are equivalent, but when these restrictions are removed the two descriptions are no longer equivalent and we have to make a choice between them. Of the alternatives we choose I and with this choice the *energy* is denoted by U. The formulation I is then a statement of the conservation of energy.

Let us now consider in greater detail the interaction between a pair of

---

* W. Thomson, Proc. Cambridge Phil. Soc. 1848 **1** 69.

systems, supposed isolated from the rest of the universe. Using superscripts ^{A}, ^{B} to relate to the two systems we have

$$dU^A + dU^B = 0 \qquad 1.10.1$$

or

$$dU^A = -dU^B \qquad 1.10.2$$

but in general this is not equal to the work w_{BA} done by B on A. In other words there can be exchange of energy between A and B of a kind other than work. Such an exchange of energy is that determined by a temperature difference and is called *heat*. If then we denote the heat flow from B to A by q_{BA}, we have the following relations

$$dU^A = w_{BA} + q_{BA} \qquad 1.10.3$$

$$dU^B = w_{AB} + q_{AB} \qquad 1.10.4$$

$$w_{AB} + w_{BA} = 0 \qquad 1.10.5$$

$$q_{AB} + q_{BA} = 0. \qquad 1.10.6$$

This set of relations together constitutes the *first law* of thermodynamics.

The sign of q is determined by the temperature difference between A and B, and the universal convention is to define the sign of a temperature difference in such a way that heat flows from the higher to the lower temperature.

The above analysis of the most general interaction between two systems can immediately be extended to the most general interaction between a given system and the rest of the universe. If we denote by U^Σ the energy of the system Σ, by q the heat flow from the surroundings to the system, and by w the work done on the system, we have

$$dU^\Sigma = q + w. \qquad 1.10.7$$

The extension of the mechanical principle of conservation of energy to include changes in thermal energy and the flow of heat was a gradual process, the earlier formulations being less rigorous than later ones. The principle is implied in a posthumous publication of Carnot[*] (died 1832) and was placed on a firm experimental basis by Joule[†] (1840–45). More explicit statements of the principle were formulated by Helmholtz[‡] (1847) and by

---

[*] Carnot, Réflexions sur la puissance motrice du feu, Bachelier, Paris, 1824. Reprinted in 1912 by Hermann, Paris and in 1953 by Blanchard, Paris, together with some notes discovered after Carnot's death in 1832 and communicated to the Académie des Sciences in 1878 by Carnot's brother.

[†] Joule, Phil. Mag. 1845 **27** 205.

[‡] Helmholtz, Über die Erhaltung der Kraft, Physik. Ges. Berlin 1847.

FUNDAMENTAL PRINCIPLES

Clausius* (1850). A completely rigorous formulation was given by Born[†] (1921).

§1.11 Conversion of work to heat

The expression *conversion of work to heat* should be used with caution if at all, since in general w and $-q$ are not numerically equal to each other. If however a system Σ is taken through a complete cycle, then since its initial and final states are identical the initial and final values of U^Σ are the same and so

$$\Delta U^\Sigma = 0 \quad w = -q \quad \text{(complete cycle).} \qquad 1.11.1$$

We may then say that in the cycle the work w done on the system is converted into the balance of heat $-q$ given off by the system during the cycle, that is to say the excess of the heat given off over the heat absorbed in various parts of the cycle.

Again if a system Σ is kept in a steady state while work is done on it, then, since its state remains unaltered, U does not change and so

$$\Delta U^\Sigma = 0 \quad w = -q \quad \text{(steady state).} \qquad 1.11.2$$

Here again we may say that in the steady state the work w done on the system is converted into the heat $-q$ given off by the system.

Except in the two special cases just mentioned, it is in general dangerous, if not meaningless, to speak of the conversion of work into heat or vice-versa. Unfortunately the expression is sometimes used incorrectly. Let us consider two simple practical examples which serve to illustrate the correct and incorrect use of the expression.

Consider as our system an ordinary *electric heater*, that is to say a resistor across which an electric potential difference E can be produced by closing a switch. Suppose that initially the resistor is in thermal equilibrium with its surroundings and the switch is open. When the switch is closed a current i flows through the resistor and the electrical work done on the resistor in an element of time dt is

$$w = Ei\, dt. \qquad 1.11.3$$

In the first instant this work produces an increase in the energy U^R of the resistor R, so that

$$w = dU^R \quad \text{(initially).} \qquad 1.11.4$$

---

* Clausius, Ann. Phys. Lpz. 1850 **79** 368, 500.
† Born, Phys. Z. 1921 **22** 218.

But immediately the temperature of the resistor becomes higher than that of its surroundings and so there is a flow of heat $-q$ from the resistor to its surroundings. Thus in a given time

$$w = \mathrm{d}U^R - q \quad \text{(generally)}. \qquad 1.11.5$$

As the temperature difference between the resistor and its surroundings increases, so $-q/w$ increases towards the value unity. Eventually a steady state is reached, the temperature of the resistor no longer increases, and we have

$$w = -q \quad \mathrm{d}U^R = 0 \quad \text{(steady state)}. \qquad 1.11.6$$

Only when this steady state has been reached, and not until then, may one correctly speak of the conversion of the work w into the heat $-q$.

Now by way of contrast consider the system consisting of the electric heater together with a fluid surrounding it, the whole being thermally insulated. The work done on the system is still given by (3). But now since the whole system Σ, consisting of resistor and fluid, is thermally insulated q is by definition zero, so that

$$w = \mathrm{d}U^\Sigma \quad q = 0 \quad \text{(thermal insulation)}. \qquad 1.11.7$$

We may now say that the work w is converted into energy; to speak of its conversion to heat would be nonsense.

§1.12 *Natural and reversible processes*

We must now consider a classification of processes due to Planck*. All the independent infinitesimal processes that might conceivably take place may be divided into three types: *natural processes, unnatural processes*, and *reversible processes*.

Natural processes are all such as actually do occur; they proceed in a direction towards equilibrium.

An *unnatural process* is one in a direction away from equilibrium; such a process never occurs.

As a limiting case between natural and unnatural processes we have *reversible processes*, which consist of the passage in either direction through a continuous series of equilibrium states. Reversible processes do not actually occur, but in whichever direction we contemplate a reversible process we can by a small change in the conditions produce a natural process differing as little as we choose from the reversible process contemplated.

* Planck, Ann. Phys. Lpz. 1887 **30** 563.

We shall illustrate the three types by an example. Consider a system consisting of a liquid together with its vapour at a pressure P. Let the equilibrium vapour pressure of the liquid be P_{eq}. Consider now the process of the evaporation of a small quantity of the liquid. If $P < P_{eq}$, this is a natural process and will in fact take place. If on the other hand $P > P_{eq}$, the process contemplated is unnatural and cannot take place; in fact the contrary process of condensation will take place. If $P = P_{eq}$ then the process contemplated and its converse are reversible, for by slightly decreasing or increasing P we can make either occur. The last case may be described in an alternative manner as follows. If $P = P_{eq} - \delta$, where $\delta > 0$, then the process of evaporation is a natural one. Now suppose δ gradually decreased. In the limit $\delta \to 0$, the process becomes reversible.

§1.13 *Reversible process and reversible change*

We have defined a reversible process as a hypothetical passage through equilibrium states. If we have a system interacting with its surroundings either through the performance of work or through the flow of heat, we shall use the term *reversible process* only if there is throughout the process equilibrium between the system and its surroundings. If we wish to refer to the hypothetical passage of the system through a sequence of internal equilibrium states, without necessarily being in equilibrium with its surroundings we shall refer to a *reversible change*. We shall illustrate this distinction by examples.

Consider a system consisting of a liquid and its vapour in mutual equilibrium in a cylinder closed by a piston opposed by a pressure equal to the equilibrium vapour pressure corresponding to the temperature of the system. Suppose now that there is a flow of heat through the walls of the cylinder, with a consequent evaporation of liquid and work done on the piston at constant temperature and pressure. The change in the system is a reversible change, but the whole process is a reversible process only if the medium surrounding the cylinder is at the same temperature as the liquid and vapour; otherwise the flow of heat through the walls of the cylinder is not reversible and so the process as a whole is not reversible, although the change in the system within the cylinder is reversible.

As a second example consider a flow of heat from one system in complete internal equilibrium to another system in complete internal equilibrium. Provided both systems remain in internal equilibrium then the change which each system undergoes is a reversible change, but the whole process of heat flow is not a reversible process unless the two systems are at the same temperature.

§1.14 *Equilibrium and reversible changes*

If a system is in complete equilibrium, any conceivable infinitesimal change in it must be reversible. For a natural process is an approach towards equilibrium, and as the system is already in equilibrium the change cannot be a natural one. Nor can it be an unnatural one, for in that case the opposite infinitesimal change would be a natural one, and this would contradict the supposition that the system is already in equilibrium. The only remaining possibility is that, for a system in complete equilibrium any conceivable infinitesimal change must be reversible.

§1.15 *Closed systems and open systems*

A system of fixed material content is called a *closed system* and a system of variable content is called an *open system*. Similarly a phase of fixed content is called a *closed phase* and a phase of variable content is called an *open phase*.

We shall often be concerned with a closed system composed of two or more open phases.

Provided a closed phase is at rest and in thermal equilibrium and provided chemical reactions are excluded, the phase is always in internal equilibrium. As already mentioned in §1.02 it has two degrees of freedom, that is to say one more than a hypothetical hydrostatic fluid having properties independent of temperature. The state of such a phase may therefore be defined by its energy U and its volume V, but other choices are possible.

§1.16 *Entropy*

There exists a function S of the state of a system called the *entropy* of the system having the following properties.

1. The entropy S^Σ of a system Σ is the sum of the entropies of its parts, $\alpha, \beta, \ldots$ so that

$$S^\Sigma = S^\alpha + S^\beta + \ldots \qquad 1.16.1$$

In this respect entropy is similar to mass, volume, and energy.

2. The entropy S^α of a closed phase α is determined by the energy U^α and the volume V^α of the phase so that

$$dS^\alpha = (\partial S^\alpha/\partial U^\alpha)_{V^\alpha} dU^\alpha + (\partial S^\alpha/\partial V^\alpha)_{U^\alpha} dV^\alpha. \qquad 1.16.2$$

3. $(\partial S^\alpha/\partial U^\alpha)_{V^\alpha}$ is always positive. $\qquad 1.16.3$

4. The entropy of an insulated closed system Σ increases in any natural

change, remains constant in any reversible change, and is a maximum at equilibrium. Hence

$$dS^\Sigma \geq 0 \quad \text{(insulated closed system)}. \qquad 1.16.4$$

5. In any reversible adiabatic change the entropy remains constant. Thus

$$dS^\Sigma = 0 \quad \text{(reversible adiabatic)}. \qquad 1.16.5$$

These properties together determine the entropy completely except for an additive constant to which any conventional value may be assigned.

§1.17 *Thermal equilibrium*

Consider a thermally insulated system composed of two closed phases each maintained at constant volume and in thermal contact with each other. Using the superscript Σ to denote the system and the superscripts α and β to denote the two phases we have

$$dV^\alpha = 0 \quad dV^\beta = 0 \qquad 1.17.1$$

$$dU^\Sigma = dU^\alpha + dU^\beta = 0 \qquad 1.17.2$$

$$dS^\Sigma = dS^\alpha + dS^\beta \geq 0 \qquad 1.17.3$$

the inequality holding for a natural heat flow and the equality for a reversible heat flow. By virtue of (1) we may rewrite (3) as

$$(\partial S^\alpha/\partial U^\alpha)_{V^\alpha} dU^\alpha + (\partial S^\beta/\partial U^\beta)_{V^\beta} dU^\beta \geq 0 \qquad 1.17.4$$

and by virtue of (2) this becomes

$$\{(\partial S^\alpha/\partial U^\alpha)_{V^\alpha} - (\partial S^\beta/\partial U^\beta)_{V^\beta}\} dU^\alpha \geq 0. \qquad 1.17.5$$

We now define a positive quantity T by

$$T = (\partial U/\partial S)_V = 1/(\partial S/\partial U)_V \qquad 1.17.6$$

and rewrite (5) as

$$(1/T^\alpha - 1/T^\beta) dU^\alpha \geq 0 \qquad 1.17.7$$

which is equivalent to

$$(T^\beta - T^\alpha) dU^\alpha \geq 0. \qquad 1.17.8$$

Hence for a natural process $dU^\alpha = -dU^\beta$ has the same sign as $T^\beta - T^\alpha$.

§1.18 *Thermodynamic temperature*

The property of T expressed by formula (1.17.8) is obviously that of a temperature and T defined by formula (1.17.6) is called *thermodynamic temperature*. This temperature is independent of any particular property of any particular substance. It will be used throughout this book and will be referred to simply as *temperature*.

§1.19 *Entropy and heat*

For a single closed phase α we have

$$dS^\alpha = (\partial S^\alpha/\partial U^\alpha)_{V^\alpha} dU^\alpha + (\partial S^\alpha/\partial V^\alpha)_{U^\alpha} dV^\alpha \qquad 1.19.1$$

and conversely if we regard U^α as a function of S^α and V^α

$$\begin{aligned} dU^\alpha &= (\partial U^\alpha/\partial S^\alpha)_{V^\alpha} dS^\alpha + (\partial U^\alpha/\partial V^\alpha)_{S^\alpha} dV^\alpha \\ &= T^\alpha dS^\alpha + (\partial U^\alpha/\partial V^\alpha)_{S^\alpha} dV^\alpha. \end{aligned} \qquad 1.19.2$$

We recall that for a reversible adiabatic change S^α remains constant and consequently

$$dU^\alpha = (\partial U^\alpha/\partial V^\alpha)_{S^\alpha} dV^\alpha = w = -P^\alpha dV^\alpha \qquad (S^\alpha = \text{const.}) \qquad 1.19.3$$

where P^α denotes the pressure of the phase α. Hence

$$(\partial U^\alpha/\partial V^\alpha)_{S^\alpha} = -P^\alpha \qquad 1.19.4$$

and substituting (4) into (2) we obtain

$$dU^\alpha = T^\alpha dS^\alpha - P^\alpha dV^\alpha. \qquad 1.19.5$$

Comparing this with the statement of the first law

$$dU^\alpha = q + w = q - P^\alpha dV^\alpha \qquad 1.19.6$$

we see that for a reversible change

$$T^\alpha dS^\alpha = q. \qquad 1.19.7$$

We shall now study the change in entropy when the system is neither thermally insulated nor in complete internal equilibrium. Let the system Σ be composed of phases α, β, ... each in internal equilibrium. If two or more parts of the system Σ have the same composition but different temperatures these are to be regarded as different phases. Now consider an infinitesimal change in Σ in which the quantities of heat gained by the phases α, β, ... are q^α, q^β, ... Evidently the changes inside the system Σ are independent of where the heat q^α, q^β, ... comes from or where the heat $-q^\alpha$, $-q^\beta$, ... goes to. We may therefore without affecting the changes inside Σ arrange for α to exchange heat only with a system α' which is in internal equilibrium and has the same temperature T^α as α; and similarly for β, ... We also arrange for the composite system $\Sigma + \alpha' + \beta' + ..$ to be thermally insulated. We accordingly have

$$dS^\Sigma + dS^{\alpha'} + dS^{\beta'} + \ldots \geq 0. \qquad 1.19.8$$

We also have

$$\begin{aligned} dS^{\alpha'} &= q^{\alpha'}/T^{\alpha'} = -q^\alpha/T^\alpha \\ dS^{\beta'} &= q^{\beta'}/T^{\beta'} = -q^\beta/T^\beta \end{aligned} \qquad 1.19.9$$

and so on. Substituting (9) into (8) we obtain

$$dS^\Sigma \geq q^\alpha/T^\alpha + q^\beta/T^\beta + \ldots \qquad 1.19.10$$

In particular for any single phase α

$$dS^\alpha \geq q^\alpha/T^\alpha \qquad 1.19.11$$

the inequality relating to a natural change and the equality to a reversible change.

The property of entropy described by (11) may alternatively be expressed as follows[*]

$$dS^\alpha = d_e S^\alpha + d_i S^\alpha \qquad 1.19.12$$

$$d_e S^\alpha = q^\alpha/T^\alpha \qquad 1.19.13$$

$$d_i S^\alpha \geq 0 \qquad 1.19.14$$

where $d_e S^\alpha$ denotes the increase in S^α associated with interaction of α with its surroundings and $d_i S^\alpha$ denotes the increase in S^α associated with a natural change occurring inside α.

§1.20 Second law

The enunciation of the properties of entropy and of thermodynamic temperature together constitute the *second law* of thermodynamics. The second law was foreshadowed by the work of Carnot[†] (1824). The first and second laws were co-ordinated by Clausius[‡] (1850) and by Kelvin[§] (1851).

---

[*] Prigogine and Defay, Chemical Thermodynamics, English translation by Everett, Longmans 1954 ch. 3.

[†] Carnot, Réflexions sur la puissance motrice du feu, Bachelier, Paris, 1824. Reprinted in 1912 by Hermann, Paris and in 1953 by Blanchard, Paris, together with some notes discovered after Carnot's death in 1832 and communicated to the Académie des Sciences in 1878 by Carnot's brother.

[‡] Clausius, Ann. Phys. Lpz. 1850 **79** 368, 500.

[§] W. Thomson, Trans. Roy. Soc. Edinb. 1853 **20** 261.

Thermodynamic temperature was introduced by Kelvin* (1848). The conception of *entropy* was first used by Clausius[†] in 1854 and the name also by Clausius[‡] in 1865. The formulation used above follows closely that of Callen[§] (1961).

§1.21 *Units*

The unit of energy in the 'Système International' is the *joule* denoted by J and defined by $J = kg\, m^2\, s^{-2}$. Another unit still widely used by physical chemists is the thermochemical calorie denoted by cal and defined by cal = 4.184 J exactly.

The unit of thermodynamic temperature in the 'Système International' is the *degree Kelvin* denoted by K and defined by the statement that the thermodynamic temperature T_{tp} of the triple point of natural water is 273.16 K exactly[∥]. The normal freezing point of water, defined as the freezing point of water saturated with air at atmospheric pressure, is within the accuracy of experiment 273.150 K. The normal boiling point of water, defined as the boiling point at a pressure of one atmosphere, is within the accuracy of experiment 373.15 K. The Celsius scale of temperature denoted by °C is defined by

$$x\,°C = (273.150 + x)K.$$

In physical chemistry the commonest unit of pressure is the atmosphere denoted by atm and defined[∥] by $atm = 1.01325 \times 10^5\, J\, m^{-3}$ exactly.

§1.22 *Extensive properties*

The mass of a system is clearly equal to the sum of the masses of its constituent phases. Any property, such as mass, whose value for the whole system is equal to the sum of its values for the separate phases is called an *extensive property*.

Important examples of extensive properties are the energy U, the entropy S, and the volume V. The energy U^Σ of a system Σ is related to the energies U^α of the separate phases α by

$$U^\Sigma = \sum_\alpha U^\alpha. \qquad 1.22.1$$

---

* W. Thomson, Proc. Cambridge Phil. Soc. 1848 **1** 69.
[†] Clausius, Ann. Phys. Lpz. 1854 **93** 481.
[‡] Clausius, Ann. Phys. Lpz. 1865 **125** 353.
[§] Callen, Thermodynamics, Wiley 1961; Guggenheim, Proc. Phys. Soc. London 1962 **79** 1079.
[∥] C. R. Conférence Générale des Poids et Mesures 1954.

Similarly, for the entropy, we have

$$S^\Sigma = \sum_\alpha S^\alpha \qquad 1.22.2$$

and for the volume

$$V^\Sigma = \sum_\alpha V^\alpha. \qquad 1.22.3$$

When we are considering a system of one phase only we may obviously omit the superscript and shall sometimes do so.

§1.23 *Intensive properties*

The density of a phase is clearly constant throughout the phase, because the phase is by definition homogeneous. Further, the density of a phase of a given kind and state is independent of the quantity of the phase. Any property of a phase with these characteristics is called an *intensive property*.

The temperature T^α and the pressure P^α of a phase α are important examples of intensive properties.

§1.24 *Chemical content of phase*

The content of a phase α is defined by the amount n_i^α of each of a finite number of *independently variable chemical species* in the phase. The unit of amount might be chosen as the amount having a given mass but this mass would not necessarily be the same mass for different chemical species. In fact, it is usually most convenient to take as unit of amount the *mole*, that is a mass proportional to that given by the accepted chemical formula of the particular species. A purely thermodynamic definition of the mole as unit of amount will be given in §3.13. In anticipation of this we shall use the mole as the unit of amount for each chemical species.

§1.25 *Chemically inert species*

We must emphasize that in the previous section we specified that the chemical species by which the chemical content of the phase is described must be *independently variable*. In the absence of any chemical reaction there is no difficulty, but if some of the species can react chemically the recipe required for selecting a set of *independently variable* species is not so simple. In order to postpone this complication we shall exclude the possibility of chemical reactions until we come to §1.43 when we revert to the subject.

§1.26 *Partial and proper quantities*

We have seen that the state of a closed phase α may be completely defined by its energy U^α and its volume V^α, or by its entropy S^α and its volume V^α. It follows that an open phase may be completely defined by U^α, V^α, and the amount n_i^α of each chemical species i, or alternatively by S^α, V^α, and the n_i^α's. But other choices are possible such as T^α, V^α, and the n_i^α's. In particular the set T^α, P^α, and the n_i^α's is especially convenient.

Now let X^α denote any extensive property of the phase α such as V^α or U^α or S^α. Then we can derive intensive properties, which we denote by X_i^α, defined by

$$X_i^\alpha = (\partial X^\alpha / \partial n_i^\alpha)_{T, P, n_j^\alpha} \qquad (j \neq i). \qquad 1.26.1$$

We shall call V_i^α the *partial volume*, U_i^α the *partial energy*, and S_i^α the *partial entropy*, and so on, of the species i in the phase α.

At given temperature and pressure we have then

$$dX^\alpha = \sum_i (\partial X^\alpha / \partial n_i^\alpha) dn_i^\alpha = \sum_i X_i^\alpha dn_i^\alpha \qquad \text{(const. } T, P\text{).} \qquad 1.26.2$$

Since X^α is homogeneous of first degree in the n_i^α's we have by Euler's theorem

$$X^\alpha = \sum_i n_i^\alpha (\partial X^\alpha / \partial n_i^\alpha) = \sum_i n_i^\alpha X_i^\alpha. \qquad 1.26.3$$

We may accordingly regard X^α as made up additively by a contribution X_i^α from each unit amount of i.

We also define another intensive property X_m^α by the formula

$$X_m^\alpha = X^\alpha / \sum_i n_i^\alpha = \sum_i n_i^\alpha X_i^\alpha / \sum_i n_i^\alpha. \qquad 1.26.4$$

We call V_m^α the *proper volume* of the phase α; we call U_m^α the *proper energy* of the phase α, and we call S_m^α the *proper entropy* of the phase α.

In the simple case of only a single chemical species i we have

$$X_i^\alpha = X_m^\alpha = X^\alpha / n_i^\alpha \qquad \text{(single species).} \qquad 1.26.5$$

We emphasize that whereas V^α, U^α, S^α, and any other X^α are extensive properties, V_i^α, U_i^α, S_i^α, and any other X_i^α are intensive properties. Since n_i^α is normally measured in moles it follows that V_i^α and V_m^α would be measured in m^3 mole^{-1} or cm^3 mole^{-1}, U_i^α and U_m^α in J mole^{-1} or cal mole^{-1}, and S_i^α and S_m^α in J K^{-1} mole^{-1} or cal K^{-1} mole^{-1}.

Corresponding to every equation homogeneous of the first degree in the extensive variables there is an analogous equation between the partial quantities and another analogous equation between the proper quantities. For example from the fundamental equation for a closed phase

$$dU^\alpha = T^\alpha dS^\alpha - P^\alpha dV^\alpha \qquad 1.26.6$$

we obtain by differentiating with respect to n_i^α

$$dU_i^\alpha = T^\alpha dS_i^\alpha - P^\alpha dV_i^\alpha \qquad 1.26.7$$

whereas by dividing by $\Sigma_i n_i^\alpha$ we obtain

$$dU_m^\alpha = T^\alpha dS_m^\alpha - P^\alpha dV_m^\alpha. \qquad 1.26.8$$

It is perhaps worth while drawing attention here to the fact that the quantity X need not be a thermodynamic property of the system. It is only required that X shall be an extensive property. We shall merely mention one example of such a non-thermodynamic property. If r denotes the refractive index of a binary mixture, we define the *total refractivity* R of the system by

$$R = (r^2 - 1)V/(r^2 + 2) \qquad 1.26.9$$

so that R is clearly an extensive property. We then define *partial refractivities* in the usual way by

$$R_1 = (\partial R/\partial n_1)_{T, P, n_2} \qquad 1.26.10$$

$$R_2 = (\partial R/\partial n_2)_{T, P, n_1} \qquad 1.26.11$$

and it then follows as usual that

$$R = n_1 R_1 + n_2 R_2. \qquad 1.26.12$$

The reason for choosing this particular example is the following. There are theoretical grounds for expecting R to be an approximately additive quantity, in which case R_1, R_2 would be independent of the composition of the mixture and have the same values as for the two pure substances. This is more or less supported by experiment. There are however theoretical grounds for expecting in certain cases deviations from simple additivity and this is also confirmed by experiment. The quantitative theoretical discussion of such deviations from simple additivity could be improved by the use of the partial refractivities defined as above.

§1.27 *Chemical potentials*

We recall formula (1.19.5) for a closed phase α at temperature T and pressure P

$$dU^\alpha = T dS^\alpha - P dV^\alpha \quad \text{(closed phase)}. \qquad 1.27.1$$

This may be extended to an open phase in the form

$$dU^\alpha = TdS^\alpha - PdV^\alpha + \sum_i \mu_i^\alpha dn_i^\alpha \quad \text{(open phase)} \qquad 1.27.2$$

where each μ_i^α is defined by

$$\mu_i^\alpha = (\partial U^\alpha / \partial n_i^\alpha)_{S^\alpha, V^\alpha, n_j^\alpha} \quad (j \neq i). \qquad 1.27.3$$

μ_i^α is called the *chemical potential* of the species i in the phase α. The dimensions of μ_i^α are energy/amount and it is therefore an intensive quantity.

§1.28 Characteristic functions. Fundamental equations

Formula (1.27.2)

$$dU^\alpha = TdS^\alpha - PdV^\alpha + \sum_i \mu_i^\alpha dn_i^\alpha \qquad 1.28.1$$

relates U^α to the independent variables S^α, V^α, and the n_i^α's, and U^α is said to be a *characteristic function* for these variables. Characteristic functions for other variables are readily obtained by the device known as a *Legendre transformation*[*]. In particular

$$d(U^\alpha - TS^\alpha) = -S^\alpha dT - PdV^\alpha + \sum_i \mu_i^\alpha dn_i^\alpha \qquad 1.28.2$$

$$d(U^\alpha + PV^\alpha) = TdS^\alpha + V^\alpha dP + \sum_i \mu_i^\alpha dn_i^\alpha \qquad 1.28.3$$

$$d(U^\alpha - TS^\alpha + PV^\alpha) = -S^\alpha dT + V^\alpha dP + \sum_i \mu_i^\alpha dn_i^\alpha. \qquad 1.28.4$$

Since these formulae are all homogeneous of first degree in U^α, S^α, V^α, n_i^α, it follows by Euler's theorem that

$$U^\alpha - TS^\alpha + PV^\alpha = \sum_i n_i^\alpha \mu_i^\alpha. \qquad 1.28.5$$

The quantity $U^\alpha - TS^\alpha$ on the left of (2) is called the *Helmholtz function* and will be denoted by F^α. The quantity $U^\alpha + PV^\alpha$ on the left of (3) is denoted by H^α and is called *enthalpy*; this name was first proposed by Kamerlingh Onnes[†]. The quantity $U^\alpha - TS^\alpha + PV^\alpha$ on the left of (4) is called the *Gibbs function* and is denoted by G^α. Using this notation we have the set of formulae

$$dU^\alpha = TdS^\alpha - PdV^\alpha + \sum_i \mu_i^\alpha dn_i^\alpha \qquad 1.28.6$$

---

[*] Courant and Hilbert, Methoden der Mathematischen Physik, Springer 1937 §1.6.
[†] Porter, Trans. Faraday Soc. 1922 **18** 140.

$$dF^\alpha = -S^\alpha dT - P dV^\alpha + \sum_i \mu_i^\alpha dn_i^\alpha \qquad 1.28.7$$

$$dH^\alpha = T dS^\alpha + V^\alpha dP + \sum_i \mu_i^\alpha dn_i^\alpha \qquad 1.28.8$$

$$dG^\alpha = -S^\alpha dT + V^\alpha dP + \sum_i \mu_i^\alpha dn_i^\alpha \qquad 1.28.9$$

$$G^\alpha = \sum_i n_i^\alpha \mu_i^\alpha. \qquad 1.28.10$$

Whereas U^α is a characteristic function for the independent variables S^α, V^α, n_i^α, we see that F^α is a characteristic function for T, V^α, n_i^α; so is H^α for S^α, P, n_i^α, and so is G^α for T, P, n_i^α.

By comparison of (9) with (1.26.1) or (10) with (1.26.3) we see that

$$\mu_i^\alpha = G_i^\alpha. \qquad 1.28.11$$

Thus in each phase the chemical potential of each species i is equal to the partial Gibbs function of this species.

The equations (6) to (9) are called *fundamental equations* for the four sets of variables S, V, n_i; T, V, n_i; S, P, n_i; T, P, n_i. The four characteristic functions U, F, H, and G were introduced by Gibbs who denoted them by ε, ψ, χ, and ζ respectively.

The characteristic functions U, F, H, and G are sufficient for all requirements. They are however not the only possible ones. For example by simple transformation of (6) we have

$$dS^\alpha = T^{-1} dU^\alpha + T^{-1} P dV^\alpha - T^{-1} \sum_i \mu_i^\alpha dn_i^\alpha \qquad 1.28.12$$

showing that S^α is a characteristic function for the variables U^α, V^α, n_i^α. Again let us define two new quantities J and Y by

$$J = S - U/T = -F/T \qquad 1.28.13$$

$$Y = S - U/T - PV/T = -G/T. \qquad 1.28.14$$

We now differentiate (13) and (14) and substitute for dS^α from (12) obtaining

$$dJ^\alpha = T^{-2} U^\alpha dT + T^{-1} P dV^\alpha - T^{-1} \sum_i \mu_i dn_i^\alpha \qquad 1.28.15$$

$$dY^\alpha = T^{-2} H^\alpha dT - T^{-1} V^\alpha dP - T^{-1} \sum_i \mu_i dn_i^\alpha \qquad 1.28.16$$

from which we see that J^α, like F^α, is a *characteristic function* for the variables T, V^α, n_i^α, and that Y^α, like G^α, is one for the variables T, P, n_i^α.

The functions J and Y were introduced by Massieu* (1869) and the latter was widely used by Planck[†]. We accordingly call J the *Massieu function* and Y the *Planck function*. It is interesting to note that these characteristic functions are six years older than $\mathcal{F}$ and G introduced by Gibbs.

By means of a *fundamental equation* all the thermodynamic functions can be expressed in terms of the characteristic function and its derivatives with respect to the corresponding independent variables. For example choosing $G^\alpha(T, P, n_i^\alpha)$ we obtain directly from (4)

$$S^\alpha = -\partial G^\alpha/\partial T \qquad 1.28.17$$

$$H^\alpha = G^\alpha - T\partial G^\alpha/\partial T \qquad 1.28.18$$

$$V^\alpha = \partial G^\alpha/\partial P \qquad 1.28.19$$

$$U^\alpha = G^\alpha - T\partial G^\alpha/\partial T - P\partial G^\alpha/\partial P \qquad 1.28.20$$

$$\mu_i^\alpha = \partial G^\alpha/\partial n_i^\alpha \qquad 1.28.21$$

$$\partial \mu_i^\alpha/\partial T = \partial^2 G^\alpha/\partial n_i^\alpha \partial T = \partial^2 G^\alpha/\partial T \partial n_i^\alpha = -\partial S^\alpha/\partial n_i^\alpha = -S_i^\alpha \qquad 1.28.22$$

$$\partial \mu_i^\alpha/\partial P = \partial^2 G^\alpha/\partial n_i^\alpha \partial P = \partial^2 G^\alpha/\partial P \partial n_i^\alpha = \partial V^\alpha/\partial n_i^\alpha = V_i^\alpha \qquad 1.28.23$$

$$\partial(\mu_i^\alpha/T)/\partial T = -S_i^\alpha/T - \mu_i^\alpha/T^2 = -H_i^\alpha/T^2. \qquad 1.28.24$$

§1.29 Mole fractions

We are often interested only in the intensive properties of a phase and not at all in the amount of the phase. It is then convenient to describe the phase entirely by intensive variables. The set of variables commonly used is T, P, x_i where x_i denotes the *mole fraction* defined by

$$x_i = n_i / \sum_k n_k \qquad 1.29.1$$

where Σ_k denotes summation over all the species.

By definition the mole fractions satisfy the identity

$$\sum_i x_i = 1. \qquad 1.29.2$$

If the number of independent species or *components* is c, then of the c+2 quantities T, P, x_i used to describe the state of the phase, apart from its amount, only c+1 are independent owing to (2). We therefore say that a single phase of c components has c+1 *degrees of freedom*.

* Massieu, C.R. Acad. Sci., Paris 1869 **69** 858.
[†] Planck, Treatise on Thermodynamics, translated by Ogg, Longmans, 3rd ed. 1927.

§1.30 Gibbs–Duhem relation

We may, if we choose, describe the state of a single phase α, apart from its size, by the set of intensive quantities T, P, μ_i^α. The number of these is $c+2$. We have however seen that the number of degrees of freedom of a single phase is only $c+1$. It follows that T, P, μ_i^α cannot be independently variable, but there must be some relation between them corresponding to the identity between mole fractions. We shall now derive such a relation.

We differentiate (1.28.10) and obtain

$$dG^\alpha = \sum_i \mu_i^\alpha dn_i^\alpha + \sum_i n_i^\alpha d\mu_i^\alpha. \qquad 1.30.1$$

From (1) we subtract (1.28.9) and obtain

$$S^\alpha dT - V^\alpha dP + \sum_i n_i^\alpha d\mu_i^\alpha = 0. \qquad 1.30.2$$

This is the sought relation between T, P, and the μ_i^α's. It is known as the *Gibbs–Duhem relation**. It is particularly useful in its application to changes at constant temperature and pressure, when it may be written

$$\sum_i x_i^\alpha d\mu_i^\alpha = 0 \qquad (T, P \text{ const.}). \qquad 1.30.3$$

§1.31 Multiphase systems

In the preceding sections most of the formulae have been written explicitly for a single phase. Corresponding formulae for a system Σ consisting of several phases are obtained by summation over all the phases. In particular from the fundamental equations in §1.28 we obtain

$$dU^\Sigma = \sum_\alpha T^\alpha dS^\alpha - \sum_\alpha P^\alpha dV^\alpha + \sum_\alpha \sum_i \mu_i^\alpha dn_i^\alpha \qquad 1.31.1$$

$$d\mathscr{F}^\Sigma = -\sum_\alpha S^\alpha dT^\alpha - \sum_\alpha P^\alpha dV^\alpha + \sum_\alpha \sum_i \mu_i^\alpha dn_i^\alpha \qquad 1.31.2$$

$$dH^\Sigma = \sum_\alpha T^\alpha dS^\alpha + \sum_\alpha V^\alpha dP^\alpha + \sum_\alpha \sum_i \mu_i^\alpha dn_i^\alpha \qquad 1.31.3$$

$$dG^\Sigma = -\sum_\alpha S^\alpha dT^\alpha + \sum_\alpha V^\alpha dP^\alpha + \sum_\alpha \sum_i \mu_i^\alpha dn_i^\alpha \qquad 1.31.4$$

where Σ_i denotes summation over the components and Σ_α denotes summation over the phases.

---

* Gibbs, Collected Works, Longmans, vol. **1**, p. 88; Duhem, Le Potentiel Thermodynamique et ses Applications 1886, p. 33. The reference given by Hildebrand and Scott, Solubility of Nonelectrolytes, Reinhold 1950, is spurious.

We are still postulating the absence of chemical reactions. This restriction will be removed in §1.43 and §1.44.

§1.32 *Adiabatic changes in closed system*

We recall that for any infinitesimal change in a closed system Σ

$$dU^\Sigma = w + q. \qquad 1.32.1$$

If the change is *adiabatic*, then by definition

$$q = 0, \quad dU^\Sigma = w \quad \text{(adiabatic)}. \qquad 1.32.2$$

All infinitesimal adiabatic changes can moreover, according to the definitions in §1.13 and §1.19, be classified as follows:

$$dU^\Sigma = w, \quad dS^\Sigma > 0 \quad \text{(natural adiabatic)} \qquad 1.32.3$$

$$dU^\Sigma = w, \quad dS^\Sigma = 0 \quad \text{(reversible adiabatic)}. \qquad 1.32.4$$

Suppose now that the whole system is enclosed by fixed rigid walls, so that $w = 0$. We then have the classification

$$dU^\Sigma = 0 \quad dV^\Sigma = 0 \quad dS^\Sigma > 0 \quad \text{(natural adiabatic)} \qquad 1.32.5$$

$$dU^\Sigma = 0 \quad dV^\Sigma = 0 \quad dS^\Sigma = 0 \quad \text{(reversible adiabatic)}. \qquad 1.32.6$$

Suppose now, instead, that each phase α is partly bounded by a piston acting against a constant pressure P^α, so that

$$w = -\sum_\alpha P^\alpha dV^\alpha = -\sum_\alpha d(P^\alpha V^\alpha) = -d(\sum_\alpha P^\alpha V^\alpha). \qquad 1.32.7$$

Then we have

$$dU^\Sigma = -d(\sum_\alpha P^\alpha V^\alpha) \qquad 1.32.8$$

$$dH^\Sigma = d(U^\Sigma + \sum_\alpha P^\alpha V^\alpha) = 0. \qquad 1.32.9$$

Consequently in this case we have the classification

$$dH^\Sigma = 0 \quad dP^\alpha = 0 \quad dS^\Sigma > 0 \quad \text{(natural adiabatic)} \qquad 1.32.10$$

$$dH^\Sigma = 0 \quad dP^\alpha = 0 \quad dS^\Sigma = 0 \quad \text{(reversible adiabatic)}. \qquad 1.32.11$$

§1.33 *Isothermal changes in closed systems*

Instead of a thermally insulated system, let us now consider a system whose temperature T is kept uniform and constant. This may be achieved by keeping the system in a temperature bath at the temperature T. Then according to the properties of entropy expounded in §1.19 and in particular formula (1.19.11) we have the classification of infinitesimal changes

$$dT=0 \quad d(TS)^\Sigma > q \quad \text{(natural isothermal)} \qquad 1.33.1$$

$$dT=0 \quad d(TS)^\Sigma = q \quad \text{(reversible isothermal)}. \qquad 1.33.2$$

We also have according to the first law of thermodynamics, in particular formula (1.10.7),

$$q = dU^\Sigma - w. \qquad 1.33.3$$

Substituting from (3) into (1) and (2) in turn we obtain

$$dT=0 \quad w > d\mathcal{F}^\Sigma \quad \text{(natural isothermal)} \qquad 1.33.4$$

$$dT=0 \quad w = d\mathcal{F}^\Sigma \quad \text{(reversible isothermal)}. \qquad 1.33.5$$

In particular if the system is enclosed by fixed rigid walls, so that $w=0$, the classification becomes

$$dT=0 \quad dV=0 \quad d\mathcal{F}^\Sigma < 0 \quad \text{(natural isothermal)} \qquad 1.33.6$$

$$dT=0 \quad dV=0 \quad d\mathcal{F}^\Sigma = 0 \quad \text{(reversible isothermal)}. \qquad 1.33.7$$

If on the other hand each phase α is partly bounded by a piston acting against a constant pressure P^α, then

$$w = -\sum_\alpha P^\alpha dV^\alpha = -\sum_\alpha d(P^\alpha V^\alpha) = -d(\sum_\alpha P^\alpha V^\alpha)$$

$$= d(\mathcal{F} - G)^\Sigma \qquad 1.33.8$$

from the definition of G. Substituting from (8) into (4) and (5), we obtain

$$dT=0 \quad dP^\alpha = 0 \quad dG^\Sigma < 0 \quad \text{(natural isothermal)} \qquad 1.33.9$$

$$dT=0 \quad dP^\alpha = 0 \quad dG^\Sigma = 0 \quad \text{(reversible isothermal)}. \qquad 1.33.10$$

§1.34 Equilibrium conditions. General form

We saw in §1.14 that if a system is in complete equilibrium then any conceivable change in it must be reversible. This enables us to put the conditions for equilibrium into various forms each of general validity.

If we first consider an infinitesimal change at constant volume, the system being thermally insulated, we have according to (1.32.6) the equilibrium conditions

$$dS^\Sigma = 0 \quad dV^\Sigma = 0 \quad dU^\Sigma = 0. \qquad 1.34.1$$

If instead we consider an infinitesimal change keeping each phase α at constant pressure P^{α}, the whole system being thermally insulated, we have according to (1.32.11) the equilibrium conditions

$$dS^{\Sigma}=0 \quad dP^{\alpha}=0 \quad dH^{\Sigma}=0. \qquad 1.34.2$$

Thirdly let us consider an infinitesimal change at constant volume and constant uniform temperature (isothermal change). We now have according to (1.33.7) the equilibrium conditions

$$dT=0 \quad dV^{\Sigma}=0 \quad d\mathcal{F}^{\Sigma}=0. \qquad 1.34.3$$

Lastly by considering an infinitesimal change keeping each phase at a constant pressure P^{α} and a constant uniform temperature T, we have according to (1.33.10) the equilibrium conditions

$$dT=0 \quad dP^{\alpha}=0 \quad dG^{\Sigma}=0. \qquad 1.34.4$$

Any one of the four sets of equilibrium conditions (1), (2), (3), (4) is sufficient by itself. They are all equivalent and each has an equal claim to be regarded as fundamental.

§1.35 *Stability and metastability*

In order to make clear what is meant by stability and instability in thermodynamic systems, we shall first discuss the significance of these expressions in a purely mechanical system. To this end, in figure 1.1 are shown in

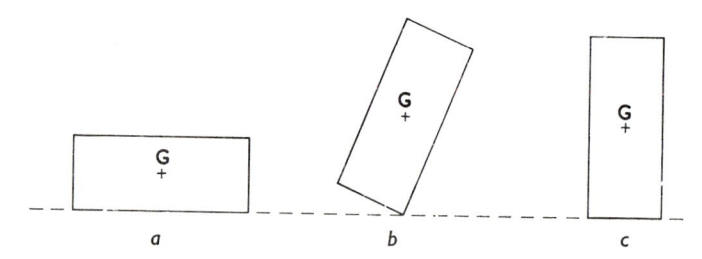

Fig. 1.1. Stable and unstable equilibrium

section three different equilibrium positions of a box on a stand. In positions *a* and *c* the centre of gravity G is lower than in any infinitesimally distant position, consistent with the box resting on the stand; the gravitational potential energy is a minimum, and the equilibrium is stable. If the position of the box be very slightly disturbed, it will of itself return to its former position. In position *b*, on the other hand, the centre of gravity G is higher than in any infinitely near position, consistent with the box resting on the stand, the gravitational potential energy is a maximum, and the equilibrium is unstable. If the position of the box be very slightly disturbed, it will of itself move right away from its original position, and finally settle in some state of stable equilibrium such as *a* or *c*. As maxima and minima of the potential energy must alternate, so must positions of stable and of unstable equilibrium. Only stable equilibria are realizable in practice since the realization of an unstable equilibrium requires the complete absence of any possible disturbing factors.

Whereas positions *a* and *c* are both stable, one may describe *a* as *more stable* than *c*. Or one may say that *a* is *absolutely stable*, while *c* is *unstable compared to a*. By this is meant that in position *c* the potential energy is less than in any position differing only infinitesimally from *c*, but is greater than the potential energy in position *a*.

Similarly, the equilibrium of a thermodynamic system may be *absolutely stable*. On the other hand it may be stable compared with all states differing only infinitesimally from the given state, but unstable compared with some other state differing finitely from the given state; such states are called *metastable*. Truly *unstable* states analogous to *b* are unrealizable in thermodynamics, just as they are in mechanics.

The fact that all thermodynamic equilibria are stable or metastable, but never unstable, is equivalent to the fact that every natural process proceeds towards an equilibrium state, never away from it. Bearing this in mind and referring to the inequalities (1.32.5), (1.32.10), (1.33.6), and (1.33.9), we obtain the following alternative conditions for equilibrium:

for given U^Σ and V^Σ that S^Σ is a *maximum*   1.35.1

for given H^Σ and P^α's that S^Σ is a *maximum*   1.35.2

for given T and V that F^Σ is a *minimum* or that J^Σ is a *maximum*   1.35.3

for given T and P^α's that G^Σ is a *minimum* or that Y^Σ is a *maximum*.   1.35.4

Since $(\partial U/\partial S)_V = (\partial H/\partial S)_P = T > 0$, we may replace the first two conditions above by two others so as to obtain the more symmetrical set of equivalent conditions

for given S^Σ and V^Σ that U^Σ is a *minimum*   1.35.5

for given S^Σ and P^α's that H^Σ is a *minimum*   1.35.6

for given T and V^Σ that F^Σ is a *minimum*   1.35.7

for given T and P^α's that G^Σ is a *minimum*.   1.35.8

Since T is a more convenient independent variable than S, the last two conditions are more useful, but nowise more fundamental, than the previous two.

Each of the above is the condition for stable equilibrium or for metastable equilibrium according as the minimum (or maximum) is absolute or only relative to neighbouring states.

§1.36 *Thermal internal stability*

Consider a closed single phase. Let its entropy be S, its volume V, and its energy U. Imagine one half of the mass of this phase to change so as to have an entropy $\frac{1}{2}(S+\delta S)$ and volume $\frac{1}{2}V$ while the other half changes so as to have an entropy $\frac{1}{2}(S-\delta S)$ and volume $\frac{1}{2}V$. According to Taylor's expansion the energy of the first half becomes

$$\tfrac{1}{2}\{U+(\partial U/\partial S)\delta S+\tfrac{1}{2}(\partial^2 U/\partial S^2)(\delta S)^2\} \qquad 1.36.1$$

when we neglect small quantities of third and higher orders; all partial differentiations in (1) are at constant V. The energy of the second half becomes similarly

$$\tfrac{1}{2}\{U-(\partial U/\partial S)\delta S+\tfrac{1}{2}(\partial^2 U/\partial S^2)(\delta S)^2\}. \qquad 1.36.2$$

Hence by addition the energy of the whole system has increased by the second order small quantity

$$\tfrac{1}{2}(\partial^2 U/\partial S^2)_V(\delta S)^2 \qquad 1.36.3$$

while the total entropy and volume remain unchanged. Now a condition for a system to be in *stable* equilibrium is that, for given values of the entropy and the volume, the energy should be a *minimum*. If then the original state of the system was *stable*, the change considered must lead to an *increase* of energy and the expression (3) must be positive. Hence we obtain as a necessary condition for stable equilibrium

$$(\partial^2 U/\partial S^2)_V > 0. \qquad 1.36.4$$

Since according to (1.17.6)

$$(\partial U/\partial S)_V = T \qquad 1.36.5$$

we can replace (4) by

$$(\partial S/\partial T)_V > 0. \qquad 1.36.6$$

The physical meaning of (6) is that when at constant volume heat is absorbed by a stable phase its temperature is raised.

§1.37 *Hydrostatic equilibrium*

Consider a system Σ of several phases in equilibrium at the temperature T. Suppose the phase α to increase in volume by an amount dV^α and the phase β to decrease by the same amount, the temperature and volume of the whole system and the composition of each phase remaining unchanged. Then, according to (1.34.3), the condition for equilibrium is

$$d\mathscr{F}^\Sigma = d\mathscr{F}^\alpha + d\mathscr{F}^\beta = 0 \qquad 1.37.1$$

or by using (1.28.7)

$$-P^\alpha dV^\alpha + P^\beta dV^\alpha = 0 \qquad 1.37.2$$

and so

$$P^\alpha = P^\beta. \qquad 1.37.3$$

That is to say that any two phases in hydrostatic equilibrium must be at the same pressure.

If we now consider two phases at the same temperature T and different pressures P^α and P^β, there will then not be hydrostatic equilibrium. There will be a tendency for the system to approach hydrostatic equilibrium by a change in which the volume of one phase, say α, increases by dV^α and that of the other phase β decreases by the same amount. Such a change is by definition a natural one. If we keep the temperature constant, we therefore have, according to (1.33.6)

$$d\mathscr{F}^\alpha + d\mathscr{F}^\beta < 0 \qquad 1.37.4$$

or using (1.28.7)

$$-P^\alpha dV^\alpha + P^\beta dV^\alpha < 0. \qquad 1.37.5$$

If we suppose dV^α to be positive, it follows that

$$P^\alpha > P^\beta. \qquad 1.37.6$$

That is to say, that the phase α with the greater pressure P^α will increase in volume at the expense of the phase β with the smaller pressure P^β.

§1.38 *Hydrostatic internal stability*

Consider again a closed single phase. Let its temperature be T, its volume V, its energy U, and its entropy S. Imagine half of the phase to change so as to have a volume $\frac{1}{2}(V+\delta V)$, and the other half to change so as to have a volume $\frac{1}{2}(V-\delta V)$, the temperature remaining uniform and unchanged. Then by an argument precisely analogous to that of the §1.36 we find that the Helmholtz function of the whole system has increased by the second order small quantity

$$\tfrac{1}{2}(\partial^2 \mathcal{F}/\partial V^2)_T (\delta V)^2 \qquad 1.38.1$$

while the temperature and total volume are unchanged. Now a condition for a system to be in *stable* equilibrium is that for given values of temperature and volume, the Helmholtz function should be a *minimum*. If then the original state of the system was *stable*, the change considered must lead to an *increase* of the Helmholtz function and the expression (1) must be positive. Hence we obtain as a necessary condition for stable equilibrium

$$(\partial^2 \mathcal{F}/\partial V^2)_T > 0. \qquad 1.38.2$$

Since according to (1.28.7)

$$(\partial \mathcal{F}/\partial V)_T = -P \qquad 1.38.3$$

we can replace (2) by

$$(\partial V/\partial P)_T < 0. \qquad 1.38.4$$

This means that when the pressure of a stable phase is increased, the volume must decrease.

§1.39 *Equilibrium distribution between phases*

Consider a system of several phases, all at the same temperature T, but not necessarily at the same pressure. Suppose a small amount dn_i^α of the species i to pass from the phase β to the phase α, the temperature of the whole system being kept constant. Then according to (1.31.2) we have

$$d\mathcal{F}^\Sigma = -\sum_\gamma P^\gamma dV^\gamma + \mu_i^\alpha dn_i^\alpha - \mu_i^\beta dn_i^\alpha \qquad 1.39.1$$

omitting the terms which obviously vanish. Since the total work w done on the whole system is $-\Sigma_\gamma P^\gamma dV^\gamma$, it follows from (1.33.4) that the process considered will be a natural one if

$$d\mathcal{F}^\Sigma < -\sum_\gamma P^\gamma dV^\gamma \qquad \text{(natural process)}. \qquad 1.39.2$$

Comparing (1) with (2) we obtain

$$(\mu_i^\alpha - \mu_i^\beta) dn_i^\alpha < 0 \qquad \text{(natural process)}. \qquad 1.39.3$$

Thus dn_i^α in a natural process has the same sign as $\mu_i^\beta - \mu_i^\alpha$. In other words each chemical species i tends to move from a phase where its potential μ_i is higher to another phase in which its potential is lower. Hence the name *potential* or *chemical potential* for μ_i.

If, instead of natural processes, we consider reversible processes we have equalities instead of inequalities; in particular instead of (3) we have

$$(\mu_i^\alpha - \mu_i^\beta)dn_i^\alpha = 0 \qquad \text{(reversible process)} \qquad 1.39.4$$

or

$$\mu_i^\alpha = \mu_i^\beta \qquad \text{(equilibrium)}. \qquad 1.39.5$$

We have obtained the important result that the condition for two phases to be in equilibrium with respect to any species is that the chemical potential of that species should have the same value in the two phases.

§1.40 Phase stability

Consider again a closed single phase. Let its temperature be T, its pressure P, and its Gibbs function G. Imagine the amount of the component i to increase in one half of the phase from $\frac{1}{2}n_i$ to $\frac{1}{2}(n_i + \delta n_i)$ while the amount in the other half of the phase changes from $\frac{1}{2}n_i$ to $\frac{1}{2}(n_i - \delta n_i)$, the temperature and pressure remaining uniform and unchanged. Then by an argument precisely analogous to those of §1.36 and §1.38 we find that the Gibbs function of the whole system has increased by the second order small quantity

$$\tfrac{1}{2}(\partial^2 G/\partial n_i^2)_{T, P, n_j}(\delta n_i)^2 \qquad (j \neq i) \qquad 1.40.1$$

while the temperature and pressure remain unchanged. Now a condition for a system to be in *stable* equilibrium is that for given values of temperature and pressure, the Gibbs function should be a *minimum*. If then the original state of the system was *stable*, the change considered must lead to an *increase* of the Gibbs function and the expression (1) must be positive. Hence we obtain as a necessary condition for stable equilibrium

$$(\partial^2 G/\partial n_i^2)_{T, P, n_j}(\delta n_i)^2 > 0 \qquad (j \neq i). \qquad 1.40.2$$

Since according to (1.28.9)

$$(\partial G/\partial n_i)_{T, P, n_j} = \mu_i \qquad (j \neq i) \qquad 1.40.3$$

we can replace (2) by

$$(\partial \mu_i/\partial n_i)_{T, P, n_j} > 0 \qquad (j \neq i). \qquad 1.40.4$$

This means that when substance i is added to a stable mixed phase the chemical potential of i is increased.

§1.41 Gibbs' phase rule

In §1.30 we mentioned that the state of a single phase α containing c independent species or *components* can, apart from its size, be completely

34   FUNDAMENTAL PRINCIPLES

described by the $c+2$ quantities $T, P, \mu_1, \mu_2, \ldots \mu_c$. Of these $c+2$ quantities only $c+1$ are independent because of the Gibbs–Duhem relation

$$S^\alpha dT - V^\alpha dP + \sum_i n_i^\alpha d\mu_i = 0. \qquad 1.41.1$$

We accordingly say that a single phase has $c+1$ *degrees of freedom*.

We shall now extend this rule to a system of c components in p phases in mutual equilibrium. We continue to use the same $c+2$ variables but there are now p Gibbs–Duhem relations, one for each phase. Consequently the number of independent variables or the number of *degrees of freedom* is $c-p+2$. This is Gibbs' phase rule*.

We have implicitly excluded chemical reaction between the species and we postpone discussion of the effect of any such complication to chapter 6.

§1.42   *Membrane equilibrium*

It is important to notice that, provided a system is at a uniform temperature, the condition for equilibrium between two phases of each chemical species is independent of that for other species and of that for hydrostatic equilibrium. If then two phases α and β are separated by a fixed wall permeable to some components *i* but not to other components *j*, the condition for the two phases to be in equilibrium as regards *i* is still

$$\mu_i^\alpha = \mu_i^\beta \qquad 1.42.1$$

but in this case in general

$$P^\alpha \neq P^\beta \qquad \mu_j^\alpha \neq \mu_j^\beta. \qquad 1.42.2$$

Such a partial equilibrium is called a *membrane equilibrium*.

§1.43   *Chemical reactions. Frozen equilibrium*

Hitherto we have explicitly excluded chemically reacting species from the system considered. We shall now explain how this restriction can be removed.

Owing to the slowness of attainment of some chemical equilibria, it can happen that the change towards chemical equilibrium is negligible during a time sufficient for other kinds of equilibrium to be observed and measured. In other cases the attainment of chemical equilibrium if not sufficiently slow for this to be the case can be made so by the addition to the system of a small quantity of a substance called an *anticatalyst* or merely by rigid

---

* Gibbs, Collected Works, Longmans, vol. 1, p. 96.

exclusion of all traces of some other substance called a *catalyst*. Even in cases where the attainment of chemical equilibrium cannot be adequately slowed down in practice it is possible and legitimate to consider the hypothetical case wherein this has been achieved.

We are thus led to consider a system not in chemical equilibrium in which however the chemical reactions leading towards its attainment have been virtually suppressed. The system is then in a special kind of metastable equilibrium called *frozen equilibrium*. The several chemical species present are then virtually independent and so we can suppose a chemical potential μ assigned to each such species.

If we now suppose the addition of a suitable catalyst so as to *thaw* the *frozen equilibrium* then generally changes of composition will take place as a result of chemical reactions; such changes are of course natural processes. In the special case that the state of frozen equilibrium corresponds to complete chemical equilibrium, then no chemical change will take place on thawing. If we imagine a virtual chemical change to take place, such a change will then be a typical reversible change. If we write down the condition for this, we therefore obtain a relation between the μ's which is a condition of chemical equilibrium.

The final result may be described as follows. Instead of choosing a set of independent chemical species or *components*, we use the set of all the chemical species present whether independent or not and then obtain restrictive relations on their behaviour. The actual form of these restrictive relations will be obtained in the next section.

§1.44 Chemical equilibrium. Affinity

We consider a system of any number of phases maintained at a constant temperature T and constant pressure P. Then according to (1.31.4)

$$dG^\Sigma = \sum_\alpha \sum_i \mu_i^\alpha dn_i^\alpha \qquad (T, P \text{ const.}) \qquad 1.44.1$$

where now, in contrast to previous practice, the species i are no longer all incapable of interacting chemically. According to (1.33.9) the condition for a natural process is

$$dG^\Sigma < 0 \qquad (T, P \text{ constant}) \qquad (\text{natural process}). \qquad 1.44.2$$

Combining (1) and (2) we obtain as the condition for a natural process

$$\sum_\alpha \sum_i \mu_i^\alpha dn_i^\alpha < 0 \qquad (\text{natural process}). \qquad 1.44.3$$

Any chemical reaction at a given temperature and pressure is described by a formula. As typical examples we quote

$$CaCO_3(s) \to CaO(s) + CO_2(g)$$
$$N_2(g) + 3H_2(g) \to 2NH_3(g)$$
$$\alpha\text{-glucose(aq)} \to \beta\text{-glucose(aq)}$$

where (s) denotes a solid phase, (g) the gaseous phase, and (aq) denotes an aqueous solution.

We can represent the most general chemical reaction symbolically by

$$\sum v_A A \to \sum v_B B \qquad 1.44.4$$

meaning that v_A moles of A and the like react together to give v_B moles of B and the like. The unit of quantity the *mole* is defined in such a way that the *stoichiometric numbers* v can all be small integers. The symbols A and B are supposed to specify not only the kind of each chemical species i but also in what phase it is present; in other words the label A implies the pair of labels i and α.

Now imagine the chemical process (4) to take place in the time dt to the extent

$$\sum v_A \mathrm{d}\xi A \to \sum v_B \mathrm{d}\xi B \qquad 1.44.5$$

where dξ denotes a small number. Then dξ/dt is called the *reaction rate* and the dn_i^α, when i relates to A, is just $-v_A \mathrm{d}\xi$. The inequality (3) thus becomes

$$\sum v_B \mu_B \mathrm{d}\xi/\mathrm{d}t < \sum v_A \mu_A \mathrm{d}\xi/\mathrm{d}t \quad \text{(natural process)} \qquad 1.44.6$$

or if we assume dξ/dt > 0

$$\sum v_B \mu_B < \sum v_A \mu_A \quad \text{(natural process)}. \qquad 1.44.7$$

Thus the chemical reaction can in fact take place from left to right only if the inequality (7) holds, and conversely.

If we replace the inequalities by equalities we obtain as the condition for the chemical change in either direction to be a reversible process

$$\sum v_A \mu_A = \sum v_B \mu_B \quad \text{(reversible process)}. \qquad 1.44.8$$

In other words the condition for equilibrium with respect to the chemical process (5) is

$$\sum v_A \mu_A = \sum v_B \mu_B \quad \text{(equilibrium)}. \qquad 1.44.9$$

It is convenient for the sake of brevity and elegance to modify our notation relating to chemical changes. We begin by rewriting the chemical reactions quoted at the beginning of this section

$$0 = CaO(s) + CO_2(g) - CaCO_3(s)$$
$$0 = 2NH_3(g) - N_2(g) - 3H_2(g)$$
$$0 = \beta\text{-glucose}(aq) - \alpha\text{-glucose}(aq)$$

and generally in place of (4) we label the reaction by

$$0 = \sum_B \nu_B B \qquad 1.44.10$$

where ν_B has negative values for the species previously denoted by A. Suppose that in a small interval of time dt the reaction takes place to the extent

$$0 = \sum_B \nu_B \, d\xi B \qquad 1.44.11$$

then the *reaction rate* is defined by $d\xi/dt$. The inequality (3) now becomes

$$\sum_B \nu_B \mu_B \, d\xi/dt < 0 \qquad \text{(natural process)}. \qquad 1.44.12$$

Thus for the reaction to take place $-\Sigma_B \nu_B \mu_B$ and $d\xi/dt$ must have the same sign. The sum $-\Sigma_B \nu_B \mu_B$ is called the *affinity* of the reaction.

It follows immediately that the condition for equilibrium is that the affinity should be zero, that is to say

$$\sum_B \nu_B \mu_B = 0 \qquad \text{(equilibrium)}. \qquad 1.44.13$$

We may combine the inequality (12) and the equation (13) into the single formula

$$\left(-\sum_B \nu_B \mu_B\right) d\xi/dt \geq 0. \qquad 1.44.14$$

The affinity is formally related to the several characteristic functions by

$$-\sum_B \nu_B \mu_B = -(\partial U^\Sigma/\partial \xi)_{S^\Sigma, V^\Sigma} = -(\partial H^\Sigma/\partial \xi)_{S^\Sigma, P}$$
$$= -(\partial F^\Sigma/\partial \xi)_{T, V^\Sigma} = -(\partial G^\Sigma/\partial \xi)_{T, P}$$
$$= T(\partial J^\Sigma/\partial \xi)_{T, V^\Sigma} = T(\partial Y^\Sigma/\partial \xi)_{T, P}$$
$$= T(\partial S^\Sigma/\partial \xi)_{U^\Sigma, V^\Sigma}. \qquad 1.44.15$$

The *affinity* was thus defined by De Donder* in 1922. Of the several equations in formula (15) the most useful and most used is

---

* For detailed references see Prigogine and Defay, Chemical Thermodynamics, English translation by Everett, Longmans 1954.

$$-\sum_B \nu_B \mu_B = -(\partial G/\partial \xi)_{T,P}. \qquad 1.44.16$$

The affinity, being a linear combination of chemical potentials, is like the chemical potentials an intensive quantity.

A different and better known notation is that introduced by G. N. Lewis*, namely

$$-\sum_B \nu_B \mu_B = -\Delta G \qquad 1.44.17$$

where the operator Δ denotes increase at constant temperature and constant pressure when ξ increases by unity. Both notations have their advantages and both will be used.

§1.45 Choice of independent variables

For practical purposes the most convenient independent variables, other than the composition, to describe any single phase are, usually, temperature and pressure. We shall therefore require to express most thermodynamic properties as functions of T, P and shall be interested in their partial derivatives with respect to T and P. In the case of gases, in contrast to liquids and solids, it is sometimes convenient to choose as independent variables T, V instead of T, P. We shall accordingly also require to express thermodynamic properties as functions of T, V and shall be interested in their partial derivatives with respect to T and V.

§1.46 Thermal expansivity and isothermal compressibility

If we regard the volume of a phase of fixed composition as a function of temperature and pressure, we have

$$dV = (\partial V/\partial T)_P dT + (\partial V/\partial P)_T dP. \qquad 1.46.$$

We define α, the *thermal expansivity*, by

$$\alpha = V^{-1}(\partial V/\partial T)_P \qquad 1.46.2$$

and κ_T, the *isothermal compressibility*, by

$$\kappa_T = -V^{-1}(\partial V/\partial P)_T. \qquad 1.46.3$$

Substituting (2) and (3) into (1) we obtain

---

* See Lewis and Randall, Thermodynamics and the Free Energy of Chemical Substances, McGraw-Hill 1923, p. 226.

FUNDAMENTAL PRINCIPLES

$$dV = \alpha V dT - \kappa_T V dP \qquad 1.46.4$$

$$d \ln V = \alpha dT - \kappa_T dP. \qquad 1.46.5$$

Alternatively if we choose to regard P as a function of T, V, we have

$$dP = \alpha \kappa_T^{-1} dT - \kappa_T^{-1} V^{-1} dV. \qquad 1.46.6$$

From (5) we deduce

$$\partial^2 \ln V / \partial T \partial P = (\partial \alpha / \partial P)_T = -(\partial \kappa_T / \partial T)_P. \qquad 1.46.7$$

§1.47 Maxwell's relations

For a closed phase α in the absence of chemical reactions we have according to (1.28.7) and (1.28.9)

$$d\mathscr{F}^\alpha = -S^\alpha dT - P dV^\alpha \qquad 1.47.1$$

$$dG^\alpha = -S^\alpha dT + V^\alpha dP. \qquad 1.47.2$$

Consequently we have

$$(\partial S^\alpha / \partial V^\alpha)_T = -\partial^2 \mathscr{F}^\alpha / \partial T \partial V^\alpha = (\partial P / \partial T)_{V^\alpha} = \alpha / \kappa_T \qquad 1.47.3$$

$$(\partial S^\alpha / \partial P)_T = -\partial^2 G / \partial T \partial P = -(\partial V^\alpha / \partial T)_P = -\alpha V^\alpha. \qquad 1.47.4$$

These two relations, due to Maxwell[*], are important since they express the dependence of entropy on volume or pressure in terms of the more readily measurable quantities α and κ_T.

§1.48 Dependence of thermodynamic functions on pressure

If, as will usually be our choice, we take as independent variables, other than the composition of each phase, the temperature T and the pressure P the relevant characteristic function is the Gibbs function G^α and according to (1.28.9) we have

$$(\partial G^\alpha / \partial P)_T = V^\alpha. \qquad 1.48.1$$

We also have Maxwell's relation (1.47.4)

$$(\partial S^\alpha / \partial P)_T = -\alpha V^\alpha. \qquad 1.48.2$$

Since the enthalpy H^α is related to G^α and S^α by

$$H^\alpha = G^\alpha + TS^\alpha \qquad 1.48.3$$

we have using (1) and (2)

[*] Maxwell, Theory of Heat, Longmans 1885 ed. p. 169.

$$(\partial H^\alpha/\partial P)_T = (\partial G^\alpha/\partial P)_T + T(\partial S^\alpha/\partial P)_T = V^\alpha(1-\alpha T). \qquad 1.48.4$$

When we use the independent variables T, P the function U is much less important than G, H. If nevertheless we should require its dependence on the pressure, it is readily derived as follows. By definition

$$U^\alpha = H^\alpha - PV^\alpha \qquad 1.48.5$$

and so by differentiation with respect to P at constant T we obtain

$$(\partial U^\alpha/\partial P)_T = (\partial H^\alpha/\partial P)_T - V^\alpha - P(\partial V^\alpha/\partial P)_T = V^\alpha(\kappa P - \alpha T). \qquad 1.48.6$$

§1.49 Gibbs–Helmholtz relation

If, as will usually be our choice, we take as independent variables, other than the composition of each phase, the temperature T and the pressure P we have for the temperature dependence of the relevant characteristic function G^α according to (1.28.9)

$$(\partial G^\alpha/\partial T)_P = -S^\alpha. \qquad 1.49.1$$

If we compare this with the definition of G^α, namely

$$G^\alpha = H^\alpha - TS^\alpha \qquad 1.49.2$$

and eliminate S^α, we obtain

$$H^\alpha = G^\alpha - T(\partial G^\alpha/\partial T)_P. \qquad 1.49.3$$

For a system Σ composed of several phases at the same pressure we obtain from (3) by addition

$$H^\Sigma = G^\Sigma - T(\partial G^\Sigma/\partial T)_P. \qquad 1.49.4$$

If we apply this relation to the final state II and to the initial state I in any isothermal process and take the difference, we obtain

$$\Delta H^\Sigma = \Delta G^\Sigma - T(\partial \Delta G^\Sigma/\partial T)_{P^\mathrm{I},\,P^\mathrm{II}} \qquad 1.49.5$$

where P^I, P^II denote the initial and final pressures respectively. Formula (5) is known as the *Gibbs–Helmholtz relation*. This name is also sometimes given to formula (4).

By simple transformation we can rewrite these formulae as

$$\{\partial(G^\Sigma/T)/\partial T\}_P = -H^\Sigma/T^2 \qquad 1.49.6$$

$$\{\partial(\Delta G^\Sigma/T)/\partial T\}_{P^\mathrm{I},\,P^\mathrm{II}} = -\Delta H^\Sigma/T^2 \qquad 1.49.7$$

or alternatively as

$$\{\partial(G^\Sigma/T)/\partial(1/T)\}_P = H^\Sigma \qquad 1.49.8$$

$$\{\partial(\Delta G^\Sigma/T)/\partial(1/T)\}_{P^{\mathrm{I}},\, P^{\mathrm{II}}} = \Delta H^\Sigma. \qquad 1.49.9$$

§1.50 Dependence of thermodynamic functions on T, V

As already stated, it is usually convenient to take T, P as independent variables. Only in the case of gases is it sometimes convenient to use instead the independent variables T, V. The dependence of the various thermodynamic functions on these variables is readily obtained and we give the chief results for a phase of fixed composition in the order in which they are conveniently derived without however giving details of the derivations.

$$d\mathscr{F}^\alpha = -S^\alpha dT - P dV^\alpha \qquad 1.50.1$$

$$dP = \alpha \kappa^{-1} dT - (\kappa V^\alpha)^{-1} dV^\alpha \qquad 1.50.2$$

$$dS^\alpha = (\partial S^\alpha/\partial T)_V dT + \alpha \kappa^{-1} dV^\alpha \qquad 1.50.3$$

$$dU^\alpha = T(\partial S^\alpha/\partial T)_V dT + (\alpha T \kappa^{-1} - P) dV^\alpha \qquad 1.50.4$$

$$dJ^\alpha = T^{-2} U^\alpha dT + T^{-1} P dV^\alpha. \qquad 1.50.5$$

§1.51 Use of Jacobians

Many thermodynamic identities, including those obtained in the preceding sections, can be obtained rapidly and elegantly by the use of Jacobians. The procedures are due to Shaw*, who has shown how to apply them to obtain a tremendous number of identities, some important, others merely amusing. We shall here give a brief sketch of the method, which we shall illustrate by a few simple examples. We would however emphasize that all the simple and most important relations are deduced in this book without using Jacobians, so that the reader not interested in their use may omit this section which does not affect the rest of the book.

We recall that Jacobians are defined by

$$\frac{\partial(x, y)}{\partial(\alpha, \beta)} \equiv -\frac{\partial(y, x)}{\partial(\alpha, \beta)} \equiv \left(\frac{\partial x}{\partial \alpha}\right)_\beta \left(\frac{\partial y}{\partial \beta}\right)_\alpha - \left(\frac{\partial x}{\partial \beta}\right)_\alpha \left(\frac{\partial y}{\partial \alpha}\right)_\beta \qquad 1.51.1$$

and that they obey the multiplicative law

$$\frac{\partial(x, y)}{\partial(u, v)} \frac{\partial(u, v)}{\partial(\alpha, \beta)} = \frac{\partial(x, y)}{\partial(\alpha, \beta)} \qquad 1.51.2$$

* Shaw, Phil. Trans. Roy. Soc. London A 1935 **234** 299.

which can be derived by simple geometrical or algebraical considerations on transformation of coordinates.

As particular cases of (1) we have

$$\left(\frac{\partial x}{\partial \alpha}\right)_\beta = \frac{\partial(x, \beta)}{\partial(\alpha, \beta)} = -\frac{\partial(\beta, x)}{\partial(\alpha, \beta)} \qquad 1.51.3$$

$$\left(\frac{\partial y}{\partial \beta}\right)_\alpha = \frac{\partial(\alpha, y)}{\partial(\alpha, \beta)} = -\frac{\partial(y, \alpha)}{\partial(\alpha, \beta)}. \qquad 1.51.4$$

Using (3) and (4) we derive from (2)

$$\left(\frac{\partial x}{\partial z}\right)_y = \frac{\partial(x, y)}{\partial(z, y)} = \frac{\partial(x, y)}{\partial(\alpha, \beta)} \bigg/ \frac{\partial(z, y)}{\partial(\alpha, \beta)}. \qquad 1.51.5$$

We now replace α, β by the pair of quantities which we regard as the usually most convenient independent variables, namely the temperature T and the pressure P. We further introduce the following new notation

$$J(x, y) = \frac{\partial(x, y)}{\partial(T, P)} = \left(\frac{\partial x}{\partial T}\right)_P \left(\frac{\partial y}{\partial P}\right)_T - \left(\frac{\partial x}{\partial P}\right)_T \left(\frac{\partial y}{\partial T}\right)_P. \qquad 1.51.6$$

In particular we have

$$\left(\frac{\partial x}{\partial T}\right)_P = J(x, P) = -J(P, x) \qquad 1.51.7$$

$$\left(\frac{\partial x}{\partial P}\right)_T = -J(x, T) = J(T, x). \qquad 1.51.8$$

Using our new notation we have instead of (5)

$$\left(\frac{\partial x}{\partial z}\right)_y = \frac{J(x, y)}{J(z, y)}. \qquad 1.51.9$$

The relations (6) and (9) together enable us to express any quantity of the type $(\partial x/\partial z)_y$ in terms of the partial differential coefficients of x, y, z with respect to T, P.

We shall illustrate by two examples, the first a useful one, the second far fetched. We have

$$\left(\frac{\partial T}{\partial P}\right)_H = \frac{J(T, H)}{J(P, H)} = -\frac{(\partial H/\partial P)_T}{(\partial H/\partial T)_P} \qquad 1.51.10$$

a relation which we shall meet again in §3.20 where it is derived more simply.

We now take a more complicated, and less useful, example:

$$\left(\frac{\partial H}{\partial G}\right)_U = \frac{J(H, U)}{J(G, U)} = \frac{(\partial H/\partial T)_P(\partial U/\partial P)_T - (\partial H/\partial P)_T(\partial U/\partial T)_P}{(\partial G/\partial T)_P(\partial U/\partial P)_T - (\partial G/\partial P)_T(\partial U/\partial T)_P}$$

$$= \frac{CV(\kappa P - \alpha T) - V(1 - \alpha T)(C - \alpha PV)}{-SV(\kappa P - \alpha T) - V(C - \alpha PV)}$$

$$= \frac{C(\kappa P - \alpha T) - (1 - \alpha T)(C - \alpha PV)}{-S(\kappa P - \alpha T) - (C - \alpha PV)} \qquad 1.51.11$$

where we have used formulae (1.48.1), (1.48.4), (1.48.6) and we have denoted $(\partial H/\partial T)_P$ by C.

These illustrative examples by no means exhaust the uses to which Jacobians can be put. The reader who is interested is referred to the original papers by Shaw.

§1.52 Reversible cycles

Suppose a system is taken through a complete cycle of states. Then as its final state is identical with its initial state, its entropy must be the same at the end as at the beginning. Thus

$$\Delta S = 0 \quad \text{(any cycle)}. \qquad 1.52.1$$

If at all stages the system is in equilibrium, so that no irreversible (natural) change takes place, then

$$\Delta S = \sum_i q_i/T_i \quad \text{(reversible changes)} \qquad 1.52.2$$

where q_i denotes the heat absorbed at the temperature T_i and the summation extends over all the temperatures through which the system passes. Substituting (2) into (1) we obtain

$$\sum_i q_i/T_i = 0 \quad \text{(reversible cycle)}. \qquad 1.52.3$$

Evidently, since T_i is always positive, some of the q_i's must be positive and some negative. It is convenient here to modify our notation so as to distinguish between the positive and negative q_i's. We accordingly replace (3) by

$$\sum_r q_r/T_r = \sum_s Q_s/T_s \quad \text{(reversible cycle)} \qquad 1.52.4$$

where each q_r is a positive quantity of heat taken in at the temperature T_r and each Q_s is a positive quantity of heat given out at the temperature T_s.

According to the first law of thermodynamics the work $-w$ done by the system during the cycle is given by

$$-w = \sum_i q_i = \sum_r q_r - \sum_s Q_s. \qquad 1.52.5$$

The ratio η defined by

$$\eta = -w/\sum_r q_r = (\sum_r q_r - \sum_s Q_s)/\sum_r q_r = 1 - \sum_s Q_s/\sum_r q_r \qquad 1.52.6$$

is called by engineers the *thermodynamic efficiency* of the cycle.

Let us suppose that there is a maximum temperature T_{max} and a minimum temperature T_{min}, between which temperatures the cycle is confined. The following question arises. Subject to this restriction on the temperatures, what is the maximum possible value of η? The answer is obviously obtained by making

$$T_r = T_{max} \qquad \text{(all } r) \qquad 1.52.7$$

$$T_s = T_{min} \qquad \text{(all } s). \qquad 1.52.8$$

This means that positive absorption of heat occurs only at the highest temperature T_{max} and positive loss of heat occurs only at the lowest temperature T_{min}. No heat is exchanged with the surroundings at any temperature intermediate between T_{max} and T_{min}. In other words the passages from T_{max} to T_{min} and the reverse are adiabatic. Thus the cycle consists entirely of isothermal absorption of heat at T_{max}, isothermal emission of heat at T_{min}, and adiabatic changes from T_{max} to T_{min} and from T_{min} to T_{max}. Such a cycle was first considered by Carnot* and is called *Carnot's cycle*.

For Carnot's cycle we have by substituting from (7) and (8) into (4)

$$\sum_r q_r/T_{max} = \sum_s Q_s/T_{min} \qquad \text{(Carnot's cycle)}. \qquad 1.52.9$$

Now substituting from (9) into (6) we obtain

$$\eta = 1 - T_{min}/T_{max} \qquad \text{(Carnot's cycle)}. \qquad 1.52.10$$

There is sometimes confusion between Carnot's cycle and reversible cycles. It will be observed that Carnot's cycle is a very special case of a reversible cycle.

A completely isothermal cycle is a special case of Carnot's cycle. For such a cycle

$$T_{max} = T_{min} = T \qquad \text{(isothermal cycle)} \qquad 1.52.11$$

---

* Carnot, Réflexions sur la puissance motrice du feu, Bachelier, Paris, 1824. Reprinted in 1912 by Hermann, Paris and in 1953 by Blanchard, Paris, together with some notes discovered after Carnot's death in 1832 and communicated to the Académie des Sciences in 1878 by Carnot's brother.

$$\sum_r q_r = \sum_s Q_s \qquad \text{(isothermal cycle)} \qquad 1.52.12$$

$$w = 0 \qquad \text{(isothermal cycle)} \qquad 1.52.13$$

$$\eta = 0 \qquad \text{(isothermal cycle).} \qquad 1.52.14$$

Formula (13) is known as *Moutier's theorem**.

We shall have no occasion to make any further reference to cycles. They are important in engineering thermodynamics for the treatment of engines and refrigerators, but these fall outside the subject-matter of this book.

§1.53 *Surface phases*

We have hitherto assumed that every system consists of one or more completely homogeneous phases bounded by sharply defined geometrical surfaces. This is an over-simplification, for the interface between any two phases will rather be a thin layer across which the physical properties vary continuously from those of the interior of one phase to those of the interior of the other. We must now consider the thermodynamic properties of these surface layers between two phases. We shall begin by considering a plane interface and shall in §1.60 extend our considerations to a curved interface.

The following treatment is essentially that of van der Waals junior and Bakker[†]. It is less abstract than the alternative treatment of Gibbs.[‡]

Figure 1.2 represents two homogeneous bulk phases, α and β, between

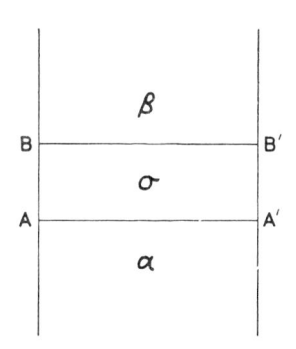

Fig. 1.2. Plane interface between two phases

---

* Moutier, Bulletin de la Société philomathique 1875 Aug. 11th.
† Van der Waals and Bakker, Handb. Experimentalphysik, 1928 vol. **6**. See also Verschaffelt, Bull. Acad. Belg. Cl. Sci. 1936 **22** No. 4, pp. 373, 390, 402; Guggenheim, Trans. Faraday Soc. 1940 **36** 398.
‡ Gibbs, Collected Works, Longmans, vol. **1**, p. 219.

which lies the surface layer σ. The boundary between σ and α is the plane AA', that between σ and β the parallel plane BB'. All properties of σ are uniform in directions parallel to AA', but not in the direction normal to AA'. At and near AA' the properties are identical with those of the phase α; at and near BB' they are identical with those of the phase β. Subject to these conditions there is freedom of choice in placing the planes AA' and BB'. It will be possible and natural though not essential, so to place the planes AA' and BB' that the uniform distance between them is submicroscopic and usually less than 10^{-6} cm, if not less than 10^{-7} cm.

§1.54 *Interfacial tension of plane interface*

Since the surface layer σ is a material system with a well-defined volume and material content, its thermodynamic properties require no special definition. We may speak of its temperature, Helmholtz function, composition, and so on just as for a homogeneous bulk phase. The only functions that call for special comment are the pressure and the interfacial tension. In any homogeneous bulk phase the force across any unit area is equal in all directions and is called the pressure. But in σ the force across unit area is not the same in all directions. If, however, we choose any plane of unit area parallel to AA' and BB', then the force normal to it has the same value for all positions of the plane whether it lie in α, β, or σ; this value of the force normal to unit area is called the pressure P. Suppose, on the other hand, we choose a plane perpendicular to AA' and extending below AA' and above BB'; let this plane have the form of a rectangle of height h (parallel to AB) and of thickness l (perpendicular to the plane of the paper). Then the force across this plane will be equal to $Phl - \gamma l$, where P is the above-defined pressure and γ is called the *interfacial tension*. If the height of this plane is chosen to extend exactly from AA' to BB', then the force across it will be equal to $P\tau l - \gamma l$ if the height AB is denoted by τ. Let the surface layer have an area A and a volume V^σ so that

$$V^\sigma = \tau A. \qquad 1.54.1$$

Suppose the area to be increased to $A + dA$, the thickness to $\tau + d\tau$, and the volume to $V^\sigma + dV^\sigma$, the material content remaining unaltered. Then the work done on σ by the forces across AA' and BB' is $-PA\,d\tau$. The work done by the forces parallel to the planes AA' and BB' is independent of the shape of the perimeter and for the sake of simplicity we may suppose the perimeter to be a rectangle. The work done by the latter forces is then evidently $-(P\tau - \gamma)dA$. The total work done on σ is therefore

$$-PA\,d\tau-(P\tau-\gamma)dA = -P(A\,d\tau+\tau dA)+\gamma dA$$
$$= -P\,dV^\sigma+\gamma dA. \qquad 1.54.2$$

This expression takes the place of $-P\,dV^\alpha$ for a homogeneous bulk phase.

§1.55 Helmholtz function of surface layer

For the most general variation of the Helmholtz function of a homogeneous bulk phase we have the fundamental equation (1.28.7)

$$dF^\alpha = -S^\alpha dT - P\,dV^\alpha + \sum_i \mu_i dn_i^\alpha. \qquad 1.55.1$$

For a surface phase σ the dependence of the Helmholtz function on the temperature and the composition will be exactly analogous to that for a bulk phase; this follows directly from the definitions of entropy and chemical potentials. But for its dependence on size and shape we must replace $-P\,dV^\alpha$ by the expression (1.54.2). We thus obtain the formula

$$dF^\sigma = -S^\sigma dT - P\,dV^\sigma + \gamma dA + \sum_i \mu_i dn_i^\sigma. \qquad 1.55.2$$

There is no need to add superscripts to T, P, μ_i because these must have values uniform throughout α, β, and σ in order that there may be thermal, hydrostatic, and physico-chemical equilibrium.

§1.56 Integrated relation. Gibbs function of surface phase

Since equation (1.55.2) is homogeneous of first degree in F^σ, S^σ, V^σ, A, and n_i^σ it follows by Euler's theorem that

$$F^\sigma + PV^\sigma - \gamma A = \sum_i n_i^\sigma \mu_i. \qquad 1.56.1$$

This formula is the analogue of

$$F^\alpha + PV^\alpha = \sum_i n_i^\alpha \mu_i \qquad 1.56.2$$

for a bulk phase.

In analogy with the definition of the Gibbs function G^α of a bulk phase

$$G^\alpha = U^\alpha - TS^\alpha + PV^\alpha = F^\alpha + PV^\alpha \qquad 1.56.3$$

we now define the Gibbs function G^σ of the surface phase by

$$G^\sigma = U^\sigma - TS^\sigma + PV^\sigma - \gamma A = F^\sigma + PV^\sigma - \gamma A \qquad 1.56.4$$

We deduce from (1.55.1) and (4)

$$dG^\sigma = -S^\sigma dT + V^\sigma dP - A\,d\gamma + \sum_i \mu_i dn_i^\sigma \qquad 1.56.5$$

$$G^\sigma = \sum_i n_i^\sigma \mu_i. \qquad 1.56.6$$

From time to time papers have been published maintaining that μ_i has a value μ_i^σ in the surface differing from its value $\mu_i^\alpha = \mu_i^\beta$ in the two bulk phases. The worst of these papers are sheer nonsense, the best of them merely confused. In the better papers the quantity denoted by μ_i^σ is a different quantity from that denoted by μ_i in the present text which follows Gibbs.

The last two formulae are the analogues of

$$dG^\alpha = -S^\alpha dT + V^\alpha dP + \sum_i \mu_i dn_i^\alpha \qquad 1.56.7$$

$$G^\alpha = \sum_i n_i^\alpha \mu_i \qquad 1.56.8$$

for a bulk phase α. From the above relations it is evident that, as in a bulk phase, the chemical potential μ_i is equal to the partial Gibbs function defined by $(\partial G/\partial n_i)_{T,P,\gamma,n_j}$.

§1.57 *Analogue of Gibbs–Duhem relation*

If we differentiate (1.56.6) we obtain

$$dG^\sigma = \sum_i \mu_i dn_i^\sigma + \sum_i n_i^\sigma d\mu_i \qquad 1.57.1$$

and subtracting (1.56.5) from this

$$S^\sigma dT - V^\sigma dP + A\,d\gamma + \sum_i n_i^\sigma d\mu_i = 0 \qquad 1.57.2$$

which is the analogue for a surface phase of the Gibbs–Duhem relation (1.30.2) for a bulk phase.

If we divide (2) throughout by A we obtain the more convenient form;

$$S_A^\sigma dT - \tau dP + d\gamma + \sum_i \Gamma_i^\sigma d\mu_i = 0 \qquad 1.57.3$$

where S_A^σ denotes S^σ/A and Γ_i denotes the amount of the species i in unit area of the surface phase σ and is thus defined by

$$\Gamma_i = n_i^\sigma/A. \qquad 1.57.4$$

We recall that τ is the thickness of the surface layer, that is to say the length AB in figure 1.2.

§1.58 Invariance of relations

We must now study what happens to the several formulae for surface layers if either of the chosen boundaries is moved in a direction normal to itself. We may regard the volume V^σ of the surface layer as defined in terms of the volume V^Σ of the whole system and the volumes V^α and V^β of the two bulk phases by the equation

$$V^\sigma = V^\Sigma - V^\alpha - V^\beta. \qquad 1.58.1$$

Similarly U^σ, S^σ, and n_i^σ are defined by

$$U^\sigma = U^\Sigma - U^\alpha - U^\beta \qquad 1.58.2$$

$$S^\sigma = S^\Sigma - S^\alpha - S^\beta \qquad 1.58.3$$

$$n_i^\sigma = n_i^\Sigma - n_i^\alpha - n_i^\beta. \qquad 1.58.4$$

If now the geometrical surface AA′ is moved so that V^α is decreased by an amount $V^\alpha \delta^\alpha$ then it is evident that V^σ becomes increased by the same amount $V^\alpha \delta^\alpha$. At the same time U^σ, S^σ, n_i^σ become increased by $U^\alpha \delta^\alpha$, $S^\alpha \delta^\alpha$, $n_i^\alpha \delta^\alpha$. It is readily verified that all the formulae of §1.54 to §1.57 remain unaltered. Exactly the same considerations apply if the geometrical surface BB′ is moved so that V^β is decreased by an amount $V^\beta \delta^\beta$. In particular the value of γA remains invariant and consequently the value of γ remains invariant. We shall see in §1.64 that for a curved surface the value of γ is not invariant.

§1.59 Gibbs geometrical surface

We have hitherto postulated that the inhomogeneous layer is completely confined between the geometrical surfaces AA′ and BB′. This restriction may be removed if we accept the possibility that some of the quantities U^σ, S^σ, n_i^σ may become negative. In particular we may make the two surfaces AA′ and BB′ coincide somewhere inside the inhomogeneous layer. This convention defined by

$$V^\sigma = 0 \qquad 1.59.1$$

was used by Gibbs. It is more elegant but more difficult to visualize than the treatment based on figure 1.2 with $V^\sigma > 0$. The single geometrical surface is called the *Gibbs geometrical surface*.

According to Gibbs' convention formulae (1.55.2), (1.56.1), (1.56.4), (1.56.5), and (1.56.6) reduce to

$$dF^\sigma = -S^\sigma dT + \gamma dA + \sum_i \mu_i dn_i^\sigma \qquad 1.59.2$$

$$U^\sigma - TS^\sigma - \gamma A = \sum_i n_i^\sigma \mu_i \qquad 1.59.3$$

$$G^\sigma = U^\sigma - TS^\sigma - \gamma A \qquad 1.59.4$$

$$dG^\sigma = -S^\sigma dT - A d\gamma + \sum_i \mu_i dn_i^\sigma \qquad 1.59.5$$

$$G^\sigma = \sum_i n_i^\sigma \mu_i \qquad 1.59.6$$

respectively. These formulae and the value of γ are all invariant with respect to the position of the Gibbs geometrical surface.

In the simplest system, namely a single substance existing as liquid + vapour with a planar boundary, it is convenient to place the Gibbs geometrical surface so that $n^\sigma = 0$. Formula (2) then reduces to

$$dF^\sigma = -S dT + \gamma dA. \qquad 1.59.7$$

§1.60 Interfacial tension of curved interface

We must now consider under what conditions the formulae already derived for plane interfaces may be applied to curved interfaces. We shall see that the formulae strictly derived for plane interfaces may be applied to curved interfaces with an accuracy sufficient for experimental purposes provided that the thickness of the inhomogeneous surface layer is small compared with its radii of curvature*.

For the sake of simplicity let us first consider a system consisting of two homogeneous bulk phases α and β connected by a surface layer σ having the form of a circular cylindrical shell. Figure 1.3 shows a cross-section of the phases α and β separated by the surface layer σ, bounded by the circular cylinders AA' and BB' with common axis O. There is complete homogeneity in the direction normal to the diagram. The properties of the surface layer σ are supposed identical at all points the same distance from the axis through O. Throughout the phase α and extending up to AA' there is a uniform pressure P^α; throughout the phase β, and extending down to BB', there is a uniform pressure P^β. Between AA' and BB' the pressure P_r parallel to the radii of the cylinders AA' and BB' varies continuously, but not necessarily monotonically, from the value P^α to the value P^β.

In the previous discussion of plane surfaces it was pointed out that the geometrical planes AA' and BB' may be placed an arbitrary distance apart.

* Guggenheim, Trans. Faraday Soc. 1940 **36** 397.

FUNDAMENTAL PRINCIPLES   51

For the present discussion of curved surfaces it is on the contrary postulated that the circular cylindrical surfaces AA' and BB' should be placed as near together as is consistent with the condition that the inhomogeneous layer be contained between them. According to this condition we may usually expect the distance AB to be about 10^{-7} cm. We shall denote by a distances measured radially from O, and in particular by a_α and a_β, the distances OA and OB respectively.

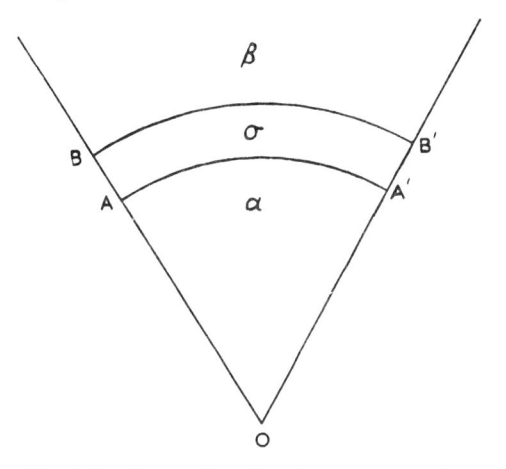

Fig. 1.3. Curved interface between two phases

Whereas the force per unit area across any element of surface inside either homogeneous phase is independent of the orientation of the element (Pascal's law), this is not the case in the inhomogeneous layer σ. It is convenient to denote the force per unit area in the direction parallel to the surface AA' and BB' by $P_r - Q$. Both P_r and Q are functions of a. Q is zero at $a = a_\alpha$ and at $a = a_\beta$, but, at least somewhere between, Q is greater than zero. It is conceivable that Q might be negative somewhere between $a = a_\alpha$ and $a = a_\beta$, but its average value in this range is unquestionably positive.

According to elementary statics the mechanical equilibrium of the matter enclosed by AA'B'B requires that for all values of a

$$d(P_r a) = (P_r - Q) da \qquad 1.60.1$$

or

$$dP_r = -(Q/a) da. \qquad 1.60.2$$

If we integrate (2) from a_β to a_α we obtain

$$P^\alpha - P^\beta = \int_{a_\alpha}^{a_\beta} (Q/a) da. \qquad 1.60.3$$

We now arbitrarily choose any length $\hat{a}$ subject only to the restriction

$$a_\alpha < \hat{a} < a_\beta \qquad 1.60.4$$

and we define a quantity $\hat{\gamma}$ by

$$\hat{\gamma} = \hat{a} \int_{a_\alpha}^{a_\beta} (Q/a) \, da. \qquad 1.60.5$$

From (3) and (5) we deduce

$$P^\alpha - P^\beta = \hat{\gamma}/\hat{a} \quad \text{(circular cylinder).} \qquad 1.60.6$$

For the sake of simplicity we have considered an interface with the form of a circular cylinder. For a spherical interface we find by similar reasoning instead of (6)

$$P^\alpha - P^\beta = 2\hat{\gamma}/\hat{a} \quad \text{(sphere).} \qquad 1.60.7$$

We may call $\hat{\gamma}$ *interfacial tension*, but its exact value depends on the choice of $\hat{a}$. We must now distinguish between the case $a_\beta - a_\alpha \ll a_\alpha$ and the case when this inequality does not hold. In the former case the distinction between a_α, a_β, and $\hat{a}$ is trivial; we may then replace (6) by

$$P^\alpha - P^\beta = \gamma/a \quad \text{(circular cylinder)} \qquad 1.60.8$$

and (7) by

$$P^\alpha - P^\beta = 2\gamma/a \quad \text{(sphere).} \qquad 1.60.9$$

For an interface of other shapes the geometry is somewhat more complicated and the general formula obtained is

$$P^\alpha - P^\beta = \gamma/\varrho_1 + \gamma/\varrho_2 \qquad 1.60.10$$

where ϱ_1, ϱ_2 are the principal radii of curvature of the interface. We shall see in §1.62 how formula (9) is the basis for measuring γ. The quantities measured are $P^\alpha - P^\beta$ and a; the value of γ is then calculated by formula (9). In the contrary case when the inequality $a_\beta - a_\alpha \ll a_\alpha$ does not hold γ can neither be uniquely defined nor accurately measured. A mathematical analysis of this situation is given in §1.64.

§1.61 *Pressure within a bubble*

Let us consider a bubble having the form of a thin spherical film of liquid of internal and external radii a_i and a_e. If P^i denotes the pressure nearer to the centre than the film, P^e the pressure further from the centre than the film, and P' the pressure in the liquid film itself, we have, according to (1.60.9)

$$P^i - P' = 2\gamma/a_i \qquad 1.61.1$$

$$P' - P^e = 2\gamma/a_e \qquad 1.61.2$$

so that

$$P^i - P^e = (2/a_i + 2/a_e)\gamma \qquad 1.61.3$$

or neglecting the difference between a_i and a_e

$$P^i - P^e = 4\gamma/a. \qquad 1.61.4$$

§1.62 Determination of interfacial tension

The commonest method of determining the value of the interfacial tension γ depends on formula (1.60.9). This method is shown diagrammatically in figure 1.4. Two fluid phases α and β are represented, the one shaded the

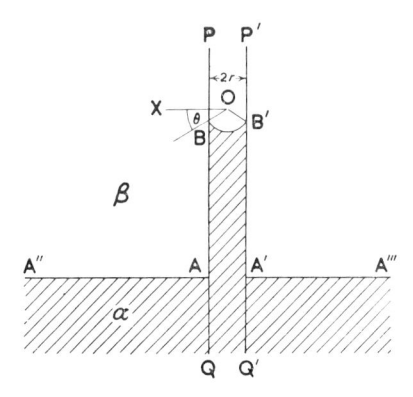

Fig. 1.4. Capillary rise due to interfacial tension

other not shaded. They are separated partly by the plane surfaces AA″ and A′A‴, and partly by the curved surface BB′ in the capillary tube PP′Q′Q of internal radius r. We may, with sufficient accuracy, regard the surface BB′ as a segment of a sphere. Let the centre of this sphere be denoted by O and let θ be the angle between OB and the horizontal OX or alternatively the angle between the tangential plane to BB′ at B and the wall of the vertical capillary tube. Then the radius of curvature of the surface BB′ is $r/\cos\theta$.

Let P^0 denote the pressure at the plane surfaces AA″ and A′A‴. It will also be the pressure inside the capillary tube at the height AA′. Let the pressures at the height BB′ be denoted by P^α in the phase α and by P^β in the phase β. Then

$$P^\alpha = P^0 - \varrho^\alpha g h \qquad 1.62.1$$

$$P^\beta = P^0 - \varrho^\beta gh \qquad 1.62.2$$

where ϱ^α, ϱ^β denote the densities of the phases α and β, g is the acceleration due to gravity, and h is the height AB. But, according to (1.60.9), since the radius of curvature is $r/\cos\theta$

$$P^\beta - P^\alpha = 2\gamma(\cos\theta)/r. \qquad 1.62.3$$

Comparing (1), (2), and (3) we obtain

$$2\gamma(\cos\theta)/r = (\varrho^\alpha - \varrho^\beta)gh. \qquad 1.62.4$$

Thus, from measurement of ϱ^α, ϱ^β, r, θ, and h we can calculate γ.

In the case that the surface BB′ is concave towards the bottom, its radius of curvature will have the opposite sign, and so h will also have the opposite sign. That is to say, BB′ will lie below AA′.

§1.63 Independence of interfacial tension of curvature

Let us now turn to the question whether the interfacial tension depends on the curvatures. We shall see that when the question is precisely defined it answers itself. In asking the question it is not sufficient to state that we vary the curvatures; we require also to state what we keep constant. For the question to be useful it should apply to the actual conditions of the experimental measurement of interfacial tension. For definiteness let us consider the capillary rise method described in the preceding section. The values of the temperature T and the chemical potentials μ_i are uniform throughout the system, and so, whatever be the size and shape of the capillary, these variables have the same values at the curved surface, where the surface tension is measured, as in the bulk phases. Hence to be useful the question should be worded: how does γ depend on ϱ_1, ϱ_2 for given values of T and the μ_i's? According to equation (1.57.3) the variation of γ under these restrictions is given by

$$d\gamma = \tau\, dP. \qquad 1.63.1$$

In its present application the ambiguity in the exact meaning of P does not matter, since it can be verified that $(P^\alpha - P^\beta)\tau$ is negligible. If now we consider a curved interface, say in a capillary, in equilibrium with a plane interface and we integrate (1) from the pressure at the plane surface to the pressure at the curved interface (either side of it) we again find that the integral of the right side is always negligible. Consequently γ has effectively the same value for the curved surface as for the plane surface with which it is in equilibrium. This is a statement of a principle usually assumed whenever an interfacial tension is measured. It is experimentally verified by the fact that within the experimental accuracy the same value is found for the interfacial tension when capillaries of different size are used, but this verification can be realized only for capillaries with diameters considerably greater than the lower bound allowed by the theory.

Gibbs summed up this situation in the words 'it will generally be easy to determine the surface tension in terms of the temperature and the chemical potentials of the several component species with considerable accuracy for plane surfaces, and extremely difficult or impossible to determine the fundamental equation more completely'.

§1.64 *Mathematical analysis of curved interface*

We have seen that it is extremely difficult if not impossible to devise any experiment which will determine the dependence of interfacial tension on curvature. This follows from the fact that the thickness of the interface is always extremely small compared with its radius of curvature; if this were not so, the interfacial tension could not be measured at all. The present section is concerned with a more exact mathematical analysis of this situation*. The reader is warned that the physical conclusions are entirely negative.

For the present discussion it is sufficient to consider only a spherical interface and it is convenient to follow Gibbs in describing the properties of the system by means of a single geometrical spherical surface lying inside the interfacial layer and concentric with it. For a chosen geometrical surface of area A the interfacial tension γ is defined by the relation

$$U^\Sigma - TS^\Sigma - \sum_i \mu_i n_i^\Sigma = -P^\alpha V^\alpha - P^\beta V^\beta + \gamma A \qquad 1.64.1$$

or by the equivalent relation

$$U^\Sigma - TS^\Sigma + P^\beta V^\Sigma - \sum_i \mu_i n_i^\Sigma = -(P^\alpha - P^\beta)V^\alpha + \gamma A. \qquad 1.64.2$$

Since all the quantities on the left, in particular U^Σ, S^Σ, V^Σ, n_i^Σ are invariant with respect to a change in the choice of the Gibbs geometrical surface, it follows that the right side is also invariant. But since a change in the Gibbs geometrical surface involves a change in V^α but not in P^α nor in P^β we may expect it also to change the value of γ. In order to investigate this change we denote the radius of the Gibbs geometrical surface by a. We have

---

* Guggenheim, Research 1957 10 478.

$$V^\alpha = 4\pi a^3/3 \qquad 1.64.3$$

$$A = 4\pi a^2 \qquad 1.64.4$$

and consequently

$$\gamma a^2 - \tfrac{1}{3}(P^\alpha - P^\beta)a^3 = \text{invariant}. \qquad 1.64.5$$

Differentiating (5) with respect to a we obtain

$$P^\alpha - P^\beta = 2\gamma/a + \mathrm{d}\gamma/\mathrm{d}a. \qquad 1.64.6$$

If we denote by a_t the value of a at which γ has a minimum value γ_t we obtain from (6)

$$P^\alpha - P^\beta = 2\gamma_t/a_t \qquad 1.64.7$$

which is of the same form as if there were a tension located exactly at the Gibbs geometrical surface of radius a_t. This surface is accordingly called the *surface of tension*. Substituting from (7) into (5) we obtain

$$\gamma a^2 - \tfrac{2}{3}\gamma_t a^3/a_t = \tfrac{1}{3}\gamma_t a_t^2 \qquad 1.64.8$$

$$\gamma/\gamma_t = \tfrac{2}{3}a/a_t + \tfrac{1}{3}a_t^2/a^2. \qquad 1.64.9$$

From the form of (9) due to Kondo* it follows that γ_t is indeed a minimum, not a maximum, and that the minimum is unique. Many relations are simplified by choosing the surface of tension as the Gibbs geometrical surface since any term containing $\mathrm{d}\gamma/\mathrm{d}a$ vanishes.

It is convenient for some purposes to define a quantity ε by

$$a^3/a_t^3 = 1 + 3\varepsilon. \qquad 1.64.10$$

It is clear that $\varepsilon \ll 1$ if the thickness of the interface is small compared with its radius of curvature. Using (10) we can rewrite (9) as

$$\gamma/\gamma_t = \tfrac{2}{3}(1+3\varepsilon)^{\frac{1}{3}} + \tfrac{1}{3}(1+3\varepsilon)^{-\frac{2}{3}}$$

$$= 1 + \varepsilon^2 + O(\varepsilon^3) \qquad 1.64.11$$

where $O(\varepsilon^3)$ denotes small terms of order ε^3.

If we denote the amount of the substance i per unit volume in the interiors of the bulk phases α and β by c_i^α and c_i^β respectively, then the surface concentration Γ_i is defined by

$$n_i^\Sigma = c_i^\alpha V^\alpha + c_i^\beta V^\beta + \Gamma_i A. \qquad 1.64.12$$

Since n_i^Σ is of course independent of the choice of the Gibbs geometrical surface it follows that the right hand side of (12) must be invariant. In terms of the radius a this implies

---

* Kondo, J. Chem. Phys. 1955 **25** 662.

FUNDAMENTAL PRINCIPLES

$$\tfrac{1}{3}(c_i^\alpha - c_i^\beta)a^3 + \Gamma_i a^2 = \text{invariant}. \qquad 1.64.13$$

For a single substance we may drop the subscript i and write

$$\tfrac{1}{3}(c^\alpha - c^\beta)a^3 + \Gamma a^2 = \text{invariant}. \qquad 1.64.14$$

In particular

$$\tfrac{1}{3}(c^\alpha - c^\beta)a^3 + \Gamma a^2 = \tfrac{1}{3}(c^\alpha - c^\beta)a_t^3 + \Gamma_t a_t^2. \qquad 1.64.15$$

We now choose a value of a to make Γ vanish. We indicate this by the subscript $_a$ and call the Gibbs geometrical surface with this radius the *auxiliary surface*. We have then

$$\Gamma_t a_t^2 = \tfrac{1}{3}(c^\alpha - c^\beta)(a_a^3 - a_t^3). \qquad 1.64.16$$

It is convenient for some purposes to rewrite (16) as

$$\Gamma_t / (c^\alpha - c^\beta) = \varepsilon_a a_t \qquad 1.64.17$$

where ε_a is defined by

$$a_a^3 / a_t^3 = 1 + 3\varepsilon_a. \qquad 1.64.18$$

Thermodynamics alone can predict neither the magnitude of ε_a nor the sign of ε_a nor the dependence of ε_a on a_t. Molecular theory indicates that $|\varepsilon_a| \ll 1$ if a_t is large compared with molecular dimensions. If we denote by γ_a the value of γ referred to the auxiliary surface, we have according to (11)

$$\gamma_a / \gamma_t = \tfrac{2}{3}(1 + 3\varepsilon_a)^{\frac{1}{3}} + \tfrac{1}{3}(1 + 3\varepsilon_a)^{-\frac{2}{3}}$$
$$= 1 + \varepsilon_a^2 + O(\varepsilon_a^3) \qquad 1.64.19$$

where $O(\varepsilon_a^3)$ denotes small terms of order ε_a^3.

Having defined γ_t uniquely we shall now study how the surface tension between a spherical portion of liquid of a single substance and its surrounding vapour depends on the radius of the sphere or, to be more precise, how γ_t depends on a_t. We have at constant temperature the thermodynamic relations

$$d\gamma_t = -\Gamma_t d\mu \qquad 1.64.20$$

$$d\mu = d\mu^\alpha = dP^\alpha / c^\alpha \qquad 1.64.21$$

$$d\mu = d\mu^\beta = dP^\beta / c^\beta \qquad 1.64.22$$

where c^α and c^β are the concentrations defined by $c = n/V$. From (21) and (22) we deduce using (7)

$$d\mu = d(P^\alpha - P^\beta)/(c^\alpha - c^\beta) = d(2\gamma_t / a_t)/(c^\alpha - c^\beta) \qquad 1.64.23$$

and substituting this into (20)

$$d\gamma_t = -\Gamma_t(c^\alpha - c^\beta)^{-1} d(2\gamma_t/a_t). \qquad 1.64.24$$

Formula (24) does not really tell us much about the dependence of γ_t on a_t because we have no means of measuring Γ_t. By using (17) we can transform (24) to

$$d\gamma_t = -\varepsilon_a a_t d(2\gamma_t/a_t) \qquad 1.64.25$$

or

$$d \ln \gamma_t / d \ln a_t = 2\varepsilon_a/(1 + 2\varepsilon_a). \qquad 1.64.26$$

The relation (26) is due to Tolman*. It is elegant but uninformative because ε_a is defined according to (18) in terms of a_a/a_t. As already mentioned thermodynamics tells us nothing about the magnitude or even the sign of ε_a and, as emphasized by Koenig[†], we have no means of measuring ε_a. Instead of claiming that (26) tells us anything about $d\gamma_t/da_t$ it would be more realistic to say that if we could measure $d\gamma_t/da_t$, that is to say the dependence of surface tension on curvature, we could then use (26) to calculate ε_a and so a_a/a_t. In fact $d\gamma_t/da_t$ is too small to be experimentally determined and this merely confirms that $|\varepsilon_a| \ll 1$.

The several experimental methods of determining surface tension are all based on the use of an equation formally resembling (7). In fact the experimental value γ_e of the surface tension is calculated from the formula

$$\gamma_e = \tfrac{1}{2}(P^\alpha - P^\beta)a_e \qquad 1.64.27$$

where a_e is the radius of the spherical interface estimated either visually or in the capillary-rise method estimated from the radius of the capillary. The most that can be said about the relation of a_e to a_t is that

$$a_e/a_t = 1 + O(\varepsilon_a). \qquad 1.64.28$$

Consequently the most that can be said about the relation of γ_e to γ_t is

$$\gamma_e/\gamma_t = 1 + O(\varepsilon_a). \qquad 1.64.29$$

§1.65 Basis of thermodynamic laws

The *zeroth law* in §1.07, the *first law* in §1.10, and the *second law* in §1.20 have all been quoted as fundamentally independent assumptions. From this point of view their justification is the empirical fact that all conclusions from

---

* Tolman, J. Chem. Phys. 1949 **17** 333.
† Koenig, J. Chem. Phys. 1950 **18** 449.

these assumptions are without exception in agreement with the experimentally observed behaviour in nature.

The form in which these laws have been enunciated is essentially that used by Born*. There are other alternative forms; some more, others less abstract, but all of an entirely empirical nature; that is to say that their justification is agreement between their implications on the one hand and experiment on the other.

It is, however, possible to obtain a deeper insight into the fundamental principles from a statistical point of view. It is in fact possible to derive these principles from our knowledge of the structure of matter including the elements of quantum theory together with a single statistical assumption of a very general form. It is a matter of taste whether to choose as a basis several empirical principles which make reference neither to atomic theory nor to quantum theory, or to choose a single principle superposed on atomic theory and quantum theory. The former choice, the one adopted in this book, is the method of *classical thermodynamics*; the latter choice corresponds to the more modern science which we call *statistical thermodynamics*.

There are however other relations of a general nature which follow naturally and directly from the statistical thermodynamic formulation, but which cannot be derived from the zeroth, first, and second laws of classical thermodynamics. The relations to which we refer are of several types concerning respectively

(a) entropy changes in highly disperse systems (i.e. gases);
(b) entropy changes in very cold systems (i.e. when $T \to 0$);
(c) entropy changes associated with mixing of very similar substances (e.g. isotopes).

The three types are of comparable importance. They resemble one another in relating to entropy changes. Their formulation in terms of classical thermodynamics is either complicated or inaccurate or else involves reference to conceptions inherently foreign to classical thermodynamics. As already mentioned they all follow naturally and directly from the statistical thermodynamic formulation.

We shall devote the following chapter to a digression on *statistical thermodynamics*, describing in general terms the methods of this science just sufficiently to give the reader an idea of the source of the relations in question without attempting to derive them in detail. The reader interested in the complete derivations must refer to a standard text-book on *statistical thermodynamics*.

* Born, Phys. Z. 1921 **22** 218.

§1.66 *Third law*

It has been customary to refer to the three types of general relations mentioned in the preceding section in three quite different ways. The relations of type (a) are referred to as the determination of *entropy constants*, those of type (b) as the *third law* of thermodynamics and those of type (c) merely as the formulae for *entropy of mixing*. This biased discrimination between types of relations of comparable importance and generality is difficult to defend. We accordingly reject this unbalanced terminology and instead choose as our *third law* the following statement.

By the standard methods of *statistical thermodynamics* it is possible to derive for certain entropy changes general formulae which cannot be derived from the *zeroth, first,* or *second laws of classical thermodynamics*. In particular one can obtain formulae for entropy changes in highly disperse systems (i.e. gases), for those in very cold systems (i.e. when $T \to 0$), and for those associated with the mixing of very similar substances (e.g. isotopes).

CHAPTER 2

DIGRESSION ON STATISTICAL THERMODYNAMICS

§2.01 *Microdescriptions and macrodescriptions of a system*

According to quantum theory the state of a system is completely specified by its eigenfunction. To each state there corresponds one eigenfunction and to each eigenfunction one state. Such a description of the system we shall call a *microdescription*.

It is often, though not always, possible to regard the system as consisting of a large number of almost independent *units* (molecules, atoms, ions, electrons) and to express each eigenfunction of the system as a linear combination of products of the eigenfunctions of all the units. According to the *symmetry restrictions*, if any, imposed on the eigenfunctions of the system, we then obtain three alternative sets of statistical formulae referred to by the names of Fermi–Dirac, Bose–Einstein, and Boltzmann, respectively. These three alternatives, however, arise only when we express the eigenfunctions of the system in terms of those of the constituent units. As long as we refer only to the eigenfunctions of the whole system, we shall not need to consider these three alternatives separately. Nor shall we do so until we reach §2.10.

When we describe the equilibrium properties of a system by thermodynamic methods, we are not interested in such a precise description as the *microdescription*, but are content with a more crude large scale description, which we shall call a *macrodescription*. For example a possible *macrodescription* of the system would be a precise statement of the energy, the volume, the chemical composition (and in special cases other quantities all measurable on a large scale) of each homogeneous part or phase. For brevity we shall confine our discussion initially to systems whose macrodescription requires a precise statement of only four quantities. The extension of the argument to more complicated systems should be obvious. Initially we shall take the first of these quantities to be the energy, the second to be the volume, the third to be the empirical composition; the nature of the fourth quantity is best indicated by some specific examples.

Example 1   Let us consider a definite quantity of hydrogen (free from deuterium) of given energy and given volume. Then we can complete the description by a statement of what fraction of it is *para*, the remaining fraction being *ortho*.

Example 2 If instead of hydrogen, we have lactic acid we can complete the description by a statement of what fraction is *dextro*, the remaining fraction being *laevo*.

Example 3 If the system consists of a given quantity of iodine of given energy and volume we can complete the description by a statement of what fraction is in the *diatomic* form I_2, the remainder being in the *monatomic* form I.

Example 4 If the system consists of a given quantity of tin of given energy and volume, we can complete the description by stating what fraction is *white*, the remainder being *grey*.

Example 5 If the system consists of a given quantity of sulphur dioxide, we can complete the description by stating what fraction is *liquid*, the remainder being *vapour*.

In the first three examples it is assumed either that the system is homogeneous or, if it consists of two phases, that we are not interested in the relative amounts, these being determined by the other conditions. Another example that might be suggested is a system of a given quantity of hydrogen of given energy and volume for which we were interested both in the ratio of *para* to *ortho* and in the ratio of *liquid* to *vapour*. Such a system, however, requires five quantities, instead of four, to complete its macrodescription and so lies outside the class which we shall discuss, although the extension of the treatment to such a system in fact offers no difficulty.

Having made clear by these examples the nature of the fourth independent variable describing the system we shall denote this variable by ξ. It corresponds closely to the quantity ξ which, following De Donder, we introduced in §1.44 and which we call the *extent of reaction* of a physico-chemical change. It is not a necessary property of ξ that one should be able completely to control its value, provided that its value can in principle be measured by *macroexperiments*.

§2.02 *System of given energy, volume, and composition*

Let us now consider in more detail a system of prescribed energy U, prescribed volume V, and containing a prescribed number N of molecules of a given kind. Let the number of independent eigenfunctions of the system consistent with the prescribed values of U, V, N and corresponding to a particular value of the parameter ξ be denoted by $\Omega(\xi)$. As long as we are not interested in distinguishing between the states of equal ξ, we may conveniently group them together.

Then the *fundamental assumption of statistical thermodynamics* is the following.

The average properties of the system for prescribed values of U, V, N can be derived statistically by averaging over all groups of states of given ξ, assigning to each group a weight $\Omega(\xi)$.

In other words it is assumed that for given U, V, N the *probability* of a particular value of ξ is

$$\Omega(\xi)/\sum_{\xi} \Omega(\xi). \qquad 2.02.1$$

It is customary to refer to the numerator $\Omega(\xi)$ in (1) as the *thermodynamic probability* of the particular value of ξ. It must be emphasized that *thermodynamic probability* thus defined is not a *probability* in the usual sense of the word. Whereas an ordinary *probability* such as (1) is a number less than or equal to unity, the *thermodynamic probability* is generally a large number.

For reasons which will appear later $\Omega(\xi)$ had better be called the *thermodynamic probability of ξ for given U, V, N* than merely the thermodynamic probability of ξ. Another name for $\Omega(\xi)$ is the *partition function for given U, V, N, ξ*. The reason for this name will also become clearer as we proceed.

We now define a quantity $S(U, V, N, \xi)$ by the relation

$$S(U, V, N, \xi) = k \ln \Omega(\xi) \qquad 2.02.2$$

where k is a universal arbitrary constant whose value will be settled later. It can then be shown as we shall see later that, in a macroscopic system, S has all the properties of the entropy of the system in the macrostate defined by U, V, N, ξ. Formula (2) is a precise formulation of the well-known relation due to Boltzmann to whom the name *thermodynamic probability* is due.

We shall see that Boltzmann's relation (2) between the *entropy* and the *thermodynamic probability* or *partition function* for given U, V, N, ξ is merely one of a number of relations of a similar type between a *characteristic function* for a particular set of variables on the one hand and the *thermodynamic probability* or *partition function* for the same set of variables on the other.

§2.03 *Characteristic of macroscopic system*

According to the *fundamental assumption of statistical thermodynamics* in a system of given U, V, N the average value $\langle \xi \rangle$ of ξ is determined by

$$\langle\xi\rangle = \sum_{\xi} \xi\Omega(\xi)/\sum_{\xi} \Omega(\xi) \qquad 2.03.1$$

and the average value $\langle\xi^2\rangle$ of ξ^2 by

$$\langle\xi^2\rangle = \sum_{\xi} \xi^2\Omega(\xi)/\sum_{\xi} \Omega(\xi). \qquad 2.03.2$$

Thus in general $\langle\xi\rangle^2$ is not the same as $\langle\xi^2\rangle$.

In other words there are fluctuations measured by

$$\langle(\xi-\langle\xi\rangle)^2\rangle = \langle\xi^2\rangle - \langle\xi\rangle^2$$
$$= \{\sum_{\xi} \Omega(\xi) \sum_{\xi} \xi^2\Omega(\xi) - \sum_{\xi} \xi\Omega(\xi) \sum_{\xi} \xi\Omega(\xi)\}/\{\sum_{\xi} \Omega(\xi)\}^2. \qquad 2.03.3$$

It can be shown generally that the larger the system the less important is this fluctuation and that for any macroscopic system the fluctuation is entirely trivial compared with $\langle\xi\rangle^2$ itself. Without attempting a proof we shall consider a little more closely how this comes about.

There is some value ξ_m of ξ for which $\Omega(\xi)$ has a maximum. Generally speaking the larger the system the sharper is this maximum and for any macroscopic system it is very sharp indeed. On each side of this maximum term $\Omega(\xi_m)$ there will be many terms almost as great as $\Omega(\xi_m)$. Then there will be a still greater number of terms appreciably smaller but not negligible; but an overwhelming majority of the terms will be entirely negligible, and this majority includes all those terms in which ξ differs appreciably from ξ_m.

As a result of such considerations it can be shown that whereas the average properties are strictly determined by attributing to each ξ the weight $\Omega(\xi)$ we may in any macroscopic system with trivial inaccuracy ignore all values of ξ other than the value ξ_m at which $\Omega(\xi)$ is maximum.

Thus for any macroscopic system we have with trivial inaccuracy

$$\langle\xi\rangle = \xi_m \qquad 2.03.4$$
$$\langle\xi^2\rangle = \xi_m^2 \qquad 2.03.5$$

and so on.

It is instructive to relate this important characteristic of a macroscopic system to the quantity $S(U, V, N, \xi)$ defined by (2.02.2), namely

$$S(U, V, N, \xi) = k \ln \Omega(\xi). \qquad 2.03.6$$

Let us now define another quantity $S(U, V, N)$ by

$$S(U, V, N) = k \ln \{\sum_{\xi} \Omega(\xi)\}. \qquad 2.03.7$$

Then by definition it is evident that

$$S(U, V, N) > S(U, V, N, \xi) \quad \text{(all values of } \xi\text{)}. \qquad 2.03.8$$

Let us now consider the ratio

$$\{\ln \sum_{\xi} \Omega(\xi) - \ln \Omega(\xi_m)\}/\ln \Omega(\xi_m). \qquad 2.03.9$$

It can be shown that roughly speaking $\Omega(\xi_m)$ is of the order $N!$ and $\Sigma_\xi(\Omega(\xi)/\Omega(\xi_m))$ is of the order N^α where α is comparable with unity. Hence the numerator in (9) is of the order $\alpha \ln N$ and the denominator of the order $N \ln N$. Thus the expression (9) is of the order α/N or near enough N^{-1}, which is entirely negligible in any macroscopic system. Hence, although the inequality (8) is strictly true by definition for all values of ξ, in any macroscopic system when ξ has the special value ξ_m we may with trivial inaccuracy replace the inequality (8) by the equality

$$S(U, V, N) = S(U, V, N, \xi_m). \qquad 2.03.10$$

We shall see in §2.05 that the functions denoted by S have in fact the properties of entropy. Anticipating this identification let us call $S(U, V, N, \xi)$ the *entropy for fixed* ξ and $S(U, V, N)$ the *entropy for equilibrium* ξ.

Consider now a system of given U, V, N with ξ *frozen*. Now suppose that by introduction of a catalyst ξ is *thawed*, so that it takes its equilibrium value. By definition the entropy changes from $S(U, V, N, \xi)$ to $S(U, V, N)$ and also by definition this is always an increase. Only in the special case that the initial value of ξ was ξ_m the entropy increase from $S(U, V, N, \xi_m)$ to $S(U, V, N)$ for any macroscopic system is trivial. In other words although $S(U, V, N)$ the entropy for *equilibrium* ξ is by definition greater than the entropy for ξ *fixed at its equilibrium value* ξ_m the difference in a macroscopic system is negligible and trivial.

We shall see later that a macroscopic system has other characteristics similar and parallel to that just formulated. These characteristics can be summed up in the single sentence that in a *macroscopic* system *fluctuations* of measurable properties are *negligible*.

§2.04 *System of given temperature, volume, and composition*

We shall now consider a system whose volume V and composition N are still prescribed, but instead of prescribing the energy we shall suppose the system to be immersed in a large temperature bath with which it can exchange energy so that the energy of the system can now take various values U_0, U_1, and so on. Let us now enumerate the eigenfunctions of the system for the prescribed values of V and N and for some definite value of ξ;

let there be Ω_r such eigenfunctions corresponding to an energy $U_r(V, N, \xi)$.

From the *fundamental assumption of statistical thermodynamics*, as stated in §2.02, without any further assumptions it can be shown that the average properties of the system in the temperature bath for the prescribed values of V and N can be derived statistically by averaging over all degenerate energy values attaching to each state r of specified ξ and U_r a weight

$$\Omega_r \exp(-\beta U_r) \qquad 2.04.1$$

where β is determined entirely by the temperature bath and so may be regarded as a temperature scale.

The fact that the parameter β is found to appear without any new assumption is the statistical thermodynamic basis of the *zeroth law of classical thermodynamics*. The *statistical thermodynamic* equivalent of the *first law of classical thermodynamics* is merely the principle of *conservation of energy* applied on the microscopic scale, that is to say applied to molecules, atoms, electrons, etc. Thus this principle is from the point of view of statistical thermodynamics not a new law but merely one item in general atomic quantum theory.

To relate the *second law of classical thermodynamics* to *statistical thermodynamics* we make certain algebraic transformations. We begin by defining a function $J(\beta, V, N, \xi)$ by

$$J(\beta, V, N, \xi) = k \ln \left\{ \sum_r \Omega_r(\xi) \exp(-\beta U_r) \right\} \qquad 2.04.2$$

where the summation is over all states of given ξ, and k is a universal arbitrary constant.

In the system with temperature specified by β there will be fluctuations of U, but the experimentally measurable U will be $\langle U \rangle$, the average value of U. Let us now consider the value of $\langle U \rangle$ for specified β, V, N, and ξ. Using the weighting factors (1) we have

$$U(\beta, V, N, \xi) = \sum_r U_r \Omega_r(\xi) \exp(-\beta U_r) / \sum_r \Omega_r(\xi) \exp(-\beta U_r)$$

$$= -\partial \ln\left\{ \sum_r \Omega_r(\xi) \exp(-\beta U_r) \right\} / \partial \beta$$

$$= k^{-1} \partial J(\beta, V, N, \xi) / \partial \beta \qquad 2.04.3$$

using (2).

Again associated with the fluctuations in U there will be fluctuations in the pressure $(-\partial U/\partial V)$ but the experimentally measured pressure P will be $\langle -\partial U/\partial V \rangle$. We accordingly have for given β, V, N, and ξ

$$P = \langle -\partial U/\partial V \rangle = -\sum_r (\partial U_r/\partial V)\Omega_r(\xi)\exp(-\beta U_r)/\sum_r \Omega_r(\xi)\exp(-\beta U_r)$$

$$= \beta^{-1}\partial \ln\{\sum_r \Omega_r(\xi)\exp(-\beta U_r)\}/\partial V$$

$$= -k^{-1}\beta^{-1}\partial J(\beta, V, N, \xi)/\partial \beta \qquad 2.04.4$$

using (2).

Let us now make the further algebraic substitution

$$T = k^{-1}\beta^{-1} \qquad 2.04.5$$

and use T as an independent variable instead of β. We now have in place of (3) and (4)

$$\langle U(T, V, N, \xi) \rangle = -\partial J(T, V, N, \xi)/\partial T^{-1}$$
$$= T^2 \partial J(T, V, N, \xi)/\partial T \qquad 2.04.6$$

$$P = T \partial J(T, V, N, \xi)/\partial V. \qquad 2.04.7$$

Combining (6) with (7) we have

$$dJ = (\langle U \rangle/T^2)dT + (P/T)dV. \qquad 2.04.8$$

Comparing (8) with (1.28.15) we see that the dependence of J, defined by (2), on T defined by (5) and on V is precisely the same as the dependence of the *Massieu function* on the thermodynamic temperature and on the volume. It can in fact be shown that T defined by (5) has all the properties of absolute temperature and J defined by (2) has all the properties of the *Massieu function*. This constitutes a brief summary of how the *second law of classical thermodynamics* follows as a natural deduction from *statistical thermodynamics*.

For the benefit of the reader not familiar with the *Massieu function* J we recall that it is defined by

$$J = S - U/T \qquad 2.04.9$$

and that either the Massieu function J or the Helmholtz function $F = -TJ$ is a characteristic function for the independent variables T, V, N.

We can now substitute from (5) into (1) and so have as a weighting factor for each energy U_r

$$\Omega_r \exp(-U_r/kT) \qquad 2.04.10$$

and this factor is called the *Boltzmann factor*. From (10) it is clear that kT has the dimensions of energy. k is a universal constant called the *Boltzmann constant*. If we use the *Kelvin scale* of thermodynamic temperature then

$$k = 1.3805 \times 10^{-23} \text{J K}^{-1}. \qquad 2.04.11$$

From (10) we see that the average properties of the system for prescribed values of T, V, N and unspecified ξ can be obtained by averaging over all ξ attaching to each ξ a weight $Q(T, \xi)$ defined by

$$Q(T, \xi) = \sum_r \Omega_r(\xi) \exp(-U_r/kT). \qquad 2.04.12$$

The function $Q(T, \xi)$ is usually called the *partition function*, but a more precise name is the *partition function for given T, V, N, ξ*. An alternative name is the *thermodynamic probability for given T, V, N, ξ*.

Substituting from (5) and (12) into (2) we obtain

$$J(T, V, N, \xi) = k \ln Q(T, \xi). \qquad 2.04.13$$

We observe that this relation between the *characteristic function J* and the statistical probability $Q(T, \xi)$ for given T, V, N, ξ is completely analogous to *Boltzmann's relation* (2.02.2) between the characteristic function S and the thermodynamic probability $\Omega(\xi)$ for given U, V, N, ξ.

§2.05 Further characteristics of macroscopic system

Let us consider the individual terms of $Q(T, \xi)$ defined by (2.04.12). Let us denote the maximum term by

$$\Omega_m \exp(-U_m/kT) \qquad 2.05.1$$

noting that this Ω_m is not the same as the $\Omega(\xi_m)$ of §2.03. Generally speaking the larger the system the sharper this maximum and for any macroscopic system it is so sharp that all terms in $Q(T, \xi)$ in which U_r differs appreciably from U_m are entirely trivial. Moreover, although the actual number of terms $Q(T, \xi)$ comparable with (1) may be great, the ratio

$$\frac{\ln Q(T, \xi) - \ln\{\Omega_m \exp(-U_m/kT)\}}{\ln\{\Omega_m \exp(-U_m/kT)\}} \qquad 2.05.2$$

is roughly of the order α/N where α is far nearer to unity than to N. Hence in any macroscopic system the ratio (2) is effectively zero and we may therefore replace the definition (2.04.2) of J by

$$J = k \ln\{\Omega_m \exp(-U_m/kT)\}. \qquad 2.05.3$$

It follows again that with an inaccuracy trivial for a macroscopic system

$$\langle U \rangle = T^2 \partial J/\partial T = U_m. \qquad 2.05.4$$

From the classical definition (1.28.13) of the *Massieu function J*, we have

$$S = J + U/T. \qquad 2.05.5$$

We accordingly in *statistical thermodynamics* define a function $S(T, V, N, \xi)$ by

$$S(T, V, N, \xi) = J(T, V, N, \xi) + \langle U \rangle/T. \qquad 2.05.6$$

Using (3), (4), and (5) we obtain from (6)

$$S(T, V, N, \xi) = k \ln \Omega_m(\xi). \qquad 2.05.7$$

Now comparing (7) with (2.02.2) we obtain the striking result

$$S(T, V, N, \xi) = S(\langle U \rangle, V, N, \xi). \qquad 2.05.8$$

Thus although the definition of entropy at a specified temperature by means of (6) together with (2.04.13) is entirely different from the definition of entropy at a specified energy by means of (2.02.2), yet for a macroscopic system the difference between the two is trivial.

This characteristic property of a macroscopic system may be described in the following instructive but less exact way. If we define S by

$$S = k \ln \Omega(\xi) \qquad 2.05.9$$

then in a system of specified energy Ω must denote the number of states having *precisely* this energy, whereas in a system of specified temperature Ω denotes the number of states of energy *nearly* equal to the average energy. The question immediately arises how nearly. The answer is that for a *macroscopic* system it just does not matter.

§2.06 System of given temperature, pressure, and composition

We now consider a system of prescribed composition surrounded by a temperature bath and enclosed by a piston subjected to a prescribed pressure P. We construct the double sum

$$W(T, P, N, \xi) = \sum_r \sum_s \Omega_{rs} \exp(-U_r/kT) \exp(-PV_s/kT) \qquad 2.06.1$$

where the summation extends over all energies U_r and all volumes V_s consistent with the prescribed value of ξ. It can then be shown without any new assumptions that we can correctly derive the average (equilibrium) properties of the system for the prescribed values of T, P, N, by averaging over all values of ξ attaching to each a weight $W(T, P, N, \xi)$.

We call $W(T, P, N, \xi)$ the *thermodynamic probability for given* T, P, N, ξ or the *partition function for given* T, P, N, ξ. It is related to the *Planck function* Y, which is a characteristic function for the independent variables T, P, N, ξ by

$$Y(T, P, N, \xi) = k \ln W(T, P, N, \xi) \qquad 2.06.2$$

analogous to (2.02.2) and (2.04.13).

For the benefit of the reader unfamiliar with the *Planck function* Y we recall its relation to the *Gibbs function* G, namely

$$Y = -G/T. \qquad 2.06.3$$

Provided the system is macroscopic we may again with only trivial inaccuracy replace W by its maximum term, say

$$\Omega_m(\xi) \exp(-U_m/kT) \exp(-PV_m/kT) \qquad 2.06.4$$

so that we may replace (2) by

$$Y(T, P, N, \xi) = -k \ln \Omega_m(\xi) - U_m/T - PV_m/T. \qquad 2.06.5$$

From (1) and (5) we immediately verify that

$$\langle U + PV \rangle = T^2 \partial Y/\partial T = U_m + PV_m \qquad 2.06.6$$

$$\langle V \rangle = -T \partial Y/\partial P = V_m \qquad 2.06.7$$

as we should expect according to (1.28.16). Furthermore comparing (5) with (1.28.14) we obtain

$$S(T, P, N, \xi) = k \ln \Omega_m(\xi) \qquad 2.06.8$$

verifying that for a macroscopic system the entropy at given T, P is indistinguishable from the entropy at given $U = U_m$ and $V = V_m$.

§2.07 *System of given temperature, pressure, and chemical potential*

To conclude we choose as independent variables the temperature T, pressure P, and chemical potential μ. An illustrative example is a gas in contact with a crystal of the same substance; the crystal is not considered as part of the system. Such a system is called *open*.

We now construct the triple sum

$$W(T, P, \mu, \xi) = \sum_r \sum_s \sum_t \Omega_{rst} \exp(-U_r/kT) \exp(-PV_s/kT) \exp(\mu N_t/LkT)$$
$$2.07.1$$

where L is a general constant called the Avogadro constant defined later in §3.13 and where Ω_{rst} denotes the number of states of energy U_r, volume V_s, and content N_t corresponding to the given value of ξ and the triple summation extends over all sets of values of U_r, V_s, N_t corresponding to the given value of

ξ. It can then be shown without any new assumptions that all the average properties of the system for the prescribed values of T, P, μ are correctly obtained by averaging over all values of ξ attaching to each a weight $W(T, P, \mu, \xi)$ this expression being the *partition function* or *thermodynamic probability* of ξ for given T, P, μ.

For a macroscopic system W can in the usual way be replaced by its maximum term say

$$\Omega_m(\xi) \exp(-U_m/kT) \exp(-PV_m/kT) \exp(\mu N_m/LkT). \qquad 2.07.2$$

If we now define a quantity $O(T, P, \mu, \xi)$ by

$$O(T, P, \mu, \xi) = k \ln W(T, P, \mu, \xi) \qquad 2.07.3$$

we may for a macroscopic system replace (3) by

$$O(T, P, \mu, \xi) = k \ln \Omega_m(\xi) - U_m/T - PV_m/T + \mu N_m/LT. \qquad 2.07.4$$

Moreover for a macroscopic system we have as usual

$$S = k \ln \Omega_m(\xi) \qquad 2.07.5$$

$$\langle U \rangle = U_m \qquad 2.07.6$$

$$\langle V \rangle = V_m \qquad 2.07.7$$

$$\langle N \rangle = N_m. \qquad 2.07.8$$

Comparing (5) to (8) with (4), dropping subscripts and replacing N/L by n we find that

$$O(T, P, N, \xi) = S - U/T - PV/T + \mu n/T = 0 \qquad 2.07.9$$

according to (1.28.5).

From the analogy between (3), (2.02.2), (2.04.13) and (2.06.2) we expect $O(T, P, \mu, \xi)$ to be a *characteristic function* for the variables T, P, and μ. According to (9) this characteristic function is identically zero. We now recall the Gibbs–Duhem relation (1.30.2)

$$S\,dT - V\,dP + \sum_i n_i\,d\mu_i = 0. \qquad 2.07.10$$

In a system of one component the sum $\Sigma_i n_i d\mu_i$ reduces to $n d\mu$ and so (10) becomes

$$0 = S\,dT - V\,dP + n\,d\mu \qquad 2.07.11$$

showing that the characteristic function for the independent variables T, P, μ is indeed zero.

§2.08 *Recapitulation*

We can now summarize the content of the several preceding sections*. For each selected set of three independent variables, other than ξ, a different kind of weighting factor w has to be attached to the microstates. The sum Σw for all microstates consistent with the prescribed values of the three chosen independent variables other than ξ and corresponding to a definite value of ξ is called the *partition function* or the *thermodynamic probability* for the prescribed values of ξ and the other three independent variables. Furthermore in each case $k \ln(\Sigma w)$ is a *characteristic function* for the chosen set of three independent variables other than ξ. These relationships are shown in table 2.1.

TABLE 2.1

Independent variables	Weighting factor for each microstate	Characteristic function equal to $k \ln (\Sigma w)$
U, V, N, ξ	1	S
T, V, N, ξ	$\exp(-U/kT)$	$J = -\mathscr{F}/T$
T, P, N, ξ	$\exp(-U/kT)\exp(-PV/kT)$	$Y = -G/T$
T, P, μ, ξ	$\exp(-U/kT)\exp(-PV/kT)\exp(\mu N/LkT)$	zero

We emphasize again that each of the listed *characteristic functions* S, J, Y, and zero is related to the corresponding *thermodynamic probability* according to

characteristic function $= k$ ln(thermodynamic probability).

The earliest and best known example of this form is Boltzmann's relation for $S(U, V, N, \xi)$, but other examples and particularly that for $J(T, V, N, \xi)$ are in fact more useful.

It is a fundamental characteristic of a macroscopic system that any partition function may with trivial inaccuracy be replaced by its maximum term. It follows that the equilibrium value of ξ is that value which maximizes the characteristic function belonging to the chosen set of independent variables. The alternative equilibrium conditions

for given U and V that S is a *maximum*   2.08.1

for given T and V that J is a *maximum*   2.08.2

for given T and P that Y is a *maximum*   2.08.3

* Guggenheim, J. Chem. Phys. 1939 **7** 103; Forh. 5te Nordiske Kemikermøde København 1939 p. 205.

thus obtained are precisely equivalent to (1.35.1), (1.35.3), and (1.35.4) respectively.

§2.09 Extension to several components. Absolute activities

We have hitherto restricted our exposition to systems of a single component purely for the sake of brevity. The extension to systems of several components is straightforward.

In particular for a system at given values of the independent variables T, P, and the μ_i's the weighting factor for each independent microstate will be

$$\exp(-U/kT)\exp(-PV/kT)\prod_i \lambda_i^{N_i} \qquad 2.09.1$$

where for brevity we have introduced quantities λ_i defined by

$$\lambda_i = \exp(\mu_i/LkT)$$

or

$$\mu_i = LkT \ln \lambda_i. \qquad 2.09.2$$

These quantities λ_i may be used instead of the μ_i and are often more convenient. λ_i is called the *absolute activity* of the species i. We shall meet these quantities again in §3.15.

§2.10 Antisymmetric and symmetric eigenfunctions

In §2.01 we mentioned that it is often, though not always, possible to regard the units (molecules, atoms, ions, electrons) composing the system as almost independent. In this case each eigenfunction of the system can be expressed as a linear combination of products of the eigenfunctions of all the units. We begin by considering the case that all the units are of the same kind. We denote the eigenfunctions of the units by ϕ and the eigenfunctions of the whole system by ψ. We have now to distinguish two cases.

If each unit is a fundamental particle (proton, neutron, or electron) or is composed of an odd number of fundamental particles, then each eigenfunction ψ of the system is constructed by forming a determinant of the eigenfunctions of the individual units. For the sake of simplicity and brevity we consider a system consisting of only three units, numbered 1, 2, 3. The symbol $\phi_\alpha(1)$ then denotes the eigenfunction of the unit 1 when in the state α. The eigenfunction is then constructed as follows

$$\psi_{\alpha\beta\gamma} = \begin{vmatrix} \phi_\alpha(1) & \phi_\beta(1) & \phi_\gamma(1) \\ \phi_\alpha(2) & \phi_\beta(2) & \phi_\gamma(2) \\ \phi_\alpha(3) & \phi_\beta(3) & \phi_\gamma(3) \end{vmatrix}. \qquad 2.10.1$$

We notice that if we interchange the states of any two units, ψ changes sign. We accordingly describe the eigenfunctions ψ as *antisymmetric* with respect to every pair of units. It follows at once that if any two of the states, α, β, γ are identical then $\psi_{\alpha\beta\gamma}$ vanishes. Thus there is one independent ψ for each combination of three $\phi_\alpha, \phi_\beta, \phi_\gamma$ provided α, β, γ are all different but none if any two of α, β, γ are the same.

If on the other hand each unit is a photon or is composed of an even number of fundamental particles (protons, neutrons, electrons), then each eigenfunction of the system is constructed from the eigenfunctions of the units by forming linear combinations called *permanents* similar to determinants, but in which all the terms are added. Thus in the case of only three units 1, 2, 3 the eigenfunction $\psi_{\alpha\beta\gamma}$ is defined by

$$\psi_{\alpha\beta\gamma} = \begin{Vmatrix} \phi_\alpha(1) & \phi_\beta(1) & \phi_\gamma(1) \\ \phi_\alpha(2) & \phi_\beta(2) & \phi_\gamma(2) \\ \phi_\alpha(3) & \phi_\beta(3) & \phi_\gamma(3) \end{Vmatrix} \qquad 2.10.2$$

which differs from (1) in that all the six terms are added. We notice that if we interchange the states of any two units, ψ remains unchanged. We accordingly describe the eigenfunction ψ as *symmetric* in all the units. It is clear that there is one independent ψ for every combination of three eigenfunctions $\phi_\alpha, \phi_\beta, \phi_\gamma$ whether or not any two or more of α, β, γ are the same.

§2.11 *Fermi–Dirac and Bose–Einstein statistics*

Let us now consider a system containing N indistinguishable units and enquire how many eigenfunctions ψ of the system can be constructed out of g eigenfunctions ϕ of the units. There are two distinct problems with different answers according as ψ is to be antisymmetric or symmetric in the units.

In the case where ψ is to be antisymmetric, to obtain any such ψ at all, g must be at least as great as N and the number of such eigenfunctions ψ is then

$$g!/N!(g-N)! \qquad \text{(antisymmetric, } g \geq N\text{).} \qquad 2.11.1$$

In the other case where ψ is to be symmetric, the number of such eigenfunctions ψ is

$$(g+N-1)!/(g-1)!N! \qquad \text{(symmetric)} \qquad 2.11.2$$

which, when $g \gg 1$, differs only trivially from the simpler expression

$$(g+N)!/g!N!. \qquad 2.11.3$$

It is of interest to note that when $g \gg N$, both (1) and (3) are nearly the

same as
$$g^N/N! \quad (g \gg N). \qquad 2.11.4$$

If now we translate the laws governing the average properties of the whole system outlined in §§2.01–2.09 into forms relating to the average distributions of the component units, we shall as a consequence of the difference between (1) and (2) find different results according as the eigenfunctions ψ are to be antisymmetric or symmetric in the units. These distribution laws take the simplest form if we choose as independent variables the temperature T, the volume V, and the absolute activity λ. We shall now state these laws without derivation.

Let ε_α denote the energy of a unit in the state α having the eigenfunction ϕ_α. Then if the unit is a fundamental particle (proton, neutron, or electron) or is composed of an odd number of fundamental particles, the eigenfunction ψ must be antisymmetric in the units and the average number N_α of units in the state α is found to be given by

$$N_\alpha/(1-N_\alpha) = \lambda \exp(-\varepsilon_\alpha/kT) \qquad 2.11.5$$

where λ denotes the *absolute activity* of the unit, T the absolute temperature, and k the Boltzmann constant. This distribution law is called that of *Fermi–Dirac statistics*.

If on the other hand the unit is a photon or is composed of an even number of fundamental particles, the eigenfunction ψ must be symmetric in the units and the average number N_α of units in the state α is found to be given by

$$N_\alpha/(1+N_\alpha) = \lambda \exp(-\varepsilon_\alpha/kT). \qquad 2.11.6$$

This distribution law is called that of *Bose–Einstein statistics*.

It is to be noted that in both the cases of *Fermi–Dirac statistics* and *Bose–Einstein statistics* the average number N_α of units in each state is related simply and explicitly to the temperature T and the absolute activity λ, which we recall is related to the chemical potential μ by (2.09.2).

§2.12 Boltzmann statistics

Let the subscript $_0$ denote the state of lowest energy ε_0 and let us consider the case that

$$\lambda \exp(-\varepsilon_0/kT) \ll 1 \qquad 2.12.1$$

so that a fortiori

$$\lambda \exp(-\varepsilon_\alpha/kT) \ll 1 \quad (\text{all } \alpha). \qquad 2.12.2$$

It then follows from either (2.11.5) or (2.11.6) that

$$N_\alpha \ll 1 \quad \text{(all } \alpha\text{)}. \qquad 2.12.3$$

We may then without loss of accuracy replace either (2.11.5) or (2.11.6) by

$$N_\alpha = \lambda \exp(-\varepsilon_\alpha/kT). \qquad 2.12.4$$

This distribution law is called that of *Boltzmann statistics*.

We now state without proof that in almost all the systems met in practice the condition (1) is satisfied. There are only two important exceptions. The first is the system of conducting electrons in a metal; these obey the Fermi–Dirac distribution law and will not be discussed in this book. The other is the system of photons forming radiation; these obey the Bose–Einstein distribution law and will be discussed in chapter 12. Boltzmann statistics are sufficient for all the other systems to be met in this book and from here onwards we shall confine our attention to these.

§2.13 *Partition functions of units and thermodynamic functions*

For any system obeying Boltzmann statistics, we have according to (2.12.4)

$$N_\alpha = \lambda \exp(-\varepsilon_\alpha/kT). \qquad 2.13.1$$

If we apply (1) to every state and add, we obtain

$$N = \lambda \sum_\alpha \exp(-\varepsilon_\alpha/kT) \qquad 2.13.2$$

so that

$$\mu/LkT = \ln \lambda = \ln\{N/\sum_\alpha \exp(-\varepsilon_\alpha/kT)\}. \qquad 2.13.3$$

The sum $\sum_\alpha \exp(-\varepsilon_\alpha/kT)$ is called the *partition function of the units*. Its structure is similar to that of the *partition function of the whole system* for the independent variables T, V, N. Formula (3) is the basis for the evaluation of the thermodynamic functions in terms of the energies of all the states of the component units.

Formula (3) is equivalent to the formula for the *Massieu function J*

$$J = -F/T = S - U/T = k \ln[\{\sum_\alpha \exp(-\varepsilon_\alpha/kT)\}^N/N!]. \qquad 2.13.4$$

If we compare (4) with (2.04.13) we see that the two are equivalent when we bear in mind that the factor $N!$ in the denominator in (4) is required to avoid counting as distinct states those obtainable from one another by a mere permutation of indistinguishable units.

The more general formula for a system containing more than one kind of units (molecules) is

$$J = -\mathcal{F}/T = S - U/T = k\sum_i \ln[\{\sum_\alpha \exp(-\varepsilon_\alpha/kT)\}^{N_i}/N_i!]. \qquad 2.13.5$$

§2.14 Separable degrees of freedom

It is often the case that there is no appreciable interaction between two or more degrees of freedom of a unit. Such degrees of freedom are said to be *separable*. Each eigenfunction ϕ_α may then be expressed as a product of the eigenfunctions for the several separable degrees of freedom, and the energy ε_α as the sum of the energies of the several separable degrees of freedom. It then follows immediately that the partition function of the unit can be expressed as the product of partition functions for its several separable degrees of freedom.

In particular the translational degrees of freedom of molecules are usually separable from the internal degrees of freedom. Among the internal degrees of freedom we here include rotational degrees of freedom as well as atomic vibrations and electronic and nuclear degrees of freedom. We may accordingly write for the partition function of a molecule

$$\sum \exp(-\varepsilon_\alpha/kT) = \sum \exp(-\varepsilon_{tr}/kT) \sum \exp(-\varepsilon_{int}/kT) \qquad 2.14.1$$

where ε_{tr} denotes the energy of the translational degrees of freedom and ε_{int} the energy of the internal degrees of freedom. Substituting (1) into (2.13.5) we obtain for the Massieu function J and the Helmholtz function $\mathcal{F}$

$$-TJ = \mathcal{F} = -kT \sum_i \ln[\{\sum \exp(-\varepsilon_{tr}/kT) \sum \exp(-\varepsilon_{int}/kT)\}^{N_i}/N_i!]. \qquad 2.14.2$$

Alternatively we may write

$$J = J_{tr} + J_{int} \qquad 2.14.3$$

$$\mathcal{F} = \mathcal{F}_{tr} + \mathcal{F}_{int} \qquad 2.14.4$$

$$\mathcal{F}_{tr} = -TJ_{tr} = -kT \sum_i \ln[\{\sum \exp(-\varepsilon_{tr}/kT)\}^{N_i}/N_i!] \qquad 2.14.5$$

$$\mathcal{F}_{int} = -TJ_{int} = -kT \sum_i N_i \ln\{\sum \exp(-\varepsilon_{int}/kT)\} \qquad 2.14.6$$

where the subscript $_{tr}$ refers throughout to contributions from the translational degrees of freedom and the subscript $_{int}$ to contributions from the internal degrees of freedom.

§2.15 Classical and unexcited degrees of freedom

It may happen that there are many energy levels less than kT. When this is the case, the sum which defines the partition function may without loss of accuracy be replaced by an integral, whose evaluation is often elementary. Such a degree of freedom is called a *classical degree of freedom*. Whether a particular degree of freedom is classical depends on the temperature. Under ordinary conditions the translational and rotational degrees of freedom of the molecules in a gas are classical.

In the opposite case it may happen that the separation between the states of lowest energy level and those of the next energy level is several times greater than kT. The partition function then reduces effectively to the terms corresponding to the lowest energy level, that is to

$$g_0 \exp(-\varepsilon_0/kT) \qquad 2.15.1$$

where ε_0 denotes the lowest energy level and g_0 denotes the number of states having this energy. Such degrees of freedom are called *unexcited degrees of freedom*. The contribution of each such unexcited degree of freedom to the Helmholtz function F is clearly

$$\varepsilon_0 - kT \ln g_0 \qquad 2.15.2$$

and the corresponding contribution to the entropy

$$k \ln g_0 \qquad 2.15.3$$

which we notice is independent of the temperature. Whether a particular degree of freedom is unexcited depends by definition on the temperature. At all the temperatures with which we are concerned all degrees of freedom internal to the atomic nucleus are unexcited. The electronic degrees of freedom of most molecules may also be regarded as unexcited at most of the temperatures which concern us; there are however a few exceptions, notably the molecule NO.

§2.16 *Translational degrees of freedom*

The translational degrees of freedom of a dilute gas may be regarded as classical. When the partition function for the translational degrees of freedom of a molecule is replaced by an integral and the integration is performed, one obtains

$$(2\pi m kT/h^2)^{\frac{3}{2}} V \qquad 2.16.1$$

where m denotes the mass of a molecule and V the volume in which it is enclosed; h denotes the Planck constant and k as usual the Boltzmann constant. Thus for a dilute gaseous mixture according to (2.14.5) we have

$$F_{tr} = -TJ_{tr} = -kT \sum_i \ln\{(2\pi m kT/h^2)^{\frac{3}{2}N_i} V^{N_i}/N_i!\} \quad \text{(dilute gas)}. \quad 2.16.2$$

Let us now consider the translational degrees of freedom in a crystal. We may regard each molecule as vibrating about an equilibrium position in the crystal lattice. Let us denote by q the partition function for a molecule attached to a given lattice position and for the moment let us imagine all the N molecules to be individually distinguishable but sufficiently alike so that any one can be interchanged with any other without destroying the crystal structure. Then the molecules can be permuted over the lattice positions in $N!$ ways, so that the partition function for the translational motion of the molecules of the whole crystal would be $N!q^N$. Actually the molecules are of course not individually distinguishable and we must consider only states whose eigenfunction is symmetric in molecules containing an even number of fundamental particles and antisymmetric in molecules containing an odd number of fundamental particles. In the simplest case when all the molecules in the crystal are of the same kind the number of states is thus reduced by a factor $N!$, which cancels the other $N!$, so that the partition function for the whole crystal becomes q^N. We thus have for a crystal of a pure substance

$$F_{tr} = -TJ_{tr} = -NkT \ln q \quad \text{(crystal)}. \qquad 2.16.3$$

Each molecule at a given lattice position usually has only one state of lowest translational energy and so at low temperatures q tends to $\exp(-\varepsilon_0/kT)$. We therefore have for a crystal of a pure substance

$$F_{tr} \to N\varepsilon_0 \quad (T \to 0) \qquad 2.16.4$$

and consequently

$$S_{tr} \to 0 \quad (T \to 0). \qquad 2.16.5$$

For a mixed crystal containing several distinguishable kinds of molecules, e.g. isotopes, the eigenfunctions have to be symmetric, or antisymmetric, only with respect to identical molecules. Hence we have to divide only by the product of all the $N_i!$ instead of by $N!$. We therefore have instead of (3)

$$F_{tr} = -TJ_{tr} = -kT \ln N! - kT \sum_i \ln\{q_i^{N_i}/N_i!\} \qquad 2.16.6$$

where $N = \Sigma_i N_i$. It has been implicitly assumed that interchanging two molecules of different kinds in the crystal does not affect the partition function q_i of either of them. This assumption is justified provided the molecules are sufficiently similar, e.g. isotopic. Since at low temperatures each q_i tends to $\exp(-\varepsilon_{i0}/kT)$ it follows that

$$(F_{tr} - \sum_i N_i \varepsilon_{i0})/kT \to -\ln N! + \sum_i \ln N_i! \quad (T \to 0) \qquad 2.16.7$$

and consequently

$$S_{\text{tr}} \to k \ln N! - k \sum_i \ln N_i! \quad (T \to 0). \qquad 2.16.8$$

§2.17 Third law of thermodynamics

After this brief and necessarily incomplete sketch of statistical thermodynamics we recall the formulation of the *third law of thermodynamics* which we adopted in §1.66.

By the standard methods of statistical thermodynamics it is possible to derive for certain entropy changes general formulae which cannot be derived from the zeroth, first, or second laws of thermodynamics. In particular we can obtain formulae for entropy changes in highly disperse systems (i.e. dilute gases), those in very cold systems (i.e. when $T \to 0$), and those associated with the mixing of very similar substances (e.g. isotopes).

We shall now briefly state these deductions from statistical thermodynamics without giving detailed derivations.

In the first place we consider the translational term in the thermodynamic functions of a highly disperse system, i.e. a dilute gas, containing N_i molecules of type i having a mass m_i. The contributions to the Helmholtz function F and to the Massieu function J are given by

$$F_{\text{tr}} = -TJ_{\text{tr}} = -kT \sum_i \ln\{(2\pi m_i kT/h^2)^{\frac{3}{2}N_i} V^{N_i}/N_i!\}. \qquad 2.17.1$$

The corresponding contribution S_{tr} to the entropy S is

$$S_{\text{tr}} = k \sum_i \ln\{(2\pi m_i kT/h^2)^{\frac{3}{2}N_i} V^{N_i}/N_i!\} + k \sum_i \tfrac{3}{2} N_i. \qquad 2.17.2$$

In particular in a gaseous single substance

$$S_{\text{tr}} = k \ln\{(2\pi mkT/h^2)^{\frac{3}{2}N} V^N/N!\} + \tfrac{3}{2} Nk. \qquad 2.17.3$$

Using Stirling's formula for large N

$$\ln N! = N \ln N - N \qquad 2.17.4$$

we can rewrite (3) as

$$S_{\text{tr}}/Nk = \ln\{(2\pi mkT/h^2)^{\frac{3}{2}} V/N\} + \tfrac{5}{2}. \qquad 2.17.5$$

Anticipating the formula given in §§ 3.13–3.14 for the pressure P of a single perfect gas

$$P = NkT/V \qquad 2.17.6$$

we can replace (5) by

$$S_{tr}/Nk = \ln\{(2\pi m/h^2)^{\frac{3}{2}}(kT)^{\frac{5}{2}}/P\} + \tfrac{5}{2}. \qquad 2.17.7$$

We shall use the equivalent of formula (7) in §3.26.

Our second example is the translational term in the entropy of a crystal of a pure substance. As the temperature tends towards zero, this contribution tends to zero. We shall return to this result in §3.51.

Finally we consider the entropy of mixtures of very similar substances such as isotopes. If several very similar substances, such as isotopes, all at the same temperature and same number of molecules per unit volume are mixed, the temperature and number of molecules per unit volume being kept unchanged, the entropy is increased by ΔS given by

$$\Delta S/k = \ln N! - \sum_i \ln N_i! \qquad 2.17.8$$

where N_i denotes the number of molecules of the species i and $N = \Sigma_i N_i$ denotes the total number of molecules of all species. Using Stirling's formula (4), we can rewrite (8) as

$$\Delta S/k = \sum_i N_i \ln(N/N_i). \qquad 2.17.9$$

This applies to solids, and incidentally to liquids, as well as to gases, provided the various species are sufficiently similar, e.g. isotopic. We shall make use of this in §3.55.

When we meet these formulae again in chapter 3, the number of molecules N_i will be replaced by the amount of substance $n_i = N_i/L$ and correspondingly the Boltzmann constant k will be replaced by the gas constant $R = Lk$.

CHAPTER 3

SYSTEMS OF A SINGLE COMPONENT

§3.01 *Single components and single phases*

The present chapter is devoted to single component systems, both single phase and multiphase. Most of the formulae of the present chapter which relate to a single closed phase are applicable also to a multicomponent closed phase. Formulae relating to an open phase or to a multiphase system are on the contrary more complicated in a multicomponent system than in a single component system. Such formulae will be derived in chapter 4.

§3.02 *Dependence of entropy on temperature*

The experimental determination of entropy and thermodynamic temperature are interlinked. We have not yet described how either can be directly or conveniently measured. In §3.12 we shall describe an especially convenient way of measuring thermodynamic temperature. Anticipating this result, that is to say assuming we have a thermometer which measures thermodynamic temperature, we shall now describe how we can determine the dependence of entropy on temperature at constant pressure.

For a single closed phase, we have according to (1.28.8)

$$dH = TdS + VdP \qquad 3.02.1$$

or if we keep the pressure constant

$$dH = TdS \quad (P\text{ const.}). \qquad 3.02.2$$

If then we supply heat q to a single component system, since the change in the system must be reversible, regardless of whether the process of supplying the heat is reversible (see §1.13), we have

$$q = dH = TdS \quad (P\text{ const.}). \qquad 3.02.3$$

Furthermore if we supply the heat by means of an electric element, the

SYSTEMS OF A SINGLE COMPONENT

heat will be equal to the electrical work done on the element. To be precise, if the potential difference across the element is E and the current flowing is i, then in a time t the heat given up by the element to the system is Eit. Since E, i, and t are all measurable we can calculate q. We see then that, apart from experimental difficulties, there is no difficulty in principle in measuring increases of H. As already mentioned we are postulating, in anticipation of §3.12, the availability of a thermometer which measures T. We thus obtain a direct experimental relationship between T and H, or rather changes in H which itself contains an arbitrary additive constant.

As an illustration we show in figure 3.1 the experimental data* for one

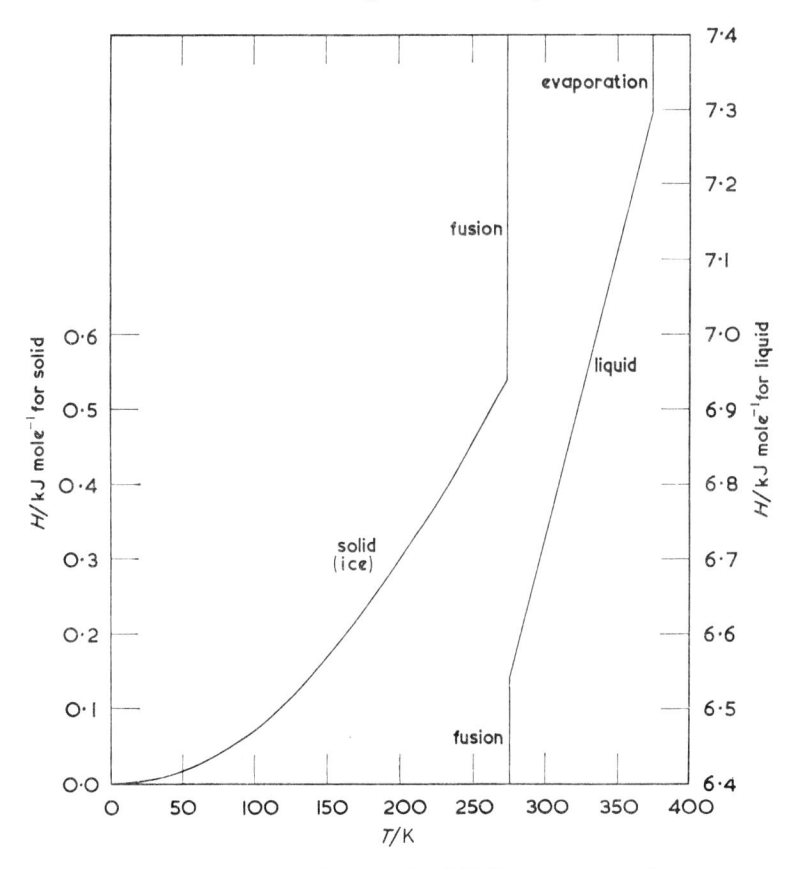

Fig. 3.1. Enthalpy of one mole of H_2O at one atmosphere

mole of H_2O at a constant pressure of one atmosphere. The first curve on the left applies to ice from 0 K to 273.15 K, at which temperature the ice

* Giauque and Stout, J. Amer. Chem. Soc. 1936 58 1144.

melts; the value of the enthalpy then rises at constant temperature by an amount equal to the *proper enthalpy of fusion*. As this change would run off the paper the scale of the curve for the liquid has been shifted downwards by 6.4 kJ mole^{-1}. The curve on the right of the figure runs from 273.15 K to 373.15 K at which temperature the water boils; the value of the enthalpy again rises at constant temperature and runs off the diagram.

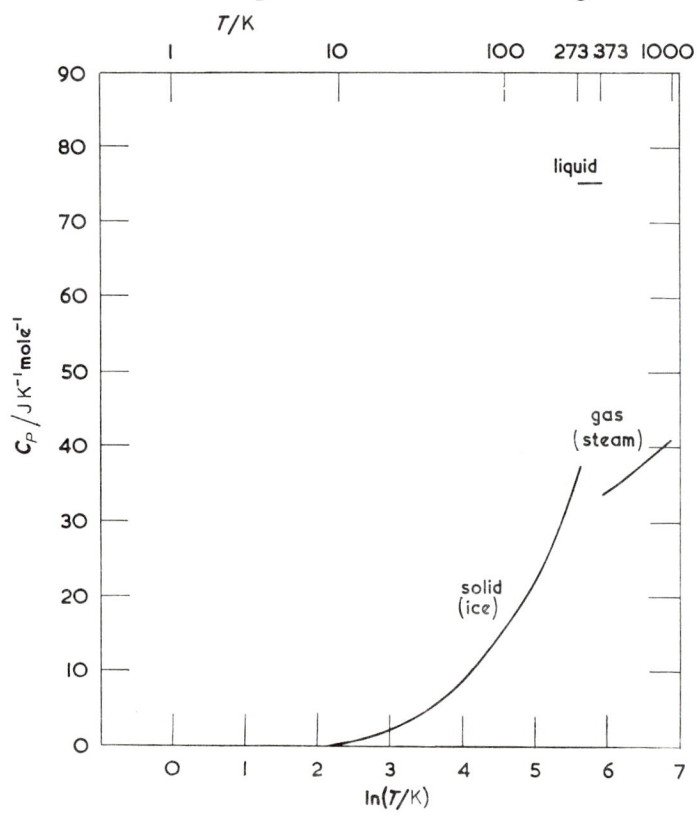

Fig. 3.2. Heat capacity of H_2O at one atmoshpere

In figure 3.2 we show the data in a somewhat different form, $(\partial H/\partial T)_P$ for unit amount or $(\partial H_m/\partial T)_P$ being now plotted against $\ln T$. The three separate curves apply to ice, liquid water, and steam respectively. From (2) we have

$$S_m = \int dS_m = \int dH_m/T = \int (\partial H_m/\partial T)_P \, d\ln T. \qquad 3.02.4$$

We see then that apart from an arbitrary constant the proper entropy of ice at a temperature T is equal to the area under the part of the curve to the left of T. In particular the proper entropy of ice at the fusion point exceeds that at 0 K by an amount corresponding to the whole area under the ice curve. This amounts to 38.09 J K^{-1} mole^{-1}.

When the ice changes to liquid water there is an increase of the proper entropy called the *proper entropy of fusion* equal to the *proper enthalpy of fusion* divided by the temperature. Thus

$$\Delta S_m = \Delta H_m/T = 6007 \text{ J mole}^{-1}/273.15 \text{ K} = 21.99 \text{ J K}^{-1} \text{ mole}^{-1}.$$

Suppose we wish to know by how much the proper entropy of steam at 1000 K and 1 atm exceeds the proper entropy of ice at 0 K. We have to sum the following contributions.

(a) Ice at 0 K → ice at 273.15 K
$\Delta S_m = 38.09$ J K^{-1} mole^{-1} (area under ice curve).

(b) Ice at 273.15 K → liquid water at 273.15 K
$\Delta S_m = \Delta H_m/T = 6007$ J mole^{-1}/273.15 K $= 21.99$ J K^{-1} mole^{-1}.

(c) Water at 273.15 K → water at 373.15 K
$\Delta S_m = 23.52$ J K^{-1} mole^{-1} (area under water curve).

(d) Water at 373.15 K → steam at 373.15 K
$\Delta S_m = \Delta H_m/T = 40656$ J mole^{-1}/373.15 K $= 108.95$ J K^{-1} mole^{-1}.

(e) Steam at 373.15 K → steam at 1000 K
$\Delta S_m = 35.8$ J K^{-1} mole^{-1} (area under steam curve).

By addition we obtain for the change
Ice at 0 K → steam at 1000 K (at 1 atm)
$\Delta S_m = 228.4$ J K^{-1} mole^{-1}.

In the case of some substances there may be several solid phases with transition temperatures at which the proper entropy increase ΔS_m is equal to the increase ΔH_m divided by T. Such transitions cause no difficulty.

We see then that the determination of changes in the entropy of any single substance through any range of temperature at constant pressure becomes straightforward provided the heat input and thermodynamic temperature can be measured.

§3.03 *Heat capacity at constant pressure*

In the previous section we saw that the determination of entropy requires us to use the relation

$$T(\partial S/\partial T)_P = (\partial H/\partial T)_P. \qquad 3.03.1$$

This quantity is called the *heat capacity at constant pressure* of the system.

The heat capacity per unit amount or the *proper heat capacity at constant pressure* will be denoted by C, or by C_P when it is desired to emphasize the contrast with another quantity C_V defined in §3.06. Thus

$$C = C_P = T(\partial S_m/\partial T)_P = (\partial H_m/\partial T)_P. \qquad 3.03.2$$

The importance of C is that it forms the connecting link between S and H. One measures directly H as a function of T and then determines S by the relation (1). Importance was in the past attached to C for a completely different, accidental, and inadequate reason, namely that for many substances at the most usual temperatures C happens to be insensitive to the temperature. For example we notice from figure 3.2 that C is nearly constant for liquid water, only roughly constant for steam, but not at all constant for ice.

The heat capacity at constant pressure per unit mass or the *specific heat capacity at constant pressure* is denoted by c_P.

§3.04 So-called mechanical equivalent of heat

Before the classical experiments of Joule, the relationship between work, heat, and energy was not understood. These experiments established that within the experimental error the work or energy input required to raise the temperature of a given mass of water through a given temperature range is independent of the particular mechanism used. The formulation of the first law of thermodynamics is largely based on these experiments and later repetitions and improvements of them. Since Joule's experiments were performed before the formulation of the first law, Joule's terminology was necessarily different from the terminology based on familiarity with the laws of thermodynamics. Joule described some of his experiments as the 'determination of the mechanical equivalent of heat'. Once the principles of thermodynamics are understood, this phrase becomes meaningless. What Joule in fact did was

(a) to establish an experimental basis for the formulation of the first law of thermodynamics;

(b) to measure the heat capacity of water.

Before the first law of thermodynamics was formulated or understood the unit of heat was the quantity of heat required to raise the temperature of one gramme of water by one degree and this unit was called the calorie. Work was however measured in mechanical units. It is found that the specific heat capacity of liquid water is approximately 4.18 J K^{-1}g^{-1} but in fact varies appreciably with the temperature. Nowadays almost all accurate thermal experiments involve measurements of volts, amperes, and seconds leading to energy values in joules. Moreover in 1948 the Conférence Générale des Poids et Mesures adopted* a recommendation of the International Union of Pure and Applied Physics that all accurate calorimetric data should be expressed in joules. It is difficult to understand why the use of the calorie as a unit persists, except as a habit. The most careful experimental workers in thermochemistry have abandoned the old definition of the calorie and have replaced it by the more satisfactory definition

$$1 \text{ calorie} = 4.184 \text{ joules exactly.}$$

The calorie thus defined is called the *thermochemical calorie*.

As already mentioned the specific heat capacity of liquid water is approximately, but by no means exactly, independent of the temperature. Its value is very near 1 cal $K^{-1}g^{-1}$ at 290 K. The best experimental values at a few other temperatures are as follows[†]:

$$
\begin{array}{rll}
\text{At} & 0\,°C & 4.2174 \text{ J K}^{-1}\text{g}^{-1} \\
& 15\,°C & 4.1855 \text{ J K}^{-1}\text{g}^{-1} \\
& 16\,°C & 4.1846 \text{ J K}^{-1}\text{g}^{-1} \\
& 17\,°C & 4.1837 \text{ J K}^{-1}\text{g}^{-1} \\
& 20\,°C & 4.1816 \text{ J K}^{-1}\text{g}^{-1} \\
& 25\,°C & 4.1793 \text{ J K}^{-1}\text{g}^{-1}.
\end{array}
$$

§3.05 *Dependence of entropy on pressure*

In §3.02 we saw how the variation of entropy with the temperature at a constant pressure is determined experimentally. In order to determine the entropy as a function of temperature and pressure, this procedure has to be supplemented by a determination of the dependence of entropy on pressure at constant temperature. This dependence is given according to Maxwell's relation (1.48.2)

$$(\partial S_m/\partial P)_T = -\alpha V_m \qquad 3.05.1$$

which when integrated becomes

$$S_m(T, P_2) - S_m(T, P_1) = -\int_{P_1}^{P_2} \alpha V_m \, dP. \qquad 3.05.2$$

If we differentiate (1) with respect to T, keeping P constant, and multiply by

---

* C.R. Conférence Générale des Poids et Mesures 1948.
† Stille, Messen und Rechnen in der Physik, Vieweg 1955, p. 289.

T we obtain

$$(\partial C/\partial P)_T = -T\{\partial(\alpha V_m)/\partial T\}_P = -\alpha^2 T V_m - T(\partial\alpha/\partial T)_P V_m. \qquad 3.05.3$$

For solids and liquids the second term on the right may be small compared with the first; for gases on the contrary the two terms are nearly equal and opposite.

§3.06 Heat capacity at constant volume

In §§3.02–3.05 we have collected the most important formulae required to determine the entropy in terms of temperature and pressure. There is an analogous set of relations for the alternative choice of temperature and volume as independent variables. Except for gases these relations are considerably less useful than those relating to the independent variables T, P. We shall refer to them briefly, without giving detailed derivations; these are in all cases analogous to those in the T, P system.

For the dependence of entropy on temperature at constant volume, we have

$$(\partial S/\partial T)_V = T^{-1}(\partial U/\partial T)_V \qquad 3.06.1$$

which when integrated becomes

$$S(T_2, V) - S(T_1, V) = \int_{T_1}^{T_2} (\partial U/\partial T)_V \, d\ln T. \qquad 3.06.2$$

Correspondingly for the dependence of entropy on volume at constant temperature, we have according to Maxwell's relation (1.47.3)

$$(\partial S/\partial V)_T = \alpha/\kappa_T \qquad 3.06.3$$

which when integrated becomes

$$S(T, V_2) - S(T, V_1) = \int_{V_1}^{V_2} (\alpha/\kappa_T) dV. \qquad 3.06.4$$

The quantity $(\partial U/\partial T)_V$ in formula (1) is called the *heat capacity at constant volume* of the system. The corresponding quantity referred to unit amount, the *proper heat capacity at constant volume*, is denoted by C_V. Thus

$$C_V = T(\partial S_m/\partial T)_V = (\partial U_m/\partial T)_V. \qquad 3.06.5$$

The heat capacity at constant volume per unit mass or the *specific heat capacity at constant volume* is denoted by c_V.

§3.07 Relation between heat capacities

According to the meaning of partial differential coefficients we have

$$(\partial S/\partial T)_P = (\partial S/\partial T)_V + (\partial S/\partial V)_T(\partial V/\partial T)_P. \qquad 3.07.1$$

Substituting from (1.46.2), (1.46.3), and from Maxwell's relation (1.47.3) into (1) we obtain

$$(\partial S/\partial T)_P = (\partial S/\partial T)_V + \alpha^2 V/\kappa_T. \qquad 3.07.2$$

Applying (2) to unit quantity, multiplying by T, and using the definitions (3.03.2) and (3.06.5) of C_P and C_V respectively we find

$$C_P = C_V + \alpha^2 T V_m/\kappa_T. \qquad 3.07.3$$

Since in a stable phase none of the quantities α^2, T, V_m, κ_T can ever be negative, it follows that C_P can never be less than C_V.

C_V is much more difficult to measure than C_P. If the value of C_V is required, it is usual to measure C_P and then calculate C_V from (3). There seems to be a widespread belief that in the comparison of a theoretical model with experimental data the most suitable quantity for the comparison is C_V. This is however a misconception. Any theoretical model susceptible to explicit analytical treatment, such as for example Debye's model of a crystal discussed in §3.33, leads to an explicit formula for the Helmholtz function and so by differentiation with respect to T to explicit formulae for the energy and the entropy, both of which are directly measurable as a function of temperature. These are clearly the most suitable quantities for comparison between a theoretical model of a crystal and experimental data. There is no reason or excuse for a further differentiation to obtain a heat capacity except in the hypothetical case that the agreement between theory and experiment is so good that a more sensitive test is required.

§3.08 Adiabatic compressibility

In §1.46 the isothermal compressibility κ_T was defined by

$$\kappa_T = -V^{-1}(\partial V/\partial P)_T. \qquad 3.08.1$$

The adiabatic compressibility κ_S is similarly defined by

$$\kappa_S = -V^{-1}(\partial V/\partial P)_S. \qquad 3.08.2$$

These two compressibilities are interrelated as follows.

$$\frac{\kappa_S}{\kappa_T} = \frac{(\partial V/\partial P)_S}{(\partial V/\partial P)_T} = \frac{(\partial S/\partial P)_V(\partial P/\partial T)_V}{(\partial S/\partial V)_P(\partial V/\partial T)_P} = \frac{(\partial S/\partial T)_V}{(\partial S/\partial T)_P} = \frac{C_V}{C_P}. \qquad 3.08.3$$

90   SYSTEMS OF A SINGLE COMPONENT

The speed a of propagation of compressional sound waves in an isotropic medium is given by

$$a^2 = V_m/M\kappa_S \qquad 3.08.4$$

where M is the proper mass. From (3), (4), and (3.07.3) we deduce

$$C_P/C_V - 1 = \alpha^2 T M a^2/C_P. \qquad 3.08.5$$

This is the most useful formula for determining C_P/C_V since all the quantities on the right, in contrast to C_V, and in solids κ_T, are readily measurable.

§3.09   Condensed phases and gases

Solids and liquids, which we shall class together under the name *condensed phases*, are under most conditions sharply distinguished from gases by a striking difference in compressibility. It is true that in the neighbourhood of the critical point, as we shall see in §3.44 the distinction between liquid and gas disappears, but at least for liquids or solids at temperatures well below the critical temperature and for gases at pressures well below the critical pressure the contrast is striking.

In a condensed phase at a given temperature the compressibility is small and practically independent of the pressure. That is to say that to a first approximation the volume is independent of the pressure and to a better approximation decreases linearly with the pressure. In a gas on the other hand the compressibility is much greater and far from independent of the pressure. In fact it is at least roughly true that the volume of a gas varies inversely as the pressure, according to Boyle's Law. In other words it is PV, not V, which to a first approximation is independent of P.

§3.10   Isothermal behaviour of a gas

It is reasonable to expect that the volume of any phase at constant temperature can be expressed as a power series in the density or the reciprocal of the proper volume. In view of what was said in the previous section, the leading term will in the case of a gas be an inverse first power. We may accordingly write

$$PV_m = A(1 + B_2/V_m + B_3/V_m^2 + B_4/V_m^3 + \ldots). \qquad 3.10.1$$

In principle the number of terms is indefinite, depending on the accuracy aimed at. Up to quite high pressures, of say a hundred atmospheres, it is often unnecessary to use more than three terms. At pressures up to a few atmospheres even the third term is often negligible, only the first two terms being required.

All the coefficients $A, B_2, B_3, \ldots$ of course depend on the temperature, but not on the volume.

B_2 is called* the second virial coefficient; B_3 is called the third virial coefficient and so on.

For the sake of simplicity and brevity we shall replace (1) by the three term expression

$$PV_m = A(1 + B_2/V_m + B_3/V_m^2). \qquad 3.10.2$$

There is in principle no difficulty in inserting further terms if required.

We can invert the series in (2) to obtain the expansion in powers of P

$$PV_m = A + B_2 P + A^{-1}(B_3 - B_2^2)P^2 + \ldots . \qquad 3.10.3$$

It is mainly a question of convenience whether one uses a formula of type (2) or of type (3). For our immediate purpose, it is more convenient to use (3). Fortunately at ordinary pressures all terms beyond the second are usually negligible and either formula then reduces to

$$V = A/P + B_2 \quad \text{(low pressures)}. \qquad 3.10.4$$

From (3) we readily obtain the proper Gibbs function G_m as a function of pressure by substituting into (3) and integrating. We thus find

$$G_m(T, P) - G_m(T, P^\ominus) = A \ln(P/P^\ominus) + B_2(P - P^\ominus) + \tfrac{1}{2}A^{-1}(B_3 - B_2^2)(P^2 - P^{\ominus 2})$$
$$3.10.5$$

where $P^\ominus$ is a standard pressure, which may be chosen arbitrarily, but in this book is always 1 atm. This does not imply that pressures must necessarily be measured in atm. In other units we have for example

$$P^\ominus = 1 \text{ atm} = 76 \text{ cmHg} = 760 \text{ mmHg} =$$
$$= 1.01325 \times 10^6 \text{ dyn cm}^{-2} = 1.01325 \times 10^5 \text{ J m}^{-3} = 1.01325 \text{ bar}.$$

We obtain for the proper enthalpy H_m by substituting (5) into (1.49.3)

$$H_m(T, P) - H_m(T, P^\ominus) = \{d(T^{-1}A)/dT^{-1}\} \ln(P/P^\ominus)$$
$$+ \{d(T^{-1}B_2)/dT^{-1}\}(P - P^\ominus)$$
$$+ \tfrac{1}{2}\{d(T^{-1}A^{-1}B_3 - T^{-1}A^{-1}B_2^2)/dT^{-1}\}(P^2 - P^{\ominus 2}). \qquad 3.10.6$$

---

* Onnes and Keesom, 'Die Zustandsgleichung'. Commun. Phys. Lab. Univ. Leiden, 11: Suppl. 23 1912; Encyk. Math. Wiss., 5: No. 10, p. 615.

§3.11 Throttling

In the previous section we set up a formula for V_m as a function of P based on experiment. From this we deduced a formula for G_m and thence a formula for H_m. We shall now consider the comparison between this formula for H_m and experiment.

The experiment which supplies the most direct information concerning the dependence of H on the pressure at constant temperature is known as *throttling*. The first experiment of this type was performed by Joule and Lord Kelvin (William Thomson); it is accordingly often called the *Joule–Thomson experiment*. In this experiment a stream of gas in a thermally insulated container is forced through a plug, the pressure being greater on the near side than on the far side and the temperatures of the gas stream approaching and leaving the plug are measured on an arbitrary scale; we denote the temperatures on this scale by θ to distinguish them sharply from thermodynamic temperatures T, which we do not yet know how to measure. Consider now the whole system in a steady state such that in a given time a certain mass of gas is pushed in at a pressure P_1 and during the same time an equal mass of gas streams away at a pressure P_2. We use the subscript 1 to denote the state of the gas being pushed in and the subscript 2 to denote that of the gas streaming away. Then during the time considered a mass of gas of pressure P_2, volume V_2, temperature θ_2, and energy U_2 is displaced by an equal mass of pressure P_1, volume V_1, temperature θ_1, and energy U_1. During this time the work done on the system is $P_1 V_1 - P_2 V_2$. Since the system is supposed thermally insulated this work must be equal to the increase in energy of the system. Thus

$$U_2 - U_1 = P_1 V_1 - P_2 V_2. \qquad 3.11.1$$

Hence according to the definition of H in §1.28, we have

$$H_2 = H_1 \qquad 3.11.2$$

or choosing θ, P as independent variables

$$H(\theta_1, P_1) = H(\theta_2, P_2). \qquad 3.11.3$$

Suppose that the effect of throttling is to cool the gas, so that θ_2 is a lower temperature than θ_1, then there is no difficulty in principle in heating the throttled gas at constant pressure so as to restore its temperature from θ_2 to θ_1. If the heat required for this purpose is measured, we then know the value of

$$H(\theta_1, P_2) - H(\theta_2, P_2) \qquad 3.11.4$$

which according to (3) is equal to

$$H(\theta_1, P_2) - H(\theta_1, P_1). \qquad 3.11.5$$

If on the contrary the effect of throttling is to warm the gas, then one must do a subsidiary experiment to determine the heat required to raise the temperature of the gas at the pressure P_2 from θ_1 to θ_2. We thus obtain an experimental value of

$$H(\theta_2, P_2) - H(\theta_1, P_2) \qquad 3.11.6$$

which according to (3) is equal to

$$H(\theta_1, P_1) - H(\theta_1, P_2). \qquad 3.11.7$$

In either case we obtain experimental values of $H(\theta_1, P_2) - H(\theta_1, P_1)$ positive in the former case, negative in the latter. It is important to notice that this experiment does not require any knowledge of how the arbitrary θ scale of temperature is related to the thermodynamic scale or to any other scale.

We shall now describe the experimental results obtained. It is found that, whatever the temperature, $H(P_1) - H(P_2)$ is at least approximately proportional to $P_1 - P_2$ and is not sensitive to the absolute magnitude of P_1. It is quite certain that at low values of P_2, the value of $H(P_1) - H(P_2)$ does not tend towards infinity, which is what one should expect from formula (3.10.6) owing to the term in $\ln P$. In short the Joule–Thomson experiment shows that the first term on the right of formula (3.10.6) is in fact missing and the linear term in P is therefore the leading one.

§3.12 *Measurement of thermodynamic temperature*

In principle to determine T, one should measure ΔH and ΔG for the same isothermal process and by comparing these obtain a differential equation for T. In particular, one can determine the coefficients A, B_2, B_3, in the formula for G simply by pressure measurements and one can obtain independent measurements of the corresponding coefficients in H, from the throttling experiment. By comparison we obtain information concerning T, but admittedly in a rather awkward form.

To our agreeable surprise the information is in a strikingly convenient form in the case of the coefficient A. The throttling experiment shows unmistakably that H contains no term tending to infinity as P tends to zero, that is to say no term in $\ln P$. Hence from (3.10.6) we conclude that

$$\mathrm{d}(T^{-1}A)/\mathrm{d}T^{-1} = 0 \qquad 3.12.1$$

which is equivalent to
$$A \propto T. \qquad 3.12.2$$

At last we have found a simple, direct, and reliable way of determining the ratio of any two thermodynamic temperatures. We use as a thermometer a fixed quantity of gas. We measure several pairs of values of P, V at the same temperature and extrapolate the product PV to $P=0$, thus obtaining the value of A. We repeat this at another temperature thus obtaining another value of A. Then the ratio of these two values of A is equal to the ratio of the two values of T. Having thus established a way of determining the ratio of any two temperatures, the numerical values are fixed by the convention described in §1.21 so that the triple point of water is 273.16 degrees Kelvin and this is called the Kelvin scale.

§3.13 *The gas constant and the mole*

We have found that the coefficient A is directly proportional to the temperature. We accordingly write
$$A = RT \qquad 3.13.1$$

where R is independent of temperature and pressure. R also becomes independent of the nature of the gas when the unit of amount is suitably chosen, e.g. by choosing the mole. From a purely thermodynamic view-point the amount of substance may be defined without any reference to molecular theory by assigning a common value of R to all gaseous substances.

The accepted definition* of the *mole* is that amount of substance which contains the same number of molecules as there are atoms in 0.012 kg of ^{12}C. In this definition 'molecules' includes ions, radicals, electrons, etc. The number of atoms in 0.012 kg of ^{12}C is 0.60225×10^{24}. Consequently the factor for transforming moles to molecules, called *the Avogadro constant L*, is
$$L = 0.60225 \times 10^{24} \text{ mole}^{-1}$$
and the factor for converting molecules to moles is
$$L^{-1} = 1.66044 \times 10^{-24} \text{ mole}.$$

It can be shown by statistical mechanics or kinetic theory that the gas constant has a common value for all gases and is related to the Boltzmann constant k introduced in chapter 2 by

---

* I.U.P.A.P. Symbols, Units and Nomenclature in Physics, 1961 p. 19, 1965 p. 25; I.U.P.A.C. Information Bulletin Number 24 1965 p. 4.

$$R = Lk$$
$$= 6.0225 \times 10^{24} \text{ mole}^{-1} \times 1.3805 \times 10^{-23} \text{ J K}^{-1}$$
$$= 8.3143 \text{ J K}^{-1} \text{ mole}^{-1}.$$

§3.14 Isothermal equation of a gas

When we replace A by RT in (3.10.2) we obtain

$$PV_m/RT = 1 + B_2/V_m + B_3/V_m^2. \qquad 3.14.1$$

It is convenient to call PV_m/RT the *compression factor*. It is sometimes denoted by Z. When the density is sufficiently low for B_2/V_m, and a fortiori B_3/V_m^2, to be negligible the gas is called a *perfect gas*.

§3.15 Absolute activity

In chapter 2 we met a quantity called the *absolute activity* which plays an important part in the statistical thermodynamics of open systems. We now give a purely thermodynamic definition of the absolute activity. This is somewhat out of place in the present chapter, but we could not give it earlier because it involves the gas constant R. We accordingly define* the absolute activity as related to the molar chemical potential μ by

$$\lambda = \exp(\mu/RT) \qquad 3.15.1$$

or

$$\mu = RT \ln \lambda. \qquad 3.15.2$$

Whereas it is not necessary to use λ as well as μ, we shall find that the absolute activity λ is often a convenient function in the study of equilibria of all kinds whether involving one species or several. In §1.44 we showed that for the most general chemical reaction represented symbolically by

$$0 = \sum_B \nu_B B \qquad 3.15.3$$

the condition for equilibrium is according to (1.44.13)

$$\sum_B \nu_B \mu_B = 0. \qquad 3.15.4$$

We now see that this condition can equally be expressed in terms of absolute activities in the form

$$\prod_B (\lambda_B)^{\nu_B} = 1. \qquad 3.15.5$$

We recall that each ν_B is negative for a reactant and positive for a product in the chemical equation for the process.

---
* Fowler and Guggenheim, Statistical Thermodynamics, Cambridge University Press 1939 p. 66.

In particular the condition for the equilibrium distribution of a species i between the phases α and β may be written

$$\lambda_i^\alpha = \lambda_i^\beta. \qquad 3.15.6$$

We can now rewrite formulae (1.28.12), (1.28.15), and (1.28.16) as

$$R^{-1}\mathrm{d}S = (RT)^{-1}\mathrm{d}U + (P/RT)\mathrm{d}V - \sum_i \ln \lambda_i \mathrm{d}n_i \qquad 3.15.7$$

$$-R^{-1}\mathrm{d}(\mathscr{F}/T) = R^{-1}\mathrm{d}J = -(U/R)\mathrm{d}(T^{-1}) + (P/RT)\mathrm{d}V - \sum_i \ln \lambda_i \mathrm{d}n_i \qquad 3.15.8$$

$$-R^{-1}\mathrm{d}(G/T) = R^{-1}\mathrm{d}Y = -(H/R)\mathrm{d}(T^{-1}) - (V/RT)\mathrm{d}P - \sum_i \ln \lambda_i \mathrm{d}n_i. \qquad 3.15.9$$

§3.16 Thermodynamic functions of a gas

When we set $A = RT$ in formula (3.10.5) for a gas we obtain

$$G_m(T, P) - G_m(T, P^\ominus) = RT \ln(P/P^\ominus) + B_2(P - P^\ominus)$$
$$+ \tfrac{1}{2}(RT)^{-1}(B_3 - B_2^2)(P^2 - P^{\ominus 2}). \qquad 3.16.1$$

When we set

$$\mu^\ominus(T) = G_m(T, P^\ominus) - B_2 P^\ominus - \tfrac{1}{2}(RT)^{-1}(B_3 - B_2^2)P^{\ominus 2} \qquad 3.16.2$$

formula (1) simplifies to

$$\mu(T, P) = G_m(T, P) = \mu^\ominus(T) + RT \ln(P/P^\ominus) + B_2 P + \tfrac{1}{2}(RT)^{-1}(B_3 - B_2^2)P^2. \qquad 3.16.3$$

Substituting (3.15.2) into (3) we obtain

$$\ln(\lambda/\lambda^\ominus) = \ln(P/P^\ominus) + B_2 P/RT + \tfrac{1}{2}(B_3 - B_2^2)(P/RT)^2 \qquad 3.16.4$$

where

$$\lambda^\ominus = \exp(\mu^\ominus/RT) \qquad 3.16.5$$

so that $\lambda^\ominus$ is a function of the temperature only.

From (3) we derive immediately

$$-S_m = \mathrm{d}\mu^\ominus/\mathrm{d}T + R\ln(P/P^\ominus) + (\mathrm{d}B_2/\mathrm{d}T)P$$
$$+ \tfrac{1}{2}P^2 \mathrm{d}(R^{-1}T^{-1}B_3 - R^{-1}T^{-1}B_2^2)/\mathrm{d}T \qquad 3.16.6$$

$$H_m = \mu^\ominus - T\mathrm{d}\mu^\ominus/\mathrm{d}T + (B_2 - T\mathrm{d}B_2/\mathrm{d}T)P$$
$$+ \tfrac{1}{2}P^2\{R^{-1}T^{-1}B_3 - R^{-1}T^{-1}B_2^2 - T\mathrm{d}(R^{-1}T^{-1}B_3 + R^{-1}T^{-1}B_2^2)/\mathrm{d}T\}. \qquad 3.16.7$$

$\mu^\ominus$ is called the *standard chemical potential*, $-\mathrm{d}\mu^\ominus/\mathrm{d}T$ the *standard proper entropy* of gas, and $\mu^\ominus - T\mathrm{d}\mu^\ominus/\mathrm{d}T$ the *standard proper enthalpy* of gas.

In general there does not exist any state in which simultaneously $\mu = \mu^\ominus$ and $\mathrm{d}\mu/\mathrm{d}T = \mathrm{d}\mu^\ominus/\mathrm{d}T$. For this reason expressions such as 'entropy in the standard state' should be avoided.

§3.17 Fugacity

It is often convenient to use a quantity p called the *fugacity* of the gas defined* by

$$\lambda/\lambda^\ominus = p/P^\ominus \qquad 3.17.1$$

$$p/P \to 1 \quad \text{as} \quad P \to 0. \qquad 3.17.2$$

An alternative equivalent definition of p is

$$\ln p = \ln P^\infty + \int_{P^\infty}^{P} V_m \, dP \qquad 3.17.3$$

where P^∞ is a pressure sufficiently small so that $p = P$. In a perfect gas the fugacity is equal to the pressure. For a real gas at moderate pressures according to (3.16.4)

$$\ln p = \ln P + B_2 P/RT + \tfrac{1}{2}(B_3 - B_2^2)(P/RT)^2. \qquad 3.17.4$$

The simplicity attained by the introduction of the fugacity is one of appearance or elegance. It leads to nothing quantitative unless we express the fugacity in terms of the pressure and then we are back where we started.

§3.18 Gases at high temperatures

By means of statistical mechanics the second, third, fourth, ... virial coefficients can be expressed in terms of integrals, called 'cluster integrals', over the position coordinates of clusters of two, three, four, ... molecules. The evaluation of these cluster integrals, except that for the second virial coefficient, is in general laborious. The effort required depends on the form of the dependence of the interaction energy w on the distance r between two molecules. The cluster integrals become much more tractable for the simplest model of non-attracting rigid spheres defined by

$$w = \infty \quad\quad r < \sigma \qquad 3.18.1$$

$$w = 0 \quad\quad r > \sigma. \qquad 3.18.2$$

For this simple model the several virial coefficients are conveniently expressed in terms of a proper volume b defined by

$$b = \tfrac{2}{3}\pi L \sigma^3 \qquad 3.18.3$$

as follows

$$B_2/b = 1 \qquad 3.18.4$$

---

* Lewis, Proc. Nat. Acad. Sci. U.S. 1901 **37** 49; Z. Phys. Chem. 1901 **38** 205.

$$B_3/b^2 = 0.625 \qquad 3.18.5$$

$$B_4/b^3 = 0.287 \qquad 3.18.6$$

$$B_5/b^4 = 0.110. \qquad 3.18.7$$

This model and these virial coefficients describe the limiting behaviour of a gas at high temperatures.

A very simple formula* giving a rough approximation to the accurate virial expansion for non-attracting rigid spheres is

$$\begin{aligned}PV_m/RT &= (1 - b/4V_m)^{-4} \\ &= 1 + 4(b/4V_m) + 10(b/4V_m)^2 + 20(b/4V_m)^3 + 35(b/4V_m)^4\end{aligned} \qquad 3.18.8$$

as compared with the accurate expansion†

$$PV_m/RT = 1 + 4(b/4V_m) + 10(b/4V_m)^2 + 18.36(b/4V_m)^3 + 28.2(b/4V_m)^4. \qquad 3.18.9$$

§3.19 Slightly imperfect gases

We shall call a gas *slightly imperfect* when the pressure or density is sufficiently low for all virial coefficients to be ignored except the second B_2. The thermodynamic properties of a slightly imperfect gas are given by the following formulae where we have dropped the subscript from B_2

$$\mu = G_m = \mu^\ominus + RT \ln(P/P^\ominus) + BP \qquad 3.19.1$$

$$\ln \lambda = \ln \lambda^\ominus + \ln(P/P^\ominus) + BP/RT \qquad 3.19.2$$

$$-S_m = d\mu^\ominus/dT + R \ln(P/P^\ominus) + (dB/dT)P \qquad 3.19.3$$

$$H_m = \mu^\ominus - T d\mu^\ominus/dT + (B - T dB/dT)P \qquad 3.19.4$$

$$PV_m = RT + PB \qquad 3.19.5$$

$$C_P = T(\partial S_m/\partial T)_P = (\partial H_m/\partial T) = -T d^2\mu^\ominus/dT^2 - TP(d^2B/dT^2) \qquad 3.19.6$$

$$\alpha V_m = (\partial V_m/\partial T)_P = R/P + dB/dT \qquad 3.19.7$$

$$\kappa_T V_m = -(\partial V_m/\partial P)_T = RT/P^2. \qquad 3.19.8$$

For the sake of brevity we shall use these formulae omitting all higher powers of P; when higher terms are required there is no difficulty in inserting them.

---

* Guggenheim, Molec. Phys. 1965 **9** 199.
† Rowlinson, Rep. Prog. Phys. 1965 **28** 180.

§3.20 Joule–Thomson coefficient

When we discussed throttling in §3.11 we stressed the fact that at that stage we could not yet measure thermodynamic temperature. Now that we know how to do this by means of a gas thermometer, it is profitable to return to a discussion of throttling. We shall generalize this discussion by including the possibility of absorption of heat by the outflowing gas. In place of formula (3.11.3) we then have

$$H(T_2, P_2) - H(T_1, P_1) = q. \qquad 3.20.1$$

We first consider the isothermal case when q is adjusted so that $T_1 = T_2$. Formula (1) then reduces to

$$H(P_2) - H(P_1) = q \qquad (\text{const. } T) \qquad 3.20.2$$

so that measurement of q leads directly to the determination of $(\partial H/\partial P)_T$.

In the adiabatic case when $q = 0$ we have

$$H(T_2, P_2) - H(T_1, P_1) = 0. \qquad 3.20.3$$

If the pressure drop is small we may usefully replace (3) by

$$dH = (\partial H/\partial T)_P dT + (\partial H/\partial P)_T dP = 0.$$

The ratio of the temperature fall to the pressure drop is called the *Joule–Thomson coefficient* given by

$$(\partial T/\partial P)_H = -(\partial H/\partial P)_T/(\partial H/\partial T)_P = -(\partial H_m/\partial P)_T/(\partial H_m/\partial T)_P$$
$$= -V_m(1-\alpha T)/C_P \qquad 3.20.4$$

by use of (1.48.4) and (3.03.2).

When we use formula (3.19.4) we obtain

$$(\partial T/\partial P)_H = (-B + T dB/dT)/C_P. \qquad 3.20.5$$

§3.21 Temperature dependence of second virial coefficient

The second virial coefficient B_2 or B is negative at low temperature, increases with temperature, and eventually becomes positive at a temperature called the *Boyle temperature* and denoted by T_B.

It is impossible to fit the quantitative dependence of B on the temperature by any two-parameter formula such as the van der Waals formula

$$B = b - aT^{-1} \qquad 3.21.1$$

or the Berthelot formula

$$B = b - cT^{-2}. \qquad 3.21.2$$

The temperature dependence of B can be fitted quantitatively by various three-parameter formulae of which the simplest is

$$B = b - aT^{-1} - cT^{-2}. \qquad 3.21.3$$

This formula is purely empirical.

To obtain a theoretical formula we have to assume a particular form for the interaction energy w between a pair of molecules as a function of their distance apart r. The simplest model, commonly called a 'square well', is described by

$$w = \infty \qquad (r < \sigma) \qquad 3.21.4$$

$$w = -\varepsilon \qquad (\sigma < r < R) \qquad 3.21.5$$

$$w = 0 \qquad (r > R) \qquad 3.21.6$$

with three adjustable parameters σ, R, and ε. This model leads to the formula for the second virial coefficient

$$B = b[1 - (R^3/\sigma^3 - 1)\{\exp(\varepsilon/kT) - 1\}] \qquad 3.21.7$$

where

$$b = \tfrac{2}{3}\pi L \sigma^3. \qquad 3.21.8$$

The application to nitrogen is shown in figure 3.3 where the curve represents formula (7) with

$$b = 40.3 \text{ cm}^3 \text{ mole}^{-1} \qquad R/\sigma = 1.50 \qquad \varepsilon/k = 116 \text{ K}. \qquad 3.21.9$$

§3.22 Boyle temperature and inversion temperature

Boyle's law $PV = f(T)$ is most nearly obeyed at the temperature at which $B = 0$. This temperature is accordingly called the *Boyle temperature* and it is denoted by T_B. According to formulae (3.21.7) to (3.21.9) for nitrogen $T_B = 330$ K. This point is shown in figure 3.3.

The Joule–Thomson coefficient for a slightly imperfect gas is positive at the lowest temperatures (cooling by throttling) but is negative at high temperatures (heating by throttling). The temperature at which the effect changes sign is called the *inversion temperature* and is denoted by T_i. According to (3.20.5) the inversion temperature is determined by

$$dB/dT = B/T. \qquad 3.22.1$$

When B is plotted against T the tangent through the origin touches the curve at $T = T_i$. For nitrogen this is shown in figure 3.3. According to formulae (3.21.7) to (3.21.9) for nitrogen $T_i = 633$ K.

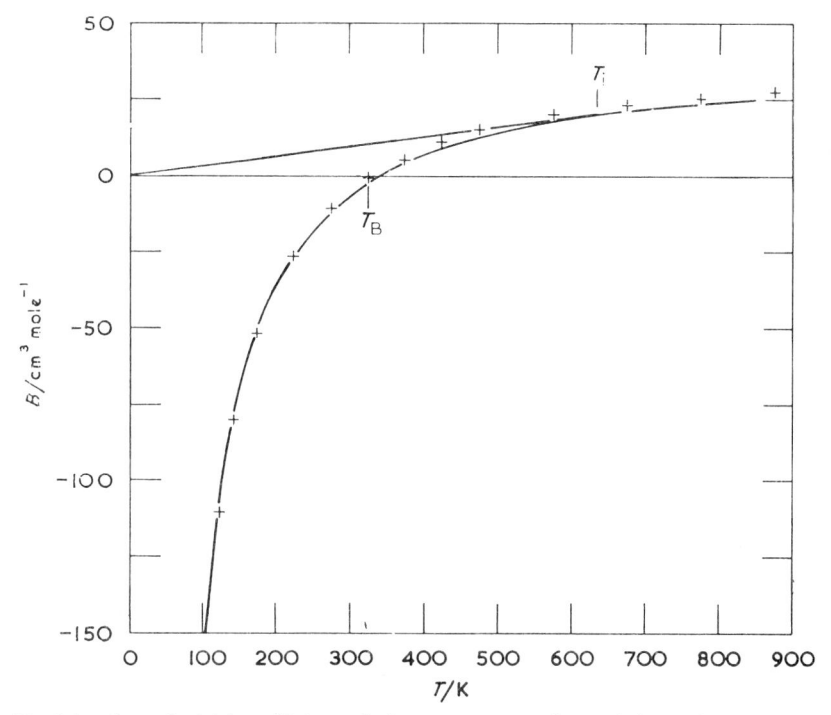

Fig. 3.3. Second virial coefficient of nitrogen, + experimental data of Holborn and Otto*, — formulae (3.21.7) to (3.21.9)

§3.23 *Relation between heat capacities of slightly imperfect gas*

We have the general relation (3.07.3)

$$C_P - C_V = \alpha^2 T V_m / \kappa_T. \qquad 3.23.1$$

Using formulae (3.19.5), (3.19.7), and (3.19.8) we deduce

$$C_P - C_V = R(1 + R^{-1} P \, dB/dT)^2. \qquad 3.23.2$$

§3.24 *Adiabatic change of a gas*

For an adiabatic change we have from (3.08.2) and (3.08.3)

$$-d \ln V/dP = \kappa_S = \kappa_T C_V / C_P \qquad (S \text{ constant}). \qquad 3.24.1$$

This differential equation for an adiabatic change cannot be integrated unless the right hand side can be expressed as an explicit function of P, V

---

* Holborn and Otto, Z. Phys. 1925 **33** 5.

and not necessarily even then. In the approximation of a perfect gas (1) becomes

$$-\mathrm{d}\ln V/\mathrm{d}\ln P = C_V/C_P \quad (S \text{ constant}). \qquad 3.24.2$$

In the special case of a gas with monatomic molecules

$$C_V/R = \tfrac{3}{2} \quad C_P/R = \tfrac{5}{2} \quad (\text{monatomic molecules}) \qquad 3.24.3$$

so that (2) becomes

$$\mathrm{d}\ln V/\mathrm{d}\ln P = -\tfrac{3}{5} \quad (\text{monatomic molecules}) \qquad 3.24.4$$

which can be integrated to

$$PV^{\tfrac{5}{3}} = \text{constant}. \qquad 3.24.5$$

In other cases (2) cannot be integrated explicitly.

§3.25 Temperature dependence of $\mu^\ominus$ and $\lambda^\ominus$

In §3.16 we have expressed all the most important thermodynamic functions of a gas in terms of $\mu^\ominus$ or $\lambda^\ominus$ each of these being a function of temperature only. We shall now consider this temperature dependence.

In the first place $\mu^\ominus$ contains an arbitrary constant term which we denote by H^0. There is a corresponding term H^0 in H_m and a corresponding factor $\exp(H^0/RT)$ in $\lambda^\ominus$. As long as chemical reactions are excluded we may fix H^0 arbitrarily for each substance, for example by setting $H^0 = 0$ at 25 °C. If on the contrary chemical processes are admissible then the values of H^0 for all substances are not independent. We may however fix H^0 arbitrarily for each element. The commonly accepted convention is $H^0 = 0$ for every element in its stable form at 25 °C.

There is also an arbitrary constant term in $-\mathrm{d}\mu^\ominus/\mathrm{d}T$ which we denote by S^0. As long as chemical reactions are excluded we may fix S^0 arbitrarily for each substance, for example by setting $S \to 0$ as $T \to 0$. If on the contrary chemical reactions are admissible then the values of S^0 for all substances are not independent. We may however fix S^0 arbitrarily for each element. The accepted convention is $S \to 0$ as $T \to 0$ for every element in its stable form. There are complications in the case of hydrogen which will be discussed in §3.56. When these conventions for H^0 and S^0 are used the chemical potential is called the *conventional chemical potential* and the entropy is called the *conventional entropy*.

In classical thermodynamics the accepted convention for S^0 is on a par with that for H^0, but statistical thermodynamics supplements the former convention in two ways. Firstly it provides a simple physical interpretation of the convention for S^0 which will be given in §3.53. Secondly by use of this interpretation of the convention it provides explicit formulae for the conventional entropy of gases. These formulae will be quoted without derivation in §§3.26–3.29 and interpreted later.

§3.26 Monatomic molecules

For gases having monatomic molecules, when we use the conventions specified in §3.25, it can be shown by statistical thermodynamics that

$$\lambda^\ominus = \exp(\mu^\ominus/RT) = \exp(H^0/RT)\, L^4 h^3 P^\ominus / g_0 (2\pi M)^{\frac{3}{2}} (RT)^{\frac{5}{2}} \qquad 3.26.1$$

where g_0 denotes the degree of degeneracy of the ground electronic level of the free atom. The value of g_0 is 1 for the noble gases He, Ne, Ar, Kr, Xe, Rn, and for Zn, Cd, Hg; its value is 2 for the alkali metals Li, Na, K, Rb, Cs.

When we use (1) in the formulae of §3.19 the conventional values of μ, λ, and S_m for a slightly imperfect gas are given by

$$\mu = G_m = H^0 - RT \ln g_0 + RT \ln\{L^4 h^3 P^\ominus / (2\pi M)^{\frac{3}{2}} (RT)^{\frac{5}{2}}\} + RT \ln(P/P^\ominus) + BP \qquad 3.26.2$$

$$\ln \lambda = H^0/RT - \ln g_0 + \ln\{L^4 h^3 P^\ominus / (2\pi M)^{\frac{3}{2}} (RT)^{\frac{5}{2}}\} + \ln(P/P^\ominus) + BP/RT \qquad 3.26.3$$

$$S_m = R \ln g_0 - R \ln\{L^4 h^3 P^\ominus / (2\pi M)^{\frac{3}{2}} (RT)^{\frac{5}{2}}\} + \tfrac{5}{2} R - R \ln(P/P^\ominus) - P(\mathrm{d}B/\mathrm{d}T). \qquad 3.26.4$$

If the gas is more than slightly imperfect it is a straightforward matter to include terms in the higher virial coefficients $B_3, B_4, \ldots$.

§3.27 Numerical values in entropy constant

We shall now insert numerical values for the constants in the formulae of §3.26 taking as our standard pressure $P^\ominus = 1$ atm. We have

$$h = 6.6256 \times 10^{-34} \text{ J s}$$
$$R = 8.3143 \text{ J K}^{-1} \text{ mole}^{-1}$$
$$L = 0.60225 \times 10^{24} \text{ mole}^{-1}$$
$$P^\ominus = 1 \text{ atm} = 1.01325 \times 10^5 \text{ J m}^{-3}.$$

With these values we obtain

$$L^4 h^3 P^\ominus/(2\pi M)^{\frac{3}{2}}(RT)^{\frac{5}{2}} = 1.236 (\text{kg/mole})^{\frac{3}{2}} (T/\text{K})^{-\frac{5}{2}}$$
$$= 39.03 (\text{g/mole})^{\frac{3}{2}} (T/\text{K})^{-\frac{5}{2}} = (\text{g/mole})^{\frac{3}{2}} (T/4.333 \text{ K})^{-\frac{5}{2}}. \qquad 3.27.1$$

Using (1) in the formulae of §3.26 we obtain

$$\ln \lambda^\ominus = \mu^\ominus/RT = H^0/RT - \ln g_0 - \tfrac{3}{2}\ln(M/\text{g mole}^{-1}) - \tfrac{5}{2}\ln(T/4.333\text{ K}) \qquad 3.27.2$$

$$\ln \lambda = \mu/RT = H^0/RT - \ln g_0 - \tfrac{3}{2}\ln(M/\text{g mole}^{-1})$$
$$\qquad - \tfrac{5}{2}\ln(T/4.333\text{ K}) + \ln(P/\text{atm}) + BP/RT \qquad 3.27.3$$

$$S_m/R = \ln g_0 + \tfrac{5}{2} + \tfrac{3}{2}\ln(M/\text{g mole}^{-1}) + \tfrac{5}{2}\ln(T/4.333\text{ K})$$
$$\qquad\qquad - \ln(P/\text{atm}) - (P/R)\mathrm{d}B/\mathrm{d}T$$

$$= \ln g_0 + \tfrac{3}{2}\ln(M/\text{g mole}^{-1}) + \tfrac{5}{2}\ln(T/1.594\text{ K})$$
$$\qquad\qquad - \ln(P/\text{atm}) - (P/R)\mathrm{d}B/\mathrm{d}T$$

$$= 10.35 + \ln g_0 + \tfrac{3}{2}\ln(M/\text{g mole}^{-1}) + \tfrac{5}{2}\ln(T/100\text{ K})$$
$$\qquad\qquad - \ln(P/\text{atm}) - (P/R)\mathrm{d}B/\mathrm{d}T \qquad 3.27.4$$

$$H_m = H^0 + \tfrac{5}{2}RT + (B - T\,\mathrm{d}B/\mathrm{d}T)P \qquad 3.27.5$$

$$C_P/R = \tfrac{5}{2} - T(\mathrm{d}^2 B/\mathrm{d}T^2)P/R. \qquad 3.27.6$$

§3.28 Linear molecules

In the formulae for monatomic molecules the electronic degrees of freedom were taken care of by g_0 while the remaining factor in $\lambda^\ominus$ or term in $\mu^\ominus$ relates to the translational degrees of freedom. In polyatomic molecules there are the same factors in $\lambda^\ominus$ and further factors to take care of the rotational and vibrational degrees of freedom. We shall describe these factors first for linear molecules and then for non-linear molecules.

For linear molecules the extra factor in $\lambda^\ominus$ due to the rotational degrees of freedom is at ordinary temperatures

$$\lambda_{\text{rot}} = (s\Theta_r/T)\{1 + \Theta_r/3T + \Theta_r^2/15T^2\}^{-1} \qquad (T \gg \Theta_r) \qquad 3.28.1$$

where Θ_r is a rotational characteristic temperature inversely proportional to the moment of inertia I of the molecule and defined by

$$\Theta_r = h^2/8\pi^2 Ik. \qquad 3.28.2$$

The factor s called *symmetry number* is 2 for a symmetrical linear molecule and 1 for an unsymmetrical molecule.

The rotational term in $\mu^\ominus$ is

$$\mu_{\text{rot}} = RT \ln \lambda_{\text{rot}} = RT \ln(s\Theta_r/T) - RT \ln\{1 + \Theta_r/3T + \Theta_r^2/15T^2\}$$
$$= RT \ln(s\Theta_r/T) - \tfrac{1}{3}R\Theta_r - R\Theta_r^2/90T. \qquad 3.28.3$$

The constant term $-\tfrac{1}{3}R\Theta_r$ may be absorbed into H^0 and (3) then becomes

$$\mu_{\text{rot}} = RT \ln(s\Theta_r/T) - R\Theta_r^2/90T. \qquad 3.28.4$$

The corresponding rotational contributions to S_m, H_m, and C are

$$S_{rot} = R - R \ln(s\Theta_r/T) + R\Theta_r^2/90T^2 \qquad 3.28.5$$

$$H_{rot} = RT - R\Theta_r^2/45T \qquad 3.28.6$$

$$C_{rot} = R + R\Theta_r^2/45T^2. \qquad 3.28.7$$

We now turn to the vibrational contributions. A linear molecule composed of a atoms has $3a-5$ normal vibrational modes each having a characteristic frequency v_i. Associated with the frequency v_i is a vibrational characteristic temperature Θ_{vi} defined by

$$k\Theta_{vi} = hv_i. \qquad 3.28.8$$

The vibrational contributions to the several thermodynamic functions are

$$\mu_{vib} = RT \ln \lambda_{vib} = RT \sum_i \ln\{1 - \exp(-\Theta_{vi}/T)\} \qquad 3.28.9$$

$$S_{vib} = R \sum_i \Theta_{vi}/T\{\exp(\Theta_{vi}/T) - 1\} - R \sum_i \ln\{1 - \exp(-\Theta_{vi}/T)\} \qquad 3.28.10$$

$$H_{vib} = R \sum_i \Theta_{vi}/\{\exp(\Theta_{vi}/T) - 1\} \qquad 3.28.11$$

$$C_{vib} = R \sum_i \{(\Theta_{vi}/2T)/\sinh(\Theta_{vi}/2T)\}^2. \qquad 3.28.12$$

We have still to discuss the electronic factor g_0. For the vast majority of linear molecules regarded as saturated g_0 is unity. The outstanding exceptions are O_2 and NO. The ground state of O_2 is $^3\Sigma$ and $g_0 = 3$. The odd molecule NO has a ground state $^2\Pi_{\frac{1}{2}}$ and an excited state $^2\Pi_{\frac{3}{2}}$ having an excitation energy ε such that ε/k is only 178 K. As a result of this the constant g_0 has to be replaced by the temperature dependent factor

$$2 + 2\exp(-178 \text{ K}/T) \qquad 3.28.13$$

having an effective value 2 when $T \ll 178$ K and 4 when $T \gg 178$ K.

Values of Θ_r, Θ_v, s, and g_0 are given* in table 3.1 for the commonest diatomic molecules and in table 3.2 for a few other typical linear molecules.

§3.29 Non-linear molecules

Whereas a linear molecule has 2 rotational degrees of freedom, a non-linear molecule has 3. For a non-linear molecule the rotational characteristic

---

* Values taken from Herzberg, Spectra of Diatomic Molecules, Van Nostrand 1950. Cf. Fowler and Guggenheim, Statistical Thermodynamics, Cambridge University Press 1939 p. 90. Cf. Slater, Introduction to Chemical Physics, McGraw-Hill 1939 p. 136, observing that their Θ_{rot} is equal to twice our Θ_r.

TABLE 3.1

Characteristic temperatures Θ_r and Θ_v, symmetry numbers s, and electronic weights g_0 for typical diatomic molecules

Formula	Θ_r/K	$10^{-2}\Theta_v/K$	s	g_0
H_2	85.3	59.8	2	1
D_2	42.7	43.0	2	1
N_2	2.88	33.5	2	1
O_2	2.07	22.4	2	3
CO	2.77	30.8	1	1
NO	2.44	27.0	1	–
HCl	15.0	41.5	1	1
HBr	12.0	36.8	1	1
HI	9.29	32.1	1	1
Cl_2	0.344	7.96	2	1
Br_2	0.116	4.62	2	1
I_2	0.0537	3.07	2	1

TABLE 3.2

Characteristic temperatures Θ_r and Θ_{vi}, symmetry numbers s, and electronic weights g_0 for typical polyatomic linear molecules

Formula	Θ_r/K	$10^{-2}\Theta_{vi}/K$	s	g_0
OCO	0.560	9.60 9.60 20.0 33.8	2	1
NNO	0.602	8.47 8.47 18.5 32.0	1	1
HCCH	1.69	8.80 8.80 10.5 10.5 28.4 47.3 48.5	2	1
HCN	2.13	10.5 10.5 28.7 49.5	1	1

temperature Θ_r is related to the 3 principal moments of inertia I_1, I_2, I_3 by

$$\Theta_r = h^2/8\pi^2(I_1 I_2 I_3)^{\frac{1}{3}}k. \qquad 3.29.1$$

The rotational factor in $\lambda^\ominus$ at ordinary temperatures is given by

$$\lambda_{\text{rot}} = s\Theta_r^{\frac{3}{2}}/\pi^{\frac{1}{2}}T^{\frac{3}{2}}. \qquad 3.29.2$$

The symmetry number s is defined as the number of indistinguishable orientations of the molecule. For example s is 1 for NOCl, 2 for OH_2, 3 for NH_3, 4 for C_2H_4, 6 for BF_3, and 12 for C_6H_6. The rotational contributions to the several thermodynamic functions are

$$\mu_{\text{rot}} = RT \ln \lambda_{\text{rot}} = RT \ln(s\Theta_r^{\frac{3}{2}}/\pi^{\frac{1}{2}}T^{\frac{3}{2}}) \qquad 3.29.3$$

$$S_{\text{rot}} = \tfrac{3}{2}R - R \ln(s\Theta_r^{\frac{3}{2}}/\pi^{\frac{1}{2}}T^{\frac{3}{2}}) \qquad 3.29.4$$

$$U_{\text{rot}} = \tfrac{3}{2}RT \qquad 3.29.5$$

$$C_{\text{rot}} = \tfrac{3}{2}R. \qquad 3.29.6$$

The vibrational contributions are exactly as for linear molecules except that there are $3a-6$ normal vibrational modes instead of $3a-5$. Thus we have

$$S_{\text{vib}} = R \sum_i \Theta_{vi}/T\{\exp(\Theta_{vi}/T)-1\} - R \sum_i \ln\{1-\exp(-\Theta_{vi}/T)\}. \qquad 3.29.7$$

The value of g_0 is unity for almost all non-linear molecules including OH_2, SH_2, NH_3, PH_3, CH_4, SO_2, and all organic molecules. Its value for free radicals such as CH_3, C_6H_5, is 2.

TABLE 3.3

Characteristic temperatures Θ_r and Θ_{vi}, symmetry numbers s, and electronic weights g_0 for typical non-linear molecules

	Formula	Θ_r/K	$10^{-2}\Theta_{vi}$/K	s	g_0
*	NOCl		4.8 8.6 25.9	1	1
	OH_2	22.3	22.9 52.5 54.0	2	1
	NH_3	12.3	13.7 23.4 23.4 48.0 49.1 49.1	3	1

* Landau and Fletcher, J. Molec. Spect. 1960 **4** 280.

Values of Θ_r, Θ_{vi}, s, and g_0 for a few non-linear molecules are given* in table 3.3.

Table 3.4 gives the vibrational contributions of a single normal mode to the several thermodynamic functions in terms of $x = \Theta_v/T$.

§3.30 Pressure dependence for condensed phases

We turn now from gases to condensed phases. Later we shall consider equilibrium between a condensed phase and a gas. As we shall see in §3.44 there are conditions of temperature and pressure called critical at which the distinction between gas and liquid disappears, but except at conditions close to the critical there is a rather sharp contrast between the properties of a gas and a liquid. The contrast between a crystal and a fluid, either gas or liquid, is always a sharp one.

Whereas the isothermal compressibility of a gas is roughly equal to the reciprocal of the pressure, the isothermal compressibility of a solid and that of a liquid, except near the critical temperature, is much smaller than that of a gas and is less dependent on the pressure. We accordingly use the approximation

$$-V^{-1}(\partial V/\partial P)_T = \kappa_T = \text{const.} \qquad 3.30.1$$

We can integrate (1) at constant temperature, obtaining

$$V_m = V^\ominus \exp\{-\kappa_T(P-P^\ominus)\} \qquad 3.30.2$$

where $V^\ominus$ is the value of V_m at the standard pressure $P^\ominus$, usually one atmosphere.

For typical liquids κ_T is about 10^{-4} atm^{-1} and for many solids is even smaller. We may therefore, without appreciable loss of accuracy replace (2) by the more convenient relation

$$V_m = V^\ominus\{1 - \kappa_T(P-P^\ominus)\}. \qquad 3.30.3$$

We can integrate again with respect to P at constant T obtaining

$$\mu = \mu^\ominus + V^\ominus\{(P-P^\ominus) - \tfrac{1}{2}\kappa_T(P-P^\ominus)^2\}. \qquad 3.30.4$$

It is sometimes convenient to rewrite (4) as

$$\mu = \mu^\ominus + (P-P^\ominus)\tfrac{1}{2}(V_m + V^\ominus). \qquad 3.30.5$$

---

* Values taken from Herzberg, Infra-red and Raman Spectra, Van Nostrand 1945. Cf. Fowler and Guggenheim, Statistical Thermodynamics, Cambridge University Press 1939 pp. 113–114.

TABLE 3.4

Contributions of a single harmonic oscillator to the several thermodynamic quantities expressed as functions of $x = h\nu/kT = \Theta_v/T$

x	$-\mu/RT$ $= -\ln\{1-\exp(-x)\}$	H_m/RT $= x/(\exp x - 1)$	S_m/R $= (H_m-\mu)/RT$	C/R $= \{\tfrac{1}{2}x/\sinh \tfrac{1}{2}x\}^2$
0.01	4.610	0.995	5.605	1.000
0.05	3.021	0.975	3.996	1.000
0.1	2.352	0.951	3.303	0.999
0.2	1.708	0.903	2.611	0.997
0.3	1.350	0.857	2.208	0.993
0.4	1.110	0.813	1.923	0.987
0.5	0.933	0.771	1.704	0.979
0.6	0.796	0.730	1.526	0.971
0.7	0.686	0.691	1.377	0.960
0.8	0.597	0.653	1.249	0.948
0.9	0.522	0.617	1.138	0.935
1.0	0.459	0.582	1.041	0.921
1.1	0.405	0.549	0.954	0.905
1.2	0.358	0.517	0.876	0.888
1.3	0.318	0.487	0.805	0.870
1.4	0.283	0.458	0.741	0.852
1.5	0.252	0.431	0.683	0.832
1.6	0.226	0.405	0.630	0.811
1.7	0.202	0.380	0.582	0.790
1.8	0.181	0.356	0.537	0.769
1.9	0.162	0.334	0.496	0.747
2.0	0.145	0.313	0.458	0.724
2.1	0.131	0.293	0.424	0.701
2.2	0.117	0.274	0.392	0.678
2.3	0.106	0.256	0.362	0.655
2.4	0.095	0.239	0.335	0.632
2.5	0.086	0.224	0.309	0.609
2.6	0.077	0.209	0.286	0.586
2.7	0.070	0.195	0.264	0.563
2.8	0.063	0.181	0.244	0.540
2.9	0.057	0.169	0.225	0.518
3.0	0.051	0.157	0.208	0.496
3.2	0.042	0.136	0.178	0.454
3.4	0.034	0.117	0.151	0.413
3.6	0.028	0.101	0.129	0.374
3.8	0.023	0.087	0.110	0.338
4.0	0.018	0.075	0.093	0.304
4.5	0.011	0.051	0.062	0.230
5.0	0.007	0.034	0.041	0.171
5.5	0.004	0.023	0.027	0.125
6.0	0.002	0.015	0.017	0.090
6.5	0.002	0.010	0.013	0.064
7.0	0.001	0.006	0.007	0.045

Differentiating (5) with respect to T at constant pressure we obtain

$$S_m = -d\mu^{\ominus}/dT - (P-P^{\ominus})\tfrac{1}{2}(\partial V_m/\partial T + \partial V^{\ominus}/\partial T)$$
$$= -d\mu^{\ominus}/dT - (P-P^{\ominus})\tfrac{1}{2}(\alpha V_m + \alpha^{\ominus} V^{\ominus}) \qquad 3.30.6$$

where α and $\alpha^{\ominus}$ are the coefficients of thermal expansion at pressures P and $P^{\ominus}$ respectively.

From (5) and (6) we derive

$$H_m = \mu + TS_m = \mu^{\ominus} - T\,d\mu^{\ominus}/dT + (P-P^{\ominus})\tfrac{1}{2}\{V_m(1-\alpha T) + V^{\ominus}(1-\alpha^{\ominus} T)\} \qquad 3.30.7$$

and

$$C = T\,d^2\mu^{\ominus}/dT^2 - (P-P^{\ominus})\tfrac{1}{2}T\{V_m\,d\alpha/dT + V^{\ominus}\,d\alpha^{\ominus}/dT\}. \qquad 3.30.8$$

§3.31 Temperature dependence for liquids

We have seen that the dependence of the thermodynamic properties of condensed phases on the pressure is simple and usually small. We have now to consider how these properties depend on the temperature.

As regards liquids there is nothing fundamental or general that can be said except that $\mu^{\ominus}$ can often be represented over quite a wide range of temperature by an empirical relation of the form

$$\mu^{\ominus} = A - (B-C)T - CT\ln T \qquad (A, B, C \text{ const.}). \qquad 3.31.1$$

From (1) we deduce

$$S^{\ominus} = -d\mu^{\ominus}/dT = B + C\ln T \qquad 3.31.2$$

$$H^{\ominus} = A + CT. \qquad 3.31.3$$

According to this empirical approximation the proper heat capacity C at the standard pressure $P^{\ominus}$ is independent of the temperature. We have already mentioned in §3.03 that for many liquids, in particular water, C is nearly independent of the temperature.

The approximate constancy of C and the consequent validity of relations of the form (1), (2), (3) also hold for many solids at ordinary and higher temperatures, but not at low temperatures. This accident has in the past caused undue importance to be attached to the heat capacity, in contrast to the enthalpy H and the entropy S. The only real importance of C is that it is the connecting link between H and S, as explained in §3.02. This link is especially simple when C is independent of T, but this occurrence, however frequent, is of no fundamental importance.

§3.32 Crystals at very low temperatures

It is predicted by statistical theory and borne out by experiment that at very low temperatures the enthalpy of a crystalline solid varies linearly with the fourth power of the thermodynamic temperature. That is to say, neglecting the small dependence on pressure,

$$H_m = H_m^0 + \tfrac{1}{4}aT^4 \quad \text{(small } T\text{)} \qquad 3.32.1$$

where a is a constant and H_m^0 is the limiting value of H_m as $T \to 0$. Correspondingly we have for the entropy

$$S_m = S_m^0 + \tfrac{1}{3}aT^3 \quad \text{(small } T\text{)} \qquad 3.32.2$$

where S_m^0 is the limiting value of S_m as $T \to 0$. The formulae (1) and (2) are not independent, but are related through the thermodynamic formula (3.02.3)

$$T dS = dH \quad \text{(const. } P\text{)}. \qquad 3.32.3$$

From (1) and (2) it immediately follows that

$$\mu = H_m^0 - TS_m^0 - \tfrac{1}{12}aT^4 \quad \text{(small } T\text{)}. \qquad 3.32.4$$

We have not stated how small T must be for these formulae to hold, nor is it possible to make any precise statement since the requirement is different for different substances. For most substances investigated these formulae appear to be at least approximately valid at temperatures below 15 K.

We shall see later that a comparison between the constant S_m^0 in (2) and the constant $-d\mu^\ominus/dT$ occurring in the formula for the proper entropy of a gas is of considerable interest. For this reason it is important to be able to extrapolate experimental data on the entropy from the lowest experimental temperature down to 0 K. For this purpose one determines a suitable value of the constant a from the relation (1) by plotting H against T^4 in the lowest temperature range in which experimental measurements have been made. This value of a is then used in (2) to give experimental values of $S(T) - S(0)$. Provided the experimental data extend below 15 K, the contribution to S from this extrapolation is usually so small that an accurate estimate of a is not required.

Actually the most important feature of the formulae of this section is not their precise form, still less the value of a, but the fact that S tends rapidly towards a constant value as T decreases. This behaviour is in striking contrast with the formulae for the entropy of gases at ordinary temperatures which contain terms in $\ln T$.

§3.33 Crystals at intermediate temperatures. Debye's approximation

In the previous section we have described the thermodynamic behaviour of crystals at very low temperatures. In §3.31 we mentioned briefly that at ordinary and higher temperatures the behaviour of many solids, as well as liquids, is represented at least approximately by the formulae of that section corresponding to a temperature-independent heat capacity. In the intermediate range the heat capacity increases with temperature, but its rate of increase falls rather rapidly. There is no precise quantitative theory except for the simplest crystals consisting of monatomic molecules. Even for these the accurate theory is so complicated as to be of little practical use and it is in fact usually replaced by a much simpler approximation due to Debye.

We shall not here describe Debye's model, still less discuss* its limitations, but shall give the formulae which follow from it. The formulae contain apart from the temperature T, two parameters namely the energy U_m^0 of the crystal at $T=0$ and a characteristic temperature Θ_D. Both these parameters U_m^0 and Θ_D are functions of the proper volume V_m, but are independent of the temperature. In considering Debye's model it is therefore expedient to regard as independent variables T, V instead of the usually more practically convenient T, P. We accordingly begin by writing down Debye's formula for the proper Helmholtz function of a crystal

$$\mathscr{F}_m = U_m - TS_m = U_m^0 - TS_m^0 + 3RT \int_0^{\Theta_D} \ln\{1 - \exp(-\theta/T)\}(3\theta^2/\Theta_D^3)d\theta \quad 3.33.1$$

wherein we repeat that U_m^0 and Θ_D are functions of V_m whereas S_m^0 on the other hand is a constant independent of V_m, as well as of T, and depends only on the arbitrary zero of entropy.

From (1) we could derive the pressure by the relation

$$P = -\partial \mathscr{F}/\partial V. \quad 3.33.2$$

We have however seen in §3.30 that the thermodynamic properties of a condensed phase, in particular a crystal, are nearly independent of the pressure; more precisely $PV_m \ll RT$. We may consequently regard the pressure as negligible and replace (2) by the condition

$$\partial \mathscr{F}/\partial V = 0 \quad 3.33.3$$

which gives an equilibrium relation between U^0 and Θ_D. From (1) and (3) we find that this relation is

$$\frac{\partial U_m^0}{\partial V_m} = 3R \frac{\partial \Theta_D}{\partial V_m} \int_0^{\Theta_D} \frac{1}{\exp(\theta/T) - 1} \frac{3\theta^3}{\Theta_D^4} d\theta. \quad 3.33.4$$

---

* Blackman, Rep. Progr. Phys. 1942 **8** 11.

From (1) we can derive formulae for the other thermodynamic functions, in particular

$$S_m - S_m^0 = -3R \int_0^{\Theta_D} \left[\ln\{1 - \exp(-\theta/T)\} - \frac{\theta/T}{\exp(\theta/T) - 1} \right] \frac{3\theta^2}{\Theta_D^3} d\theta \qquad 3.33.5$$

$$U_m - U_m^0 = 3R \int_0^{\Theta_D} \frac{1}{\exp(\theta/T) - 1} \frac{3\theta^3}{\Theta_D^3} d\theta \qquad 3.33.6$$

$$C_V = 3R \int_0^{\Theta_D} \left\{ \frac{\theta/2T}{\sinh(\theta/2T)} \right\}^2 \frac{3\theta^2}{\Theta_D^3} d\theta. \qquad 3.33.7$$

We may note that at very low temperatures, $T \ll \Theta_D$ and we may without sensible error replace the upper limits of integration in the above formulae by ∞. We thus obtain

$$U_m - U_m^0 = 3R \frac{T^4}{\Theta_D^3} \int_0^\infty \frac{3\xi^3 d\xi}{\exp \xi - 1} = \frac{3\pi^4 R T^4}{5 \Theta_D^3} \qquad 3.33.8$$

which, in view of the negligible difference between U_m and H_m is in agreement with (3.32.1) with a given by

$$\tfrac{1}{4} a = \frac{3\pi^4 R}{5 \Theta_D^3}. \qquad 3.33.9$$

While we shall not here discuss the extent of agreement or disagreement to be expected between these formulae and the behaviour of real crystals, we shall however devote some space to the consideration of how the comparison can most directly be made. Let us therefore consider which quantities are most directly measurable, bearing in mind that with all condensed phases it is convenient to make measurements at constant pressure but extremely difficult to make measurements at constant volume.

The usual calorimetric measurements determine directly how H depends on T. Provided these measurements have been carried to a low enough temperature, the extrapolation to $T=0$ can be performed as described in §3.32 so that we know $H_m(T) - H_m^0$ as a function of T. Then by using the relation (3.32.3) we can without any further experimental data compute $S_m(T) - S_m^0$. We can now compare this experimental quantity with the right side of (5), which is tabulated as a function of Θ_D/T. We thus obtain for each temperature T a value of Θ_D fitting the experimental value of $S_m - S_m^0$. These values of Θ_D will be constant neither in practice, nor according to Debye's model. For we are considering data at constant pressure, consequently at varying volume, and, as the volume varies, so Θ_D varies. In fact

TABLE 3.5
Comparison of proper entropy of gold with Debye's formula

T/K	$(S_m - S_m^0)/R$ expt.	Θ_D/T	Θ_D/K
15	0.06	10.90	163
20	0.14	8.25	165
30	0.385	5.57	167
40	0.705	4.225	167
50	1.05	3.39	170
60	1.40	2.87	172
70	1.73	2.45	171
80	2.03	2.15	172
90	2.32	1.91	172
100	2.58	1.72	172
120	3.07	1.44	173
140	3.49	1.23	172
160	3.87	1.075	172
180	4.22	0.95	171
200	4.53	0.855	171
300	5.77	0.555	167

TABLE 3.6
Comparison of proper entropy of magnesium with Debye's formula

T/K	$(S_m - S_m^0)/R$ expt.	Θ_D/T	Θ_D/K
20	0.01	18.0	360
30	0.05	11.5	345
40	0.13	8.38	335
50	0.26	6.53	326
60	0.41	5.41	324
80	0.77	4.03	322
100	1.15	3.22	322
120	1.52	2.68	322
140	1.87	2.30	322
160	2.20	2.00	320
180	2.50	1.78	320
200	2.76	1.61	322
300	7.77	1.06	318

as the volume increases, theory predicts that Θ_D should steadily decrease. If then it is found that as T increases, the value of Θ_D, determined as described above, slowly but steadily decreases then we may say that at least there is no contradiction between the experimental data and the model. If on the other hand as T increases, the value of Θ_D thus determined increases or fluctuates, then we may say with certainty that the experimental data are in disagreement with the model.

We give typical illustrations of this method of comparison in tables 3.5 and 3.6 for gold* and magnesium[†] respectively. We observe that for gold Θ_D rises from 163 K to 173 K and then falls again to 167 K. For magnesium Θ_D decreases steadily from 360 K to 318 K. In a few cases, such as copper and lead, Θ_D varies even less than in the case of gold. In other cases, notably lithium, Θ_D varies by nearly 20 %.

We must emphasize that the entropy is the only simple thermodynamic function for which we have both a closed formula and an experimental value obtainable from a single set of calorimetric measurements performed at constant pressure. In spite of the directness and simplicity of the above method of comparison, it is not generally used. The usual procedure is, from the experimental measurements of H_m as a function of T, first to compute $dH/dT = C_P$; then by measured, or estimated values of α and κ_T to use formula (3.07.3) or (3.08.5) to compute C_V from C_P; lastly to compare the C_V so calculated with formula (7). There are two objections to this procedure as compared with that recommended here. In the first place the computation of C_P from H involves a differentiation and so increases any experimental errors whereas in the computation of S from H the integration helps to smooth out the errors introduced by the differentiation. In the second place the computation of C_V from C_P by (3.07.3) or (3.08.5) requires either several other pieces of difficult experimental measurement or else some guess work, neither of which is required if one makes comparisons of entropy. When the value of C_V, thus computed or estimated, is compared with formula (7) we can calculate at each temperature a value of Θ_D which fits. Just as in the comparison of entropies, these values of Θ_D should, if the model is good, decrease slowly and steadily as the temperature, and so the volume, increases. There appears to be a widespread, but mistaken, belief that Θ_D should be independent of temperature in spite of the thermal expansion.

Quite apart from the change in Θ_D due to thermal expansion, variations of Θ_D with temperature are to be expected owing to the limitations of Debye's model. In view of all the complications in the lattice theory, Debye's theory is remarkable not in the extent of its failure, but rather in the extent of its success[‡].

---

* Clusius and Harteck, Z. Phys. Chem. 1928 **134** 243.
† Clusius and Vaughen, J. Amer. Chem. Soc. 1930 **52** 4686.
‡ Blackman, Rep. Progr. Phys. 1942 **8** 11.

§3.34 *Corresponding temperatures of crystals*

We have seen that Debye's model is only an approximate representation of a simple crystal of monatomic molecules and further that even if it were an accurate representation, the characteristic temperature Θ_D should still vary with temperature owing to thermal expansion. Nevertheless it is an experimental fact that Debye's formulae with constant Θ_D do give a remarkably good approximate representation over a wide temperature range of the actual behaviour of many simple crystals, especially metals crystallizing in the cubic system. For such substances the values of $S_m - S_m^0$, of $(H_m - H_m^0)/T$, and consequently of $G_m - G_m^0$ are universal functions of T/Θ_D. Thus several important thermodynamic properties of different crystals have the same value when T/Θ_D has the same value. Temperatures of different substances such that T/Θ_D has the same value are called *corresponding temperatures*. The principle that certain thermodynamic properties have equal values for different substances at corresponding temperatures is called a *principle of corresponding temperatures*. It is to be observed that this principle for simple crystals makes no reference to the pressure, which is tacitly assumed to be low and to have no appreciable effect on the values of the properties under discussion. In §3.48 we shall discuss a more interesting principle of corresponding temperatures and corresponding pressures for liquids and gases.

§3.35 *Comparison of Debye's functions with Einstein's functions*

Debye's model was preceded by a simpler model due to Einstein leading to the simpler formulae

$$U_m - TS_m = U_m^0 - TS_m^0 + 3RT \ln\{1 - \exp(-\Theta_E/T)\} \qquad 3.35.1$$

$$U_m = U_m^0 + 3R\Theta_E/\{\exp(\Theta_E/T) - 1\} \qquad 3.35.2$$

$$S_m = S_m^0 - 3R[\ln\{1 - \exp(\Theta_E/T)\} - \Theta_E/T\{\exp(\Theta_E/T) - 1\}] \qquad 3.35.3$$

$$C_V = 3R\{(\Theta_E/2T)/\sinh(\Theta_E/2T)\}^2 \qquad 3.35.4$$

where Θ_E is Einstein's characteristic temperature.

By comparing Debye's formulae with Einstein's we observe that the former contain integrals from zero to Θ_D where the latter contain merely simple functions of Θ_E. Thus Θ_E in a sense represents an average Θ covering the range from 0 to Θ_D. Thus at any given temperature the value of Θ_E which fits is always smaller than the value of Θ_D which fits.

If one tries to fit the experimental data by Einstein's formulae with a constant Θ_E one fails completely at the lowest temperatures, but at higher temperatures there is little to choose between Einstein's formulae and Debye's provided the value chosen for Θ_E is suitably adjusted. In fact when $T > \frac{1}{3}\Theta_D$ the values of $U_m - U_m^0$ calculated from Debye's formula do not differ appreciably from the values calculated from Einstein's formula provided one uses for Θ_E the value given by $\Theta_E = 0.73\Theta_D$. Similarly when $T > \frac{1}{3}\Theta_D$ the values of $S_m - S_m^0$ calculated from Debye's formula do not differ appreciably from the values calculated from Einstein's formula provided one takes $\Theta_E = 0.71\Theta_D$. The comparison is shown in table 3.7. The slight difference of

TABLE 3.7

Comparison of Einstein's formulae with Debye's assuming
$\Theta_E = 0.73\Theta_D$ for energies
and $\Theta_E = 0.71\Theta_D$ for entropies

Θ_D/T	$(U_m - U_m^0)/3RT$		$(S_m - S_m^0)/3R$	
	Debye	Einstein	Debye	Einstein
0.1	0.964	0.963	3.64	3.64
0.2	0.929	0.929	2.945	2.95
0.4	0.860	0.861	2.26	2.26
0.6	0.794	0.797	1.85	1.86
0.8	0.733	0.736	1.575	1.58
1.0	0.675	0.679	1.36	1.37
1.2	0.620	0.625	1.19	1.19
1.4	0.571	0.575	1.045	1.045
1.6	0.525	0.527	0.925	0.925
1.8	0.482	0.483	0.825	0.820
2.0	0.442	0.442	0.735	0.730
2.2	0.405	0.403	0.657	0.650
2.4	0.371	0.368	0.590	0.580
2.6	0.339	0.334	0.529	0.518
2.8	0.310	0.304	0.476	0.463
3.0	0.284	0.276	0.429	0.414

about 2% between the best values of Θ_E corresponding to a given Θ_D in the cases of the energy and the entropy is a measure of the accuracy lost by the substitution. Since the experimental data cannot be fitted exactly by a constant value of Θ_D considerable simplification can often be attained without significant loss of accuracy by using Einstein's formulae rather than Debye's provided one is concerned only with temperatures greater than $\frac{1}{3}\Theta_D$. At lower temperatures Debye's formulae should be used in preference to Einstein's.

§3.36 *Equilibrium between two phases*

Having discussed the thermodynamic properties of a single phase, we now turn to consider two phases in equilibrium. If we denote the two phases by superscripts α and β, the condition for equilibrium between the two phases is according to (1.39.5)

$$\mu^\alpha = \mu^\beta \qquad 3.36.1$$

or according to (3.15.6)

$$\lambda^\alpha = \lambda^\beta. \qquad 3.36.2$$

Since in any single phase of a pure substance the temperature T and pressure P may be varied independently and μ or λ may be regarded as a function of T, P, we may therefore regard (1) or (2) as expressing a relation between T and P for equilibrium between the two phases. It follows that when the two phases are in equilibrium, the temperature T and pressure P are not independently variable but either determines the other. We accordingly say that a single phase of one component has two degrees of freedom but a pair of phases of one component has only one degree of freedom.

§3.37 *Relation between temperature and pressure for two-phase equilibrium*

We now proceed to determine how the equilibrium pressure between two phases α and β depends on the temperature T. Differentiating (3.36.1) we have

$$d\mu^\alpha = d\mu^\beta \qquad 3.37.1$$

or

$$(\partial\mu^\alpha/\partial T)dT + (\partial\mu^\alpha/\partial P)dP = (\partial\mu^\beta/\partial T)dT + (\partial\mu^\beta/\partial P)dP. \qquad 3.37.2$$

Using (1.28.22) and (1.28.23), we obtain

$$(V_m^\beta - V_m^\alpha)dP = (S_m^\beta - S_m^\alpha)dT. \qquad 3.37.3$$

Formula (3) can also be obtained more directly from Maxwell's relation

$$(\partial P/\partial T)_V = (\partial S/\partial V)_T. \qquad 3.37.4$$

We apply this relation to a system consisting of the two phases α and β in equilibrium with each other. Since for this equilibrium to persist P is completely determined by T and is independent of V, we may replace the partial differential coefficient $(\partial P/\partial T)_V$ by dP/dT. Moreover at constant temperature and incidentally also constant pressure, S and V can only change through some amount of substance passing from the phase α to the phase β or conversely. Thus the ratio of the changes in S and in V is independent of the amount transferred from the one phase to the other. If then we denote by the symbol Δ the increase of any property when unit amount passes from the phase α to the phase β, we have

$$(\partial S/\partial V)_T = \Delta S/\Delta V \qquad 3.37.5$$

and so (4) becomes

$$dP/dT = \Delta S/\Delta V \qquad 3.37.6$$

which is formula (3) in different notation.

Since we may rewrite (3.36.1) as

$$H_m^\alpha - TS_m^\alpha = H_m^\beta - TS_m^\beta \qquad 3.37.7$$

it follows immediately that

$$T\Delta S = T(S_m^\beta - S_m^\alpha) = H_m^\beta - H_m^\alpha = \Delta H. \qquad 3.37.8$$

This relation has an obvious physical meaning, the same as that of (3.02.2). If unit amount passes isothermally from the phase α to the phase β, the heat q absorbed is equal to ΔH because the process occurs at constant pressure and it is also equal to $T\Delta S$ because, the system being in equilibrium throughout, the change is reversible.

If we now substitute from (8) into (6), we obtain

$$dP/dT = \Delta H/T\Delta V = (H_m^\beta - H_m^\alpha)/T(V_m^\beta - V_m^\alpha) \qquad 3.37.9$$

which is known as *Clapeyron's relation*. This can also be obtained more directly by starting from

$$\mu^\alpha/T = \mu^\beta/T \qquad 3.37.10$$

instead of (3.36.1). Differentiating (10) we obtain

$$\{\partial(\mu^\alpha/T)/\partial T\}dT + T^{-1}(\partial\mu^\alpha/\partial P)dP = \{\partial(\mu^\beta/T)/\partial T\}dT + T^{-1}(\partial\mu^\beta/\partial P)dP \qquad 3.37.11$$

and so using (1.28.24) and (1.28.23)

$$-(H_m^\alpha/T)dT + (V_m^\alpha/T)dP = -(H_m^\beta/T)dT + (V_m^\beta/T)dP \qquad 3.37.12$$

whence (9) follows immediately. We have given these alternative derivations of (9) because of its great importance, as the prototype of other similar formulae in systems of more than one component.

§3.38 *Clapeyron's relation applied to two condensed phases*

Let us consider the application of Clapeyron's relation to the equilibrium between a solid and a liquid. Using the superscripts ^{S} and ^{L} to denote these two phases, we have for the variation of the equilibrium pressure with the equilibrium temperature according to (3.37.9)

$$dP/dT = (H_m^L - H_m^S)/T(V_m^L - V_m^S) = \Delta_f H/T(V_m^L - V_m^S) = \Delta_f S/(V_m^L - V_m^S) \qquad 3.38.1$$

where $\Delta_f H$ is the proper enthalpy of fusion and $\Delta_f S$ is the proper entropy of fusion. Since fusion is always an endothermic process, the numerator of (1) is always positive, but the denominator may have either sign. It is negative for water, but positive for most substances. Thus the melting point of ice is decreased by increase of pressure, but that of most solids is increased.

The application of Clapeyron's relation to the equilibrium between two solid phases is analogous. In (1) we need only make the superscript ^{L} denote the phase stable at the higher temperature and ^{S} the phase stable at the lower temperature, so that $H_m^L - H_m^S$ is positive. The sign of dP/dT will then be the same as that of $V_m^L - V_m^S$.

For condensed phases, both V_m^L and V_m^S are small and their difference is much smaller. Usually a pressure of some hundred atmospheres is required to change the freezing point by a single degree. Let us take water as an illustrative example. We have

$$-dP/dT = 22 \text{ J K}^{-1} \text{ mole}^{-1}/(19.6-18.0) \text{ cm}^3 \text{ mole}^{-1} = 22 \text{ J K}^{-1}/1.6 \text{ cm}^3$$
$$= 220 \text{ atm}/1.6 \text{ K} = 1.4 \times 10^2 \text{ atm/K.} \qquad 3.38.2$$

As a second example, let us take sodium. We have

$$dP/dT = 7.1 \text{ J K}^{-1} \text{ mole}^{-1}/(24.6-24.2) \text{ cm}^3 \text{ mole}^{-1} = 7.1 \text{ J K}^{-1}/0.4 \text{ cm}^3$$
$$= 71 \text{ atm}/0.4 \text{ K} = 1.8 \times 10^2 \text{ atm/K.} \qquad 3.38.3$$

Hence as long as the pressure does not exceed a few atmospheres, the freezing point may for many purposes be regarded as unaffected by the pressure.

§3.39 *Clapeyron's relation applied to saturated vapour*

Let us now consider the equilibrium between a liquid and a gaseous phase. Using the superscripts ^{L} for the liquid and ^{G} for the gas we have according to (3.37.9)

$$dP/dT = (H_m^G - H_m^L)/T(V_m^G - V_m^L). \qquad 3.39.1$$

This exact relation can be transformed by making two approximations. In the first place we neglect the proper volume of the liquid compared with that of the vapour. In the second place we neglect the virial coefficients of the gas and treat it as perfect. With these approximations we have

SYSTEMS OF A SINGLE COMPONENT

$$V_m^G - V_m^L \approx RT/P \qquad 3.39.2$$

Substituting (2) into (1), we obtain

$$d \ln P/dT = (H_m^G - H_m^L)/RT^2 = \Delta_e H/RT^2 \qquad 3.39.3$$

where $\Delta_e H$ is the proper enthalpy of evaporation.

We denote equilibrium or saturated vapour pressure of a condensed phase by P_s and accordingly write in place of (3)

$$d \ln P_s/dT = \Delta_e H/RT^2 \qquad 3.39.4$$

or

$$d \ln P_s/d(1/T) = -\Delta_e H/R. \qquad 3.39.5$$

It follows from (5) that if we plot $\ln P_s$ against $1/T$ the curve so obtained has at each point a slope equal to $-\Delta H_e/R$. Actually $\Delta_e H$ varies so slowly with the temperature that this curve is nearly a straight line.

Formula (5) incidentally provides a method, rarely if ever mentioned, for determining the proper mass in the vapour. For by measuring P_s at several known temperatures we can use (5) to calculate $\Delta_e H$. We can then make direct calorimetric measurements to determine what mass of liquid is converted to vapour when a quantity of heat equal to $\Delta_e H$ is absorbed. This mass is then the proper mass of vapour.

The treatment of equilibrium between a solid and its vapour is precisely analogous. The saturated vapour pressure P_s of the solid is related to the temperature by

$$d \ln P_s/d(1/T) = -\Delta_s H/R \qquad 3.39.6$$

where $\Delta_s H$ is the proper enthalpy of sublimation.

§3.40 *Heat capacities of two phases in equilibrium*

Consider two phases of a single component in mutual equilibrium. Suppose now that we isolate unit amount of either of these phases and change its temperature, not at constant pressure, but adjusting the pressure to the value corresponding to two-phase equilibrium at each temperature. The quantity of heat absorbed in this phase will evidently, for a small temperature increase dT, be proportional to dT. We may therefore write for either of the two phases

$$q = C_{eq} dT. \qquad 3.40.1$$

C_{eq} is the heat capacity at two-phase equilibrium. Since moreover the change is reversible we may write instead of (1)

$$dS_m = C_{eq} dT/T. \qquad 3.40.2$$

But for the change in question

$$dS_m = \{(\partial S_m/\partial T)_P + (\partial S_m/\partial P)_T dP/dT\} dT. \qquad 3.40.3$$

Comparing (2) with (3) we see that

$$C_{eq} = T\{(\partial S_m/\partial T)_P + (\partial S_m/\partial P) dP/dT\} = C_P - \alpha V_m T dP/dT \qquad 3.40.4$$

using the definition (3.03.2) of C_P and Maxwell's relation (1.47.4). Now substituting from (3.37.9) into (4) we obtain

$$C_{eq} = C_P - \alpha V_m \Delta H / \Delta V \qquad 3.40.5$$

where Δ denotes the increase when unit amount passes isothermally from the one phase to the other; as regards sign the same convention must of course be used for ΔH and ΔV.

§3.41 Heat capacities at saturation

The most important application of the formulae of the previous section is to the equilibrium between a liquid and its vapour. The quantities C_{eq} are then called the heat capacities at saturation and are denoted by C_{sat}. If we neglect the second virial coefficient of the gas and also neglect the proper volume of the liquid compared with that of the gas, formula (3.40.5) becomes

$$C_{sat} = C_P - \alpha \Delta_e H P V_m / RT \qquad 3.41.1$$

where $\Delta_e H$ is the proper enthalpy of evaporation.

Formula (1) is applicable either to the vapour or to the liquid, but the importance of the second term on the right is very different in the two cases. For the vapour we have, still neglecting the second virial coefficient,

$$\alpha = T^{-1} \qquad PV_m = RT \qquad 3.41.2$$

so that, using the superscript ^{G} for the gas, we obtain

$$C_{sat}^G = C_P^G - \Delta_e H / T = C_P^G - \Delta_e S. \qquad 3.41.3$$

The second term on the right may be numerically greater than the first, in which case C_{sat}^G is negative. For example for steam at its normal boiling point

$$C_P = 34 \text{ J K}^{-1} \text{ mole}^{-1}$$
$$\Delta_e S = \Delta_e H / T = 40.6 \text{ kJ mole}^{-1} / 373 \text{ K} = 109 \text{ J K}^{-1} \text{ mole}^{-1}$$

so that

$$C_{sat}^G = (34 - 109) \text{ J K}^{-1} \text{ mole}^{-1} = -75 \text{ J K}^{-1} \text{ mole}^{-1}$$

SYSTEMS OF A SINGLE COMPONENT 123

and we see that the heat capacity of steam at saturation is negative.

For the liquid phase on the other hand the second term on the right of (1) is much smaller than for the gas because V_m is smaller by a factor of something like 10^{-3} or less. Consequently for the liquid phase we may neglect this term and replace (1) by

$$C_{\text{sat}}^L = C_P^L \qquad 3.41.4$$

where the superscript ^{L} denotes the liquid phase.

The formulae of this section may also be applied to the equilibrium conditions between solid and vapour. Formula (3) is then applicable to the vapour and formula (4) to the solid.

§3.42 Temperature dependence of enthalpies of evaporation and of fusion

Consider any phase change such as evaporation or fusion and let the symbol Δ denote the increase in any property when unit amount passes isothermally from the one phase to the other in the direction such that ΔH is positive, i.e. from liquid to gas or from solid to liquid. Then we have

$$\Delta H/T = \Delta S. \qquad 3.42.1$$

Differentiating with respect to T, varying P so as to maintain equilibrium, we have

$$\mathrm{d}(\Delta H/T) = \mathrm{d}(\Delta S)/\mathrm{d}T = \Delta(\mathrm{d}S/\mathrm{d}T) = \Delta C_{\text{eq}}/T \qquad 3.42.2$$

or

$$\mathrm{d}(\Delta H)/\mathrm{d}T - \Delta H/T = \Delta C_{\text{eq}}. \qquad 3.42.3$$

For equilibrium between liquid and vapour, $C_{\text{eq}} = C_{\text{sat}}$ is given by (3.41.3) for the vapour and by (3.41.4) for the liquid. Substituting these into (3) we obtain

$$\mathrm{d}(\Delta_e H)/\mathrm{d}T = C_P^G - C_P^L \qquad 3.42.4$$

the terms $\Delta_e H/T$ on either side cancelling. Formula (4) involves the several approximations mentioned in §3.41. It is formally similar to the exact formula for a process taking place between pressure limits independent of the temperature.

To obtain the temperature coefficient of an enthalpy of fusion, we have to go back to (3.40.5), which we rewrite in the form

$$C_{\text{eq}} = C_P - (\partial V_m/\partial T)_P (\Delta_f H/\Delta_f V) \qquad 3.42.5$$

where Δ_f denotes the increase of a proper quantity on fusion. Substituting

(5) into (3), we obtain

$$d(\Delta_f H)/dT = \Delta_f C_P + \Delta_f H/T - (\Delta_f H/T)(\partial \ln \Delta_f V/\partial \ln T)_P \qquad 3.42.6$$

a formula due to Planck*. The magnitude of the last term on the right is usually unknown and it is often neglected. Formula (6) then reduces to

$$d(\Delta_f H)/dT = \Delta_f C_P + \Delta_f H/T. \qquad 3.42.7$$

Of the two terms on the right, either may be numerically greater. We thus have a formula not even approximately of the same form as the formula for a process taking place between pressure limits independent of the temperature.

Evidently the formulae of this section may mutatis mutandis be applied to the equilibrium between two solid phases.

§3.43 *Triple point*

We have seen that the equilibrium condition for a single component between two phases α and β

$$\mu^\alpha(T, P) = \mu^\beta(T, P) \qquad 3.43.1$$

is equivalent to a relation between P and T which can be represented by a curve on a P–T diagram. Similarly the equilibrium between the phases α and γ can be represented by a curve on a P–T diagram. If these two curves cut, we shall have at the point of intersection

$$\mu^\alpha(T, P) = \mu^\beta(T, P) = \mu^\gamma(T, P) \qquad 3.43.2$$

and the three phases α, β, γ will be in mutual equilibrium. This point of intersection is called a *triple point* and the values of T and P at the triple point are called the *triple-point temperature* and the *triple-point pressure*.

We have seen that a single component in one phase has two degrees of freedom since temperature and pressure can be varied independently and that two phases in mutual equilibrium have only one degree of freedom since temperature and pressure are mutually dependent. We now see that three phases can exist in mutual equilibrium only at a particular temperature and particular pressure. Thus three phases of a single component in mutual equilibrium have no degree of freedom.

In figure 3.4 the conditions of mutual equilibrium for H_2O are shown[†] on the P–T diagram.

---
* Planck, Ann. Phys. Lpz. 1887 **30** 574.
† From Landolt–Börnstein Tables.

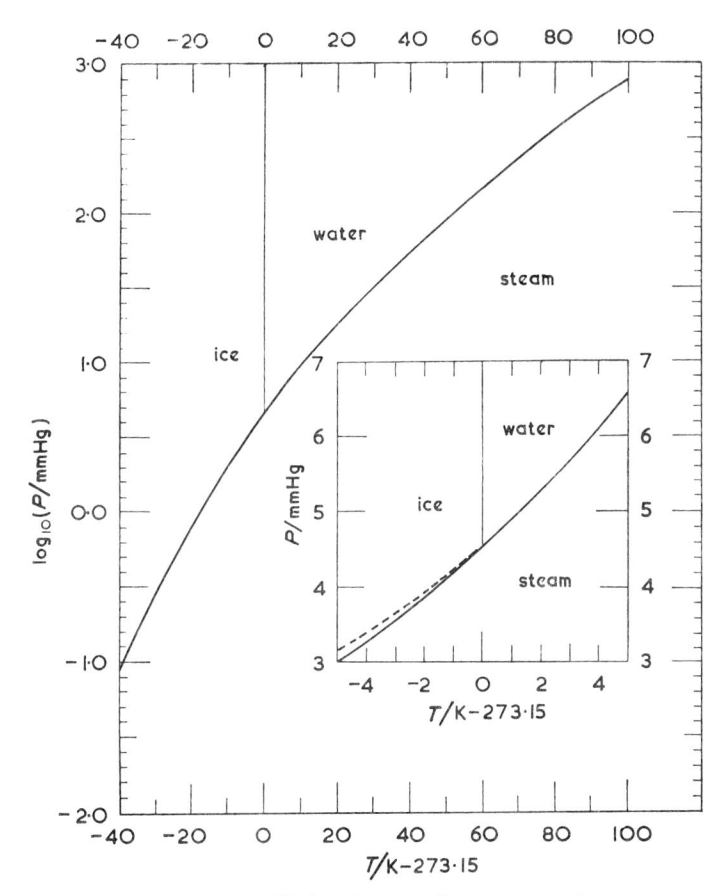

Fig. 3.4. Equilibrium between ice, water, and steam

Triple points can also exist for two solid phases and one liquid phase or for two solid phases and a vapour phase or for three solid phases. More rarely we may have two liquid phases and a vapour phase or a solid phase. A triple point can occur in a region where all three phases are metastable. The conditions of equilibrium for sulphur are shown* in figure 3.5. There are three stable triple points

T_1: equilibrium between monoclinic, liquid, and vapour
T_2: equilibrium between rhombic, monoclinic, and liquid
T_3: equilibrium between rhombic, monoclinic, and vapour

* From Landolt–Börnstein Tables.

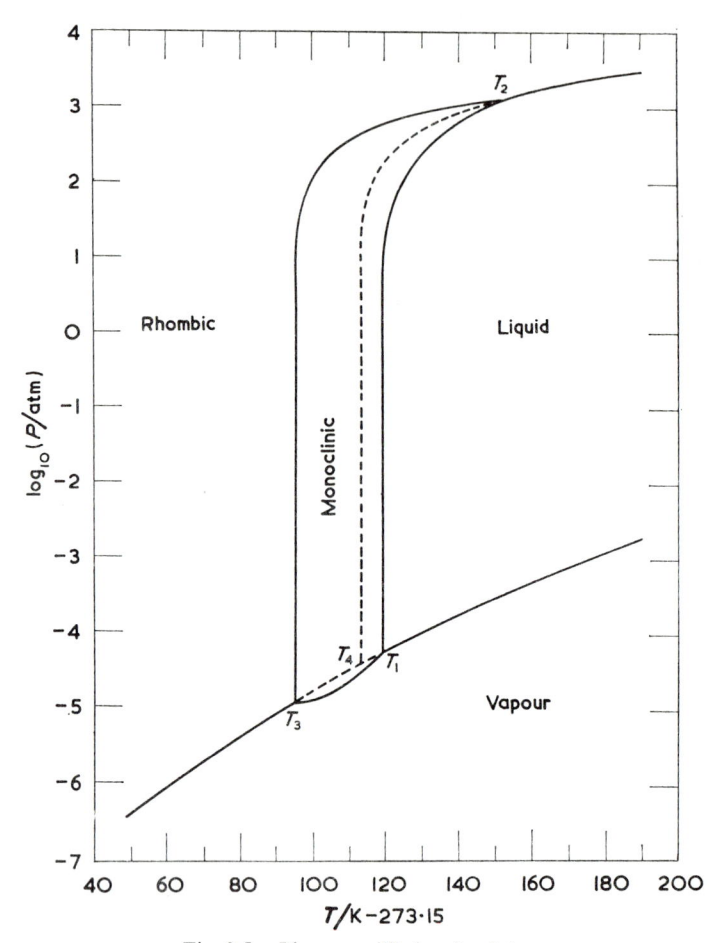

Fig. 3.5. Phase equilibria of sulphur

and one metastable triple point

T_4: equilibrium between rhombic, liquid, and vapour, all three phases being metastable and the monoclinic being the stable form.

§3.44 Critical points

The P–V_m isotherms of all pure substances fall into two classes according as the temperature lies above or below a *critical temperature* T_c. Examples of each class are shown in figure 3.6 for carbon dioxide* and in figure 3.7 for xenon[†].

* Michels, Blaisse, and Michels, Proc. Roy. Soc. A 1937 **160** 367.
† Habgood and Schneider, Can. J. Chem. 1954 **32** 98.

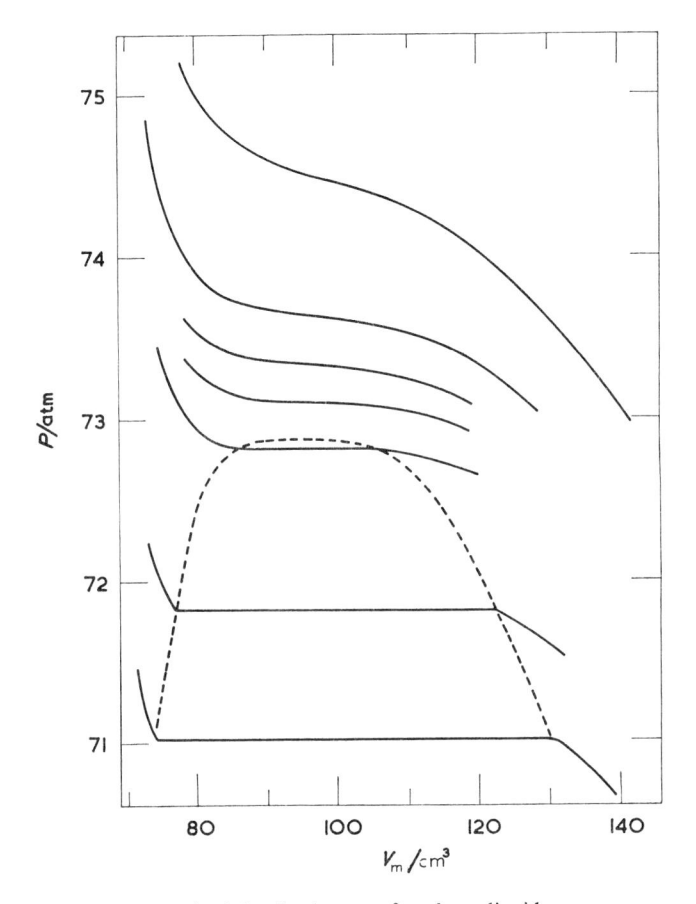

Fig. 3.6. Isotherms of carbon dioxide

When the proper volume is sufficiently large both classes approximate to the rectangular hyperbolae $PV_m = RT$ of a perfect gas. As the proper volume diminishes, the form of the two classes is quite different. At temperatures greater than the critical, there is a smooth regular variation along the whole isotherm, which can be expressed mathematically by saying that it is a single analytic curve or expressed physically by saying that throughout the isotherm there is a single fluid phase. At temperatures below the critical on the other hand, the isotherm consists of three analytically distinct parts separated by discontinuities of the slope. The middle portion is a straight line parallel to the V_m axis. These parts represent respectively the pure gas, the saturated vapour in equilibrium with the liquid, and the pure liquid. The isothermal curve for the critical temperature T_c is the borderline between

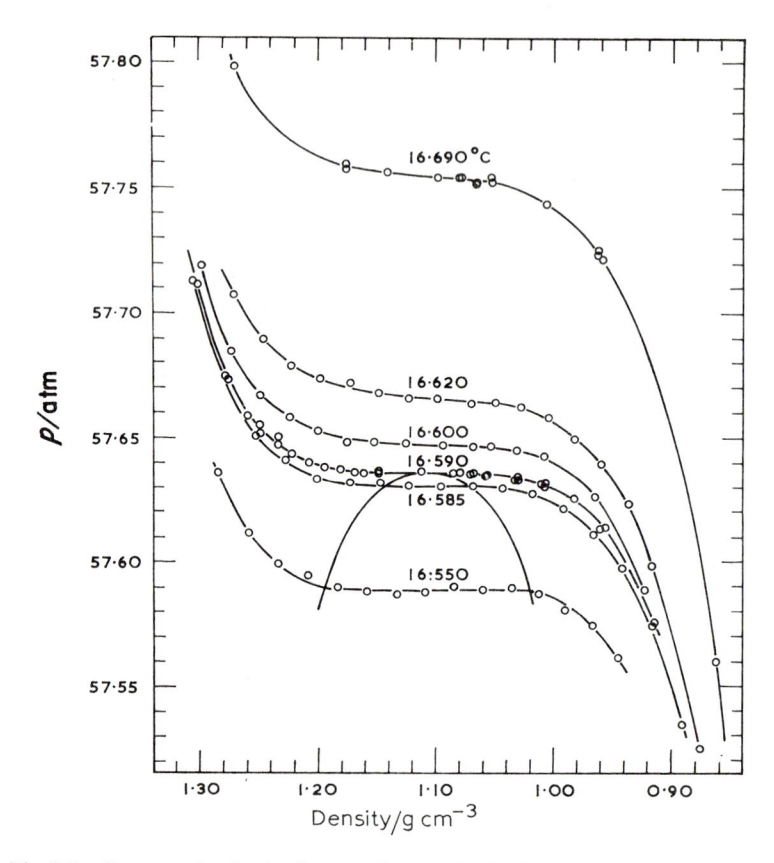

Fig. 3.7. Pressure–density isotherms of xenon in the immediate neighbourhood of the critical temperature

the two classes of isotherms. In this isotherm the horizontal portion is reduced to a single point of horizontal inflexion.

Both diagrams show the locus of the points representing on the left the liquid phase under the pressure of its vapour and on the right the locus of the points representing the saturated vapour. As the temperature increases the proper volume of the liquid at the pressure of its vapour increases, while the proper volume of the saturated vapour decreases. At the critical temperature the isotherm has a point of horizontal inflexion where the liquid and vapour phases cease to be distinguishable. The state represented by this point is called the *critical state*; the pressure and proper volume in the critical state are called the *critical pressure* P_c and the *critical volume* V_c respectively.

SYSTEMS OF A SINGLE COMPONENT 129

To recapitulate, above the critical temperature the substance can exist in only one fluid state. Below the critical temperature it can exist in two states, the liquid with a proper volume less than the critical volume and the gas with a proper volume greater than the critical volume. The equilibrium pressure between the two phases, liquid and vapour, can have values up to but not exceeding the critical pressure.

§3.45 *Continuity of state*

The relation between pressure P and proper volume V_m of a single component at a temperature below the critical temperature is shown diagrammatically in figure 3.8. The portion KL represents the liquid state, the portion VW the gaseous state, and the straight portion LV the two-phase system liquid + saturated vapour.

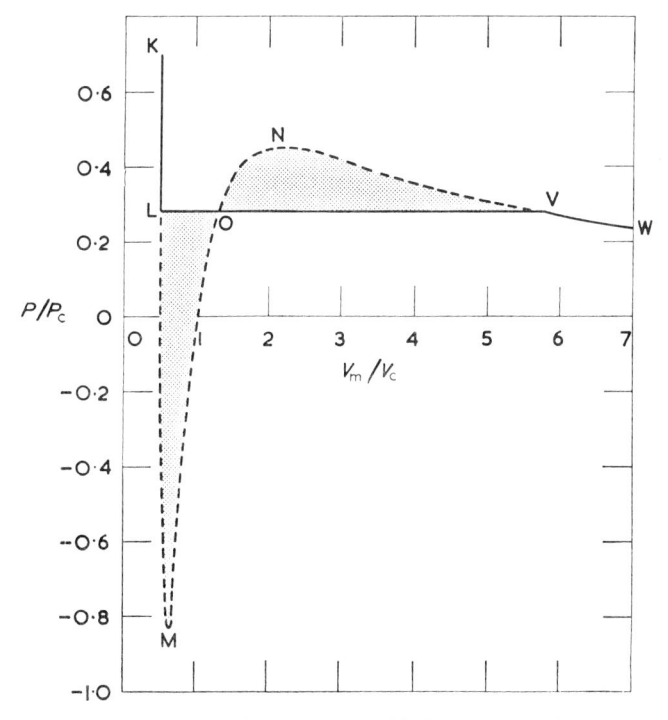

Fig. 3.8. Continuity between liquid and gas phases

At the given temperature the substance can be brought from the liquid state to the gaseous state, or conversely, only by a change during part of which two separate phases will be present. By varying the temperature, however, it is possible to bring the substance from the gaseous state represented by W to the liquid state represented by K by a continuous change throughout which there is never more than one phase present. It is only necessary to raise the temperature above the critical temperature, keeping the volume sufficiently greater than the critical volume, then compress the fluid to a volume below the critical volume, keeping the temperature above the critical temperature, and finally cool the liquid to its original temperature, keeping the volume sufficiently below the critical volume. This possibility of continuity between the liquid and gaseous states was first realized by James Thomson[*], and he suggested that the portions KL and VW of the isotherm are actually parts of one smooth curve, such as KLMONVW. In point of fact, states corresponding to the portion VN are realizable as supersaturated vapour, and under certain circumstances the same may be true of the portion LM representing superheated liquid. Each of these portions represents states stable with respect to infinitesimal variations, but metastable relative to the two-phase system liquid + saturated vapour. The portion of the curve MON, on the other hand, represents states absolutely unstable, since here

$$(\partial V_m/\partial P)_T > 0 \qquad 3.45.1$$

and, according to (1.38.4), such states are never realizable.

Although the states represented by points on the curve LMONV are either metastable or unstable, they have been treated[†] as equilibrium states. It follows that the sequence of states represented by the curve LMONV corresponds to a reversible process. The change in the chemical potential μ of the fluid in passing through this sequence of states is, according to (1.28.23) given by

$$\mu^G - \mu^L = \int_L^G (\partial \mu/\partial P)_T \, dP = \int_L^G V_m \, dP \qquad 3.45.2$$

where the integrals are to be evaluated along the curve LMONV. But, since the two states represented by ^{L} and ^{G} can exist in equilibrium with each other, we have

$$\mu^G = \mu^L. \qquad 3.45.3$$

From (2) and (3) we deduce

$$\int_L^G V_m \, dP = 0 \qquad 3.45.4$$

where the integral is to be evaluated along the curve LMONV. The geo-

[*] J. Thomson, Proc. Roy. Soc. 1871 **20** 1.
[†] Maxwell, Nature 1875 **11** 357.

metrical significance of (4) is that the two shaded surfaces LMO and ONV are of equal area. This condition is due to Maxwell*.

It is instructive to reconsider continuity of state in terms of the Helmholtz function $\mathscr{F}$. Imagine this to be plotted as vertical coordinate against T and V as horizontal Cartesian coordinates. The resulting locus is a curved surface. Consider now cross-sections of this surface by planes $T=$const. Examples of these are shown diagrammatically in figure 3.9 and since

$$(\partial \mathscr{F}/\partial V)_T = -P \qquad 3.45.5$$

the slope of each curve at each point is equal to $-P$.

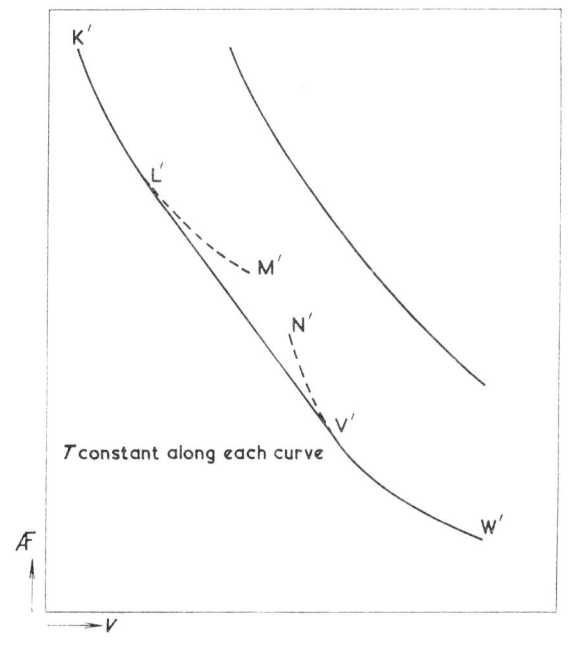

Fig. 3.9. Stable and metastable isotherms

In the upper curve we see that as V increases, the negative slope steadily decreases numerically and so P decreases steadily. This is typical of any temperature above the critical.

In the lower curve we see that there are two portions K′L′ and V′W′ in which the negative slope decreases steadily as V increases and these are joined by a straight line L′V′ touching K′L′ at L′ and touching V′W′ at V′. These three portions correspond to liquid, to gas, and to a two-phase liquid–vapour system. This is typical of a temperature below the critical. The

* Maxwell, Nature 1875 **11** 357.

broken portion of curve L'M' represents superheated liquid and the broken portion V'N' represents supersaturated vapour. We see immediately that all states represented by these portions of curve are metastable, for any point on either of them lies above a point of the same volume V on the straight line L'V'. This means that the Helmholtz function of the superheated liquid or supersaturated vapour is greater than in a system of the same volume consisting of a mixture of liquid L' and saturated vapour V'.

The portions of curve L'M' and N'V' have curvature concave upwards so that
$$\partial P/\partial V = -\partial^2 \mathcal{F}/\partial V^2 < 0. \qquad 3.45.6$$

Hence according to (1.38.4) they represent states internally stable, though metastable with respect to a two-phase mixture. If however we wish to unite these two portions into a single smooth curve, the middle portion would necessarily have a curvature concave downwards. This would correspond to a positive value of $\partial P/\partial V$ and so to unstable states and we saw in §1.38 that such states are never realizable. It may therefore be argued that no physical significance could be attached to this part of the curve. Nevertheless, if the realizable parts K'L'M' and N'V'W' of the surface could be represented by the same analytical function, it would be reasonable from a mathematical point of view to consider the complete surface. Having constructed such a surface and considering a section corresponding to a particular temperature below the critical, we could then plot $P = -\partial \mathcal{F}/\partial V$ against V and so construct a curve such as that in figure 3.8. From this construction it follows of necessity that in figure 3.8 the area below the broken curve LMONV and the area below the straight line LV are both equal to the height of L' above V' in figure 3.9. Consequently these two areas are equal. From this it follows immediately that the two shaded areas are equal as already proved. Since the portion MON of the curve cannot be realized experimentally, instead of saying that the two-phase equilibrium is determined by the condition of equality of the two shaded areas, it is perhaps more correct to say that, L and V being known, if the connecting portion of the curve were sketched in such a manner as to make the two shaded areas unequal it would be nonsensical for then $-P$ would not be the slope of any conceivable curve in the plot of $\mathcal{F}$ against V.

§3.46 *Two phases at different pressures*

In our previous considerations of equilibrium between two phases of one component, we have assumed the equilibrium to be complete so that the two phases were at the same pressure. The distribution equilibrium of one

SYSTEMS OF A SINGLE COMPONENT   133

component between two phases at different pressures is also of interest. Let us denote the two phases by the superscripts $^\alpha$ and $^\beta$. Then the equilibrium condition determining the change from the one phase to the other is according to (1.42.1)

$$\mu^\alpha = \mu^\beta. \qquad 3.46.1$$

If we vary the common temperature T of the two phases and the pressures P^α and P^β of the two phases, the condition for maintenance of equilibrium is

$$d\mu^\alpha = d\mu^\beta \qquad 3.46.2$$

or

$$(\partial\mu^\alpha/\partial T)dT + (\partial\mu^\alpha/\partial P^\alpha)dP^\alpha = (\partial\mu^\beta/\partial T)dT + (\partial\mu^\beta/\partial P^\beta)dP^\beta. \qquad 3.46.3$$

Substituting from (1.28.22) and (1.28.23) we obtain

$$-S_m^\alpha dT + V_m^\alpha dP^\alpha = -S_m^\beta dT + V_m^\beta dP^\beta \qquad 3.46.4$$

or

$$V_m^\beta dP^\beta - V_m^\alpha dP^\alpha = (S_m^\beta - S_m^\alpha)dT = \underset{\alpha\to\beta}{\Delta S}\, dT \qquad 3.46.5$$

where $\Delta_{\alpha\to\beta}$ is used to denote the increase of a quantity when unit amount passes from the phase α to the phase β.

Since we may rewrite (1) as

$$H_m^\alpha - TS_m^\alpha = H_m^\beta - TS_m^\beta \qquad 3.46.6$$

it follows immediately that

$$T\underset{\alpha\to\beta}{\Delta S} = \underset{\alpha\to\beta}{\Delta H} \qquad 3.46.7$$

just as for two phases at the same pressure. In fact formula (3.37.8) is a special example of (7) and the physical significance is the same in both cases.

If we now substitute from (7) into (5) we obtain

$$V_m^\beta dP^\beta - V_m^\alpha dP^\alpha = \underset{\alpha\to\beta}{\Delta H}\, dT/T. \qquad 3.46.8$$

It is evident that two of the three quantities T, P^α, P^β are independent and so the system has two degrees of freedom. The most important application of these formulae is to the equilibrium between a liquid and its vapour. We then use the superscript ^{L} for the liquid and ^{G} for the vapour. In this notation (8) becomes

$$V_m^G dP^G - V_m^L dP^L = (\Delta_e H/T)dT \qquad 3.46.9$$

where $\Delta_e H$ is the proper enthalpy of evaporation. According to the definition

(3.17.1) of fugacity p, we may replace (9) by

$$RT\,\mathrm{d}\ln p - V_\mathrm{m}^\mathrm{L}\mathrm{d}P^\mathrm{L} = (\Delta_\mathrm{e} H/T)\mathrm{d}T. \qquad 3.46.10$$

In particular at constant temperature we have for the dependence of the fugacity of the gas on the external pressure P on the liquid

$$\mathrm{d}\ln p/\mathrm{d}P = V_\mathrm{m}^\mathrm{L}/RT \qquad (T\text{ const.}). \qquad 3.46.11$$

If we treat the vapour as a perfect gas, we may replace p by P^G.

If, on the other hand, we maintain constant the pressure P on the liquid, we obtain from (10) for the dependence of the gas fugacity on the temperature

$$\mathrm{d}\ln p/\mathrm{d}T = \Delta_\mathrm{e} H/RT^2 \qquad (P\text{ const.}) \qquad 3.46.12$$

or if we treat the vapour as a perfect gas

$$\mathrm{d}\ln P^\mathrm{G}/\mathrm{d}T = \Delta_\mathrm{e} H/RT^2 \qquad 3.46.13$$

or

$$\mathrm{d}\ln P^\mathrm{G}/\mathrm{d}(1/T) = -\Delta_\mathrm{e} H/R. \qquad 3.46.14$$

It is instructive to compare (14) with (3.39.5). The latter involves neglecting the proper volume of the liquid compared with that of the vapour, but the former involves no such approximation. The difference between the exact formula (14) and the approximate formula (3.39.5) is usually negligible owing to the fact that in order to affect the saturated vapour pressure P_s appreciably by change of the hydrostatic pressure P^L at constant temperature, one requires according to (11) pressures considerably greater than the vapour pressure itself.

The direct experimental application of these formulae would require the separation of the liquid from the vapour by a membrane permeable to the vapour, but not to the liquid. This is difficult to achieve, though not impossible. Consequently the formulae have not much direct practical application. They have nevertheless a real importance, which will become clear when we consider systems of two or more components. We shall find that these formulae remain true in the presence of another component gas insoluble in the liquid, provided we interpret P^G as the partial pressure of the vapour when mixed with the inert gas. We cannot profitably say more at this stage, but we shall return to this point in §4.13.

§3.47 *Fugacity of a condensed phase*

In §3.17 we defined the fugacity p of a gaseous pure substance in terms of its absolute activity λ. We now define the *fugacity of a pure substance* in any condensed phase as being equal to the fugacity in the gas phase with which it is in equilibrium. Evidently when two condensed phases are in equilibrium with each other the fugacities must be equal in the two phases.

With this extended definition we may regard formula (3.46.10), namely

$$d \ln p = (\Delta_e H/RT^2)dT + (V_m^L/RT)dP \qquad 3.47.1$$

as expressing the dependence of p, the *fugacity of a liquid*, on the temperature T and the external pressure P. A precisely analogous relation applies to a solid.

§3.48 *Corresponding states of fluids*

The principle of *corresponding states* asserts that for a group of similar substances the equation of state can be written in the form

$$P/P_c = \phi\{(T/T_c), (V_m/V_c)\} \qquad 3.48.1$$

where ϕ is the same function for all the substances of the group.

Whereas it is not possible to express the equation of state in any simple analytical form, the principle of corresponding states is obeyed with a useful degree of accuracy by a considerable number of substances. It is in fact obeyed within the accuracy of experiment by the three inert elements Ar, Kr, Xe and to a high degree of accuracy by these substances together with Ne, N_2, O_2, CO, and CH_4. It would be misleading to try to divide substances sharply into two groups, those which do and those which do not obey the principle. It is obeyed more or less accurately by a great variety of substances. Deviations occur due to any one or several of the following causes:
(a) quantal effects in the lightest molecules, especially H_2, He, and to a much smaller extent Ne;
(b) polarity of the molecule or presence of strong polar groups even though the resultant dipole moment vanishes as in CO_2, SF_6;
(c) large departures of shape from rough spherical symmetry as in the higher alkanes and alkenes.

The principle is not obeyed at all by substances whose molecules form hydrogen bonds, especially those containing hydroxyl or amino groups, nor those such as NO_2 whose molecules associate.

We shall now review* briefly some of the experimental data which show directly or indirectly how well certain substances especially Ne, Ar, Kr, Xe, N_2, O_2, CO, CH_4 obey an equation of state of the common form (1).

---

* Guggenheim, J. Chem. Phys. 1945 **13** 253; cf. Pitzer, J. Chem. Phys. 1939 **7** 583.

TABLE 3.8

Corresponding states of gases and liquids

	Formula	Ne	Ar	Kr	Xe	N_2	O_2	CO	CH_4
1	M/g mole^{-1}	20.18	39.94	83.7	131.3	28.02	32.00	28.00	16.03
2	T_c/K	44.8	150.7	209.4	289.8	126.0	154.3	133.0	190.3
3	V_c/cm^3 mole^{-1}	41.7	75.3	92.1	118.8	90.2	74.5	93.2	98.8
4	P_c/atm	26.9	48.0	54.1	57.6	33.5	49.7	34.5	45.7
5	$P_c V_c/RT_c$	0.305	0.292	0.290	0.288	0.292	0.292	0.294	0.289
6	T_B/K	121	411.5			327		345	491
7	T_B/T_c	2.70	2.73			2.59		2.6	2.58
8	T_s/K ($P_s = P_c/50$)	25.2	86.9	122.0	167.9	74.1	90.1	78.9	110.5
9	T_s/T_c	0.563	0.577	0.582	0.580	0.588	0.583	0.593	0.581
10	$\Delta_e H/R$ K	224	785	1086	1520	671	820	727	1023
11	$\Delta_e H/RT_s$	8.9	9.04	8.91	9.06	9.06	9.11	9.22	9.26
12	V_m/cm^3 mole^{-1}		228.1	34.1	42.7				
13	V_m/V_c		0.374	0.371	0.376				

In table 3.8 the first row gives the proper mass M, the next three rows the critical temperature T_c, critical volume V_c, and critical pressure P_c. The fifth row gives values of $P_c V_c/RT_c$ which according to the principle should have a universal value. All the values lie close to 0.29. It is of interest to note that the value for xenon based on recent measurements is closer to 0.29 than the best experimental value 0.278 quoted in 1945.

In figure 3.10 the experimental data on the second virial coefficients of Ar, Kr, Xe, and CH_4 are shown in the form of B/V_c plotted against T/T_c. The data for the four substances were shown by McGlashan and Potter* to be well fitted from high values of T/T_c down to $T/T_c = 0.6$ by the empirical formula

$$B/V_c = 0.430 - 0.886(T_c/T) - 0.694(T_c/T)^2. \qquad 3.48.2$$

They are also well fitted from high values of T/T_c down to 0.5 by the curve in figure 3.10 which represents the formula[†]

$$B/V_c = 0.440 + 1.40\{1 - \exp(0.75 T_c/T)\} \qquad 3.48.3$$

which can be derived theoretically from an interaction energy w between a pair of molecules distant r apart of the 'square-well' form given by

$$\begin{array}{ll} r < \sigma & w = \infty \\ \sigma < r < 1.5\sigma & w = -\varepsilon \\ r > 1.5\sigma & w = 0 \end{array} \qquad 3.48.4$$

* McGlashan and Potter, Proc. Roy. Soc. A 1962 **267** 478.

† Guggenheim, Applications of Statistical Mechanics, Clarendon Press 1966 p. 36.

when the parameters σ and ε have the values given by
$$\tfrac{2}{3}\pi L\sigma^3 = 0.447 V_c$$
$$\varepsilon = 0.936 k T_c.$$
3.48.5

The Boyle temperature T_B at which the second virial coefficient changes sign is given in the sixth row of table 3.8. In the seventh row are given values of T_B/T_c and all these values lie near 2.7.

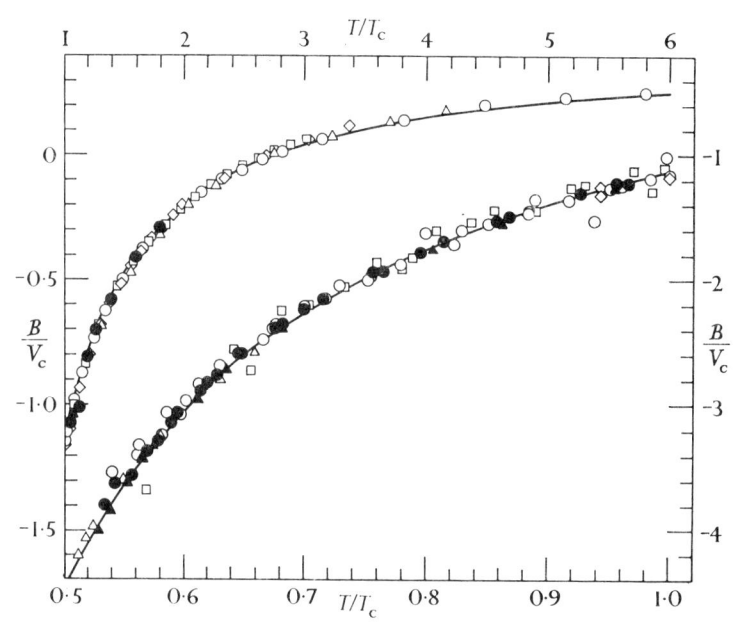

Fig. 3.10. Reduced second virial coefficients
The left-hand and upper scales relate to the upper curve.
The right-hand and lower scales relate to the lower curve.
○●, Ar;   △▲, Kr;   ◇, Xe;   □, CH_4.

● Recent measurements communicated privately to the author by Rowlinson and by Staveley.

▲ Recent measurements communicated privately to the author by Rowlinson.

If ϱ^L denotes the density of the liquid and ϱ^G that of the vapour in mutual equilibrium at the temperature T, while ϱ_c denotes the density at the critical point, then according to the principle of corresponding states we should expect ϱ^L/ϱ_c and ϱ^G/ϱ_c to be common functions of T/T_c. How nearly this is the case is shown in figure 3.11. The curve in the diagram is drawn according to the empirical formulae

$$(\varrho^L + \varrho^G)/2\varrho_c = 1 + \tfrac{3}{4}(1 - T/T_c) \qquad 3.48.6$$
$$(\varrho^L - \varrho^G)/\varrho_c = \tfrac{7}{2}(1 - T/T_c)^{\frac{1}{3}}. \qquad 3.48.7$$

It is a pure accident that the data can be represented by formulae with such simple numerical coefficients. These formulae as displayed above are of high relative accuracy, but if used to compute ϱ^G the percentage inaccuracy increases with decrease of temperature and becomes serious below $T \approx 0.65 T_c$. It is therefore not recommended to use these formulae for computing values of ϱ^G. There are however occasions when we require relatively accurate

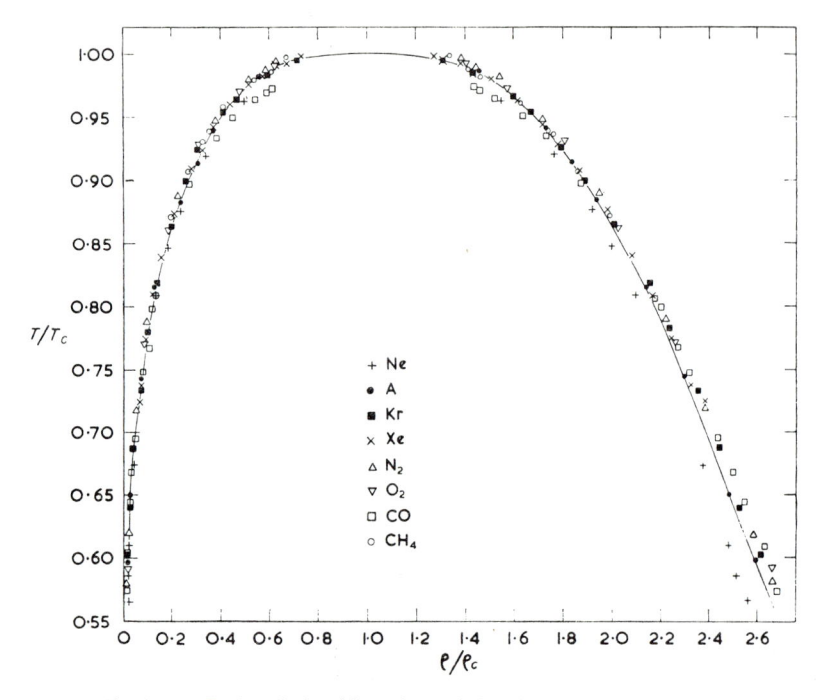

Fig. 3.11. Reduced densities of coexisting liquid and gas phases

values, not of ϱ^G itself, but of $(\varrho^L - \varrho^G)/\varrho_c$; on such occasions formula (9) in view of its extreme simplicity and surprisingly high accuracy, has much to recommend it. An example of its use will occur in §3.65.

At temperatures considerably below the critical temperature say $T < 0.65 T_c$, it is more useful to consider the saturated vapour pressure P_s instead of ϱ^G. According to the principle of corresponding states we should expect P_s/P_c to be a common function of T/T_c. That this is approximately the case is seen from figure 3.12, where $\ln(P_s/P_c)$ is plotted against T_c/T for several substances. It is clear that the relation is nearly linear, so that we may write

$$\ln(P_s/P_c) = A - BT_c/T \qquad 3.48.8$$

where A, B are constants having nearly the same values for the several substances. In the diagram the straight line which best fits the data for argon, krypton, and xenon has been drawn. For this line

$$A = 5.29 \qquad B = 5.31 \qquad \text{(triple point to critical point)}. \qquad 3.48.9$$

The fact that A is nearly but not exactly equal to B, means that the straight line goes near to but not through the critical point. A formula of the

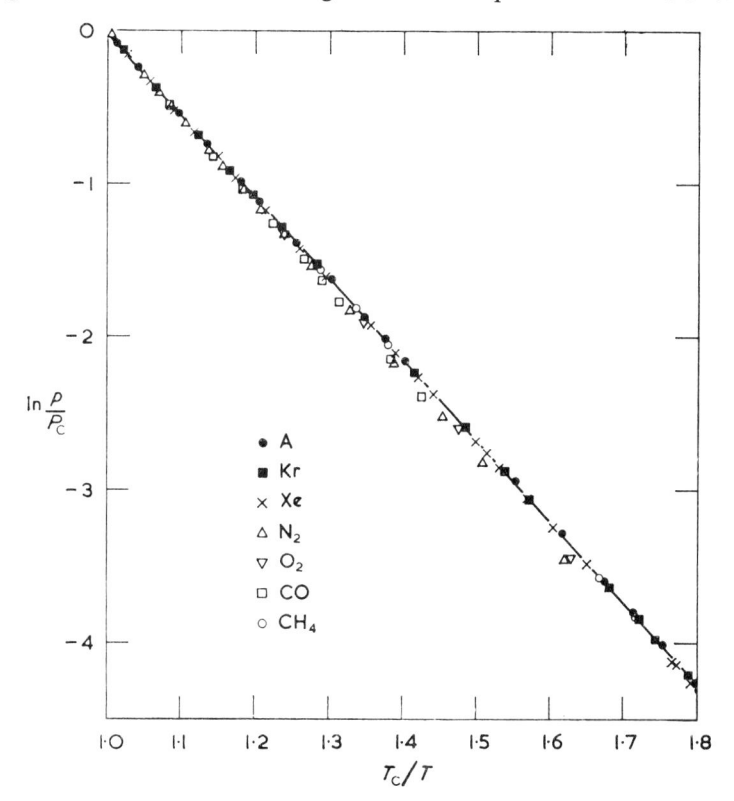

Fig. 3.12. Relation between vapour pressure and temperature

type (8) has a theoretical basis at low temperatures, where the vapour does not differ significantly from a perfect gas and the proper enthalpy of evaporation is nearly independent of the temperature. Under these conditions $\Delta_e H/R = BT_c$. At higher temperatures where the vapour pressure is greater, neither of these conditions holds; the vapour deviates appreciably from a perfect gas and $\Delta_e H$ decreases, becoming zero at $T = T_c$. At such temperatures formula (8) is empirical, but remains surprisingly accurate owing to a compensation between the two deviations.

In the temperature range between the triple point and the normal boiling point a formula of the type (8) becomes almost if not quite as accurate as the experimental data, but the best value for the constants A, B over this temperature range are not quite the same as the best values for the whole range from triple point to critical point. For argon an excellent fit of the experimental vapour pressures between the triple point and the normal boiling point and of the calorimetrically determined enthalpy of evaporation is attained with the values

$$A = 5.13 \qquad B = 5.21 \qquad \text{(temperatures below n.b.p.)} \qquad 3.48.10$$

$$\Delta_e H = 5.21 RT_c. \qquad 3.48.11$$

In the eighth row of table 3.8 are given the temperatures T_s at which the vapour pressure has a value one fiftieth of the critical pressure. In the ninth row are given values of the ratio T_s/T_c. These are all close to 0.58.

In the tenth row of this table are given values of the proper enthalpy of evaporation in the low temperature range where it is nearly independent of the temperature. In the eleventh row are given values of $\Delta_e H/RT_s$. All these values lie near to 9.0. Since $\Delta_e H/T_s$ is the entropy of evaporation, this aspect of the principle of corresponding states may be formulated thus: the entropy of evaporation at corresponding temperatures, e.g. temperatures at which the vapour pressure is one fiftieth the critical pressure, has a common value. The older *rule of Trouton* that substances should have the same entropy of evaporation at their normal boiling points is not in accord with the principle of corresponding states and is in somewhat less good agreement with the facts.

In the twelfth row of the table are given values of V_m the proper volume of the liquid at temperatures just above the triple point and in the thirteenth row values of the ratio V_m/V_c. These values are all near to 0.375.

§3.49 *Corresponding states of solids*

The principle of corresponding states has a much more restricted applicability to solids. It however applies with high accuracy to the group of the inert elements Ne, Ar, Kr, Xe. The relevant data for comparison are given in table 3.9. In the first three rows are given values of T_c, V_c, and P_c.

In the fourth row are given values of the triple point temperature T_t and in the fifth row values of the ratio T_t/T_c. All these values are near to 0.555.

In the sixth row are given values of the proper enthalpy of fusion $\Delta_f H$ divided by R and in the seventh row values of the entropy of fusion $\Delta_f S$ divided by R. These are all near to 1.69.

In the eighth row are given values of P_t, the triple point pressure, and in the ninth row values of the ratio $100P_t/P_c$. These are all near to 1.4.

Finally in the tenth and eleventh rows are given the proper volumes V^L and V^S of the liquid and solid respectively both at the triple point. In the twelfth row are given the ratios V^L/V^S, all near to 1.15.

TABLE 3.9

Corresponding states of solids

	Formula	Ne	Ar	Kr	Xe
1	T_c/K	44.8	150.7	209.4	289.8
2	V_c/cm³ mole⁻¹	41.7	75.3	92.1	118.8
3	P_c/atm	26.9	48.0	54.1	58.0
4	T_t/K	24.6	83.8	116.0	161.3
5	T_t/T_c	0.549	0.557	0.553	0.557
6	$\Delta_f H/R$ K	40.3	141.3	196.2	276
7	$\Delta_f H/RT_t$	1.64	1.69	1.69	1.71
8	P_t/atm	0.425	0.682	0.721	0.810
9	$100P_t/P_c$	1.58	1.42	1.33	1.40
10	V^L/cm³ mole⁻¹		28.14	34.13	42.68
11	V^S/cm³ mole⁻¹		24.61	29.65	37.09
12	V^L/V^S		1.144	1.151	1.151

§3.50 Two simple equations of state

Many attempts have been made in the past to represent the equation of state of gas and liquid throughout the whole P–V–T domain by an analytical formula. It is now known that it is not possible so to represent the experimental data accurately except by complicated and unwieldy formulae of little interest. On the other hand the distinction between liquid and gas and the existence of a critical point can be deduced qualitatively from various quite simple equations of state. Of these we shall mention only two of the simplest.

The earliest attempt to describe semi-quantitatively the behaviour of a real fluid was made by van der Waals. His well-known formula is

$$(P+a/V_m^2)(V_m-b)=RT \qquad 3.50.1$$

but in the present context it is more convenient to write it as

$$PV_m=RT(1-4y)^{-1}-a/V_m \qquad 3.50.2$$

where

$$y=b/4V_m. \qquad 3.50.3$$

Van der Waals assumed a model of spherical molecules of volume $b/4L$ or yV_m/L. He also assumed that the proper attractive potential energy could be expressed as $-a/V_m$.

From (1) or (2) we derive for the proper total energy U_m

$$U_m = -a/V_m \qquad 3.50.4$$

and for the proper entropy S_m

$$-S_m/R = \ln\{y/(1-4y)\}. \qquad 3.50.5$$

In (4) the zero of energy is that at zero density. In (5) the arbitrary constant in the entropy is chosen so that in the limit of low density $S_m/R \to -\ln y$.

It is now known that formula (5) for the entropy is valid only for densities so low that y^2 is negligible; in other words it leads to a correct contribution to the second virial coefficient, but very inaccurate contributions to all higher virial coefficients. Formula (4) for the energy by contrast is inaccurate at low densities but is a useful approximation at high densities.

We shall compare and contrast equation (2) with the equally simple formula*

$$PV_m = RT(1-y)^{-4} - a/V_m \qquad 3.50.6$$

from which follows

$$U_m = -a/V_m \qquad 3.50.7$$

$$-S_m/R = \ln\{y/(1-y)\} + 3y(1-\tfrac{1}{2}y)/(1-y)^2 + y^3/3(1-y)^3. \qquad 3.50.8$$

In the limit of high temperatures the term $-a/V_m$ in (6) becomes unimportant compared with the term proportional to T. It is known that the latter term is correct up to y^3 in contrast to the van der Waals term $RT(1-4y)$ which is correct only up to y. In the limit of high temperatures we have the virial expansions according to van der Waals

$$PV_m/RT = 1 + 4y + 16y^2 + 64y^3 + 256y^4 + 1024y^5 + \ldots \qquad 3.50.9$$

and according to (6)

$$PV_m/RT = 1 + 4y + 10y^2 + 20y^3 + 35y^4 + 56y^5 + \ldots \qquad 3.50.10$$

whereas the accurate expansion for non-attracting rigid spheres is known to be

$$PV_m/RT = 1 + 4y + 10y^2 + 18.36y^3 + 29.4y^4 + \ldots \qquad 3.50.11$$

---

* Guggenheim, Molec. Phys. 1965 **9** 199; Longuet-Higgins and Widom, Molec. Phys. 1964 **8** 549.

We shall now compare other consequences of formulae (2), (4), (5) on the one hand and (6), (7), (8) on the other. Where possible we shall also compare with experimental data on argon. The complete comparison is given in table 3.10. We first derive formulae relating to the critical point. These are determined by the simultaneous conditions

$$\partial P/\partial y = 0 \qquad \partial^2 P/\partial y^2 = 0 \qquad 3.50.12$$

which lead to the values of y_c and of V_c/b and of a/RT_cV_c given in the first three rows.

TABLE 3.10
Comparison of equations of state

		Van der Waals	Modified equation	Experiment
1	y_c	0.083	0.126	
2	V_c/b	3	1.98	
3	a/RT_cV_c	1.125	1.37	
4	$T_B/T_c = a/RT_cb$	3.38	2.72	2.73
5	$(PV/RT)_c$	0.375	0.33	0.29
6	$a/RT_tV_m^L$	(8.56)	(8.56)	8.56
7	y^L	0.221	0.416	
8	V_m^L/b	1.13	0.600	
9	V_m^L/V_c	0.377	0.303	0.374
10	$\ln(PV_m^L/RT)$	-7.40	-6.01	-5.89
11	P_cV^L/RT_c	0.141	0.109	0.108

We next obtain the Boyle temperature T_B given by

$$RT_B = a/b \qquad 3.50.13$$

and the values of the ratio T_B/T_c are given in the fourth row.

The fifth row gives values of $(PV/RT)_c$ obtained from the values in the first and third rows by means of the equation

$$(PV/RT)_c = 1 - 4y_c - a/RT_cV_c \qquad 3.50.14$$

or the equation

$$(PV/RT)_c = (1 - y_c)^4 - a/RT_cV_c. \qquad 3.50.15$$

Hitherto we have not assumed any experimental values. To obtain quantitative results concerning the liquid denoted by the superscript ^{L} at or near the triple point denoted by the subscript $_t$ we equate $a/RT_tV_m^L$ to the experimental value for argon, or indeed any substance conforming to corresponding states with respect to argon, of $\Delta_e U/RT_t$ where $\Delta_e U$ is the energy of evaporation. This is shown in the sixth row. We now put $PV_m^L = 0$ in (2) and in (6) to obtain the values of y^L in the seventh row and thence the values of V_m^L/b in the eighth row. By combining the figures in the second and eighth rows we obtain the values of V_m^L/V_c in the ninth row.

For the equilibrium between liquid and vapour at the triple point we have

$$\mu^G = \mu^L. \qquad 3.50.16$$

If we treat the vapour as a perfect gas and set $PV_m^L/RT = 0$ we have the equations

$$\ln y^G - 1 = a/RT_t V_m^L + \ln\{y^L/(1-4y^L)\} \qquad 3.50.17$$

or alternatively

$$\ln y^G - 1 = a/RT_t V_m^L + \ln\{y^L/(1-y^L)\} + 3y^L(1-\tfrac{1}{2}y^L)/(1-y^L)^2$$
$$+ y^{L3}/3(1-y^L)^3. \qquad 3.50.18$$

Using

$$y^G/y^L = V_m^L/V_m^G = PV_m^L/RT \qquad 3.50.19$$

we can rewrite (17) as

$$\ln(PV_m^L/RT_t) = 1 + a/RT_t V_m^L - \ln(1-4y^L) \qquad 3.50.20$$

and (18) as

$$\ln(PV_m^L/RT_t) = 1 + a/RT_t V_m^L - \ln(1-y^L) + 3y^{L3}(1-\tfrac{1}{2}y^L)/(1-y^L)^2$$
$$+ y^{L3}/3(1-y^L)^3. \qquad 3.50.21$$

Using the values of $a/RT_t V_m^L$ in the sixth row and of y^L in the seventh row we obtain the values of $\ln(PV_m^L/RT_t)$ in the tenth row.

We observe that the equation of state (6) in contrast to the van der Waals equation (2) leads to remarkably good agreement with experiment except for expressions containing V_c. This is not surprising. Because $\partial V/\partial P \to \infty$ at the critical point a small inexactitude in the P-V curve may affect P_c only slightly but will have a pronounced effect on V_c. This is borne out by multiplying $(PV/RT)_c$ by V_t/V_c and obtaining the values of $P_c V_t/RT_c$ in the last line of the table.

§3.51 Zero-temperature entropy in crystals

In §3.26 we gave formula (3.26.4) for the conventional entropy of a slightly imperfect gaseous element composed of monatomic molecules. This formula is composed additively of contributions from
  (a) translational degrees of freedom
  (b) electronic degrees of freedom
  (c) gas imperfection.

In §3.28 we gave, for the conventional entropy of a slightly imperfect gas composed of linear molecules, in formulae (3.28.5) and (3.28.10) respectively, the further contributions from rotational and vibrational degrees of freedom.

In §3.29 we gave formula (3.29.4) for the rotational contribution to the entropy of a slightly imperfect gas composed of non-linear molecules and formula (3.29.7) for the vibrational degrees of freedom.

Thus all the conventional formulae for a slightly imperfect gas include contributions from

(a) translational degrees of freedom
(b) electronic degrees of freedom, if any
(c) gas imperfection
(d) rotational degrees of freedom, if any
(e) vibrational degrees of freedom, if any

All other possible contributions are excluded, in particular

(f) intranuclear degrees of freedom
(g) mixing of isotopes.

There are two good reasons for disregarding the contribution of intranuclear degrees of freedom. In the first place they are in many cases not known. In the second place under terrestrial conditions the nuclear contribution of each nuclidic species is a constant independent of temperature, pressure, phase, composition, and chemical change.

The contribution due to isotopic mixing is ignored because it remains constant as long as the isotopic composition remains unchanged. Variations in isotopic composition will be discussed in §3.55.

We have now reinterpreted what we mean by the conventional proper entropy $S_m^G(T, P)$ of a slightly imperfect single gas at a chosen temperature and pressure. Since any entropy change can be determined by calorimetric and related measurements, we can in particular determine

$$S_m^G(T, P) - S_m^S(T', P') \qquad 3.51.1$$

where the superscript ^{S} denotes the solid crystalline phase. If T' is sufficiently small we can use Debye's approximation (3.32.2) to extrapolate $T' \to 0$ so as to obtain a value of $S_m^S(0, P')$. The dependence of S_m^S on P is negligible, and in fact vanishes as $T \to 0$, and we therefore abbreviate $S_m^S(0, P)$ to $S_m^S(0)$. We thus have a calorimetric value of

$$S_m^G(T, P) - S_m^S(0). \qquad 3.51.2$$

The quantity specified in (2) is often called the *calorimetric entropy* of the gas at the given temperature and pressure, whereas the conventional entropy of a gas $S_m^G(T, P)$ is often called the *spectroscopic entropy* of the gas at the given temperature and pressure.

It is found experimentally that for all elements the calorimetric entropy is equal to the spectroscopic entropy. (The exceptional behaviour of hydrogen will be discussed in §3.56.) Thus

$$S_m^G(T, P) - S_m^S(0) = S_m^G(T, P) \quad \text{(element)} \qquad 3.51.3$$

and consequently

$$S_m^S(0) = 0 \quad \text{(element)}. \qquad 3.51.4$$

Formula (4) expresses our earlier definition of conventional entropy given in §3.25. We have now verified that the convention used in §3.25 is equivalent to the convention described in the present section.

It is found experimentally that the equations (3) and (4) also hold for most compounds, but there are about half a dozen well-established exceptions.

§3.52 Two numerical examples

We shall now illustrate the content of §3.51 by two numerical examples. We choose N_2 and CO.

To calculate the conventional or spectroscopic entropy of nitrogen we use the following data. The proper mass is 28.02 g mole^{-1}. The rotational characteristic temperature Θ_r is 2.87 K and the vibrational characteristic temperature Θ_v is 3.35×10^3 K. The symmetry number s is 2. At the boiling point 77.32 K and a pressure of 1 atm we have, using (3.27.4) and (3.28.5)

$$S_m^G(T_b)/R = \tfrac{7}{2} + \tfrac{5}{2} \ln(T_b/4.333 \text{ K}) + \tfrac{3}{2} \ln(M/\text{g mole}^{-1}) + \ln(T_b/s\Theta_r)$$
$$= \tfrac{7}{2} + \tfrac{5}{2} \ln(77.32/4.333) + \tfrac{3}{2} \ln 28.02 + \ln\{77.32/(2 \times 2.87)\}$$
$$= 3.50 + 7.20 + 5.00 + 2.60 = 18.30. \qquad 3.52.1$$

The contributions from the vibrational degree of freedom and from gas imperfection are negligible.

We next calculate the calorimetric entropy. As usual we use superscripts ^{S} for solid, ^{L} for liquid, and ^{G} for gas. We use subscripts $_{tr}$ to denote a transition, $_f$ to denote fusion, and $_e$ to denote evaporation. We use the following data

$$\int_0^{T_{tr}} C^S \, d \ln T = 6.49 \text{ cal K}^{-1} \text{ mole}^{-1}$$

including an extrapolation from 15 K to 0 K,

$$\int_{T_{tr}}^{T_f} C^S \mathrm{d}\ln T = 5.59 \text{ cal K}^{-1} \text{ mole}^{-1}$$

$$\int_{T_f}^{T_b} C^L \mathrm{d}\ln T = 2.73 \text{ cal K}^{-1} \text{ mole}^{-1}$$

$T_{tr} = 35.61$ K $\quad \Delta_{tr} H = 54.71$ cal mole^{-1}
$T_f = 63.14$ K $\quad \Delta_f H = 172.3$ cal mole^{-1}
$T_b = 77.32$ K $\quad \Delta_e H = 1332.9$ cal mole^{-1}.

Hence the calorimetric entropy at the boiling point and at one atmosphere

$S_m^G(T_b) - S_m^S(0)$
$= (6.49 + 54.71/35.61 + 5.59 + 172.3/63.14 + 2.73 + 1332.9/77.32)$ cal K^{-1} mole^{-1}
$= (6.49 + 1.54 + 5.59 + 2.73 + 2.73 + 17.24)$ cal K^{-1} mole^{-1}
$= 36.32$ cal K^{-1} mole^{-1}

and consequently

$$\{S_m^G(T_b) - S_m^S(0)\}/R = 18.3. \qquad 3.52.2$$

From (1) and (2) we conclude that the conventional entropy $S_m^S(0)$ of the solid at 0 K is zero.

When we do precisely analogous calculations for carbon monoxide we obtain at 1 atm

$S_m^G(T_b)/R = \tfrac{7}{2} + \tfrac{5}{2} \ln(81.61/4.333) + \tfrac{3}{2} \ln 28.01 + \ln(81.61/2.77)$
$\qquad\qquad = 3.50 + 7.34 + 5.00 + 3.38 = 19.22 \qquad 3.52.3$

$S_m^G(T_b) - S_m^S(0)$
$= (10.09 + 151.3/61.55 + 1.23 + 199.7/68.09$
$\qquad\qquad + 2.61 + 1443.6/81.61)$ cal K^{-1} mole^{-1}
$= (10.09 + 2.46 + 1.23 + 2.93 + 2.61 + 17.69)$ cal K^{-1} mole^{-1}
$= 37.01$ cal K^{-1} mole^{-1}

and consequently

$$\{S_m^G(T_b) - S_m^S(0)\}/R = 18.6. \qquad 3.52.4$$

Comparing (3) and (4) we find for the conventional entropy of the crystal at 0 K

$$S_m^S(0)/R = 0.6 = \ln 2 \qquad 3.52.5$$

within the experimental accuracy.

§3.53 *Statistical-mechanical interpretation*

We recall that by definition the vanishing of the conventional entropy of a crystal extrapolated to $T \to 0$

$$S_m^S(0) = 0 \qquad 3.53.1$$

holds for all elements. The special behaviour of hydrogen is discussed in §3.56. Formula (1) also holds for the vast majority of compounds. The best established exceptions are CO, NNO, NO, H_2O.

Classical thermodynamics has nothing to add to this statement. It is however instructive and interesting to discuss the statistical-mechanical interpretation. The interpretation of (1) is that the lowest energy level of the crystal is non-degenerate. This implies that the structure of the crystal is perfectly regular showing no kind of randomness.

The condition (1) found for most crystals states that the conventional zero-temperature entropy of the crystal is zero. This means that the contributions to the entropy from the translational, electronic, rotational, and internal vibrational degrees of freedom are all zero. In other words disregarding intranuclear degrees of freedom and isotopic composition, we may say that no other degrees of freedom contribute anything to the entropy. Statistical theory tells us that this corresponds to the crystal being in a perfectly ordered state, provided we disregard intranuclear degrees of freedom and isotopic composition. Thus a combination of statistical theory with experimental data tells us that as the temperature decreases, most crystals tend towards a state of perfect order apart from intranuclear phenomena and isotopic composition. More strictly we should say that this is how the crystal appears to behave judged by the experimental data in the region of 0 K.

§3.54 *Simple typical exceptions*

We shall now consider exceptions to the general rule $S_m^S(0) = 0$. For this purpose it is convenient to define a number o by

$$S_m^S(0) = R \ln o \qquad 3.54.1$$

so that usually $o = 1$. The two simplest exceptions are CO and NNO. In both cases within the experimental accuracy $o = 2$. The statistical interpretation of the value 2 for o is that instead of perfect order in the crystal, there are two possible orientations for each molecule and the molecules are randomly distributed between these two orientations. This is what we should expect to happen in the case of a linear molecule whose field of force is nearly but not quite symmetrical so that the molecule can be reversed end for end without an appreciable energy change. Statistical theory tells us that the equilibrium distribution of directions will remain random down to temperatures at which kT is comparable with the energy difference in the two orientations. At temperatures where kT is much smaller than the energy difference between the two orientations, then only one orientation will be stable, but at such low temperatures it may well be that the molecules have not sufficient energy to turn round. In simple words when the crystal is so cold that the molecules have a preference for one orientation they have too little energy (are 'too cold') to change their orientations. Such a crystal at the lowest temperatures will remain in a state with $o=2$ and this state is metastable with respect to the ideal unrealizable state of ordered orientation with $o=1$. It is believed that this is a true description of the behaviour of crystalline CO and NNO at the lowest temperatures. It is interesting to note that the SCO molecule is not sufficiently symmetrical to behave in this way and the experimental data are consistent with $o=1$.

The case of NO is somewhat more complicated. It is suggested that at the lowest temperatures the molecular unit is $_{ON}^{NO}$ and that owing to the similarity between N and O atoms the two orientations $_{ON}^{NO}$ and $_{NO}^{ON}$ have nearly equal energies. There would then be a random distribution over these two orientations. This would lead to a value of $o=2$ for the molecular unit N_2O_2; the corresponding value of o expressed in terms of the molecule NO is $2^{\frac{1}{2}}$ and this value is in agreement with experiment within the estimated accuracy.

The other well established case of $o>1$ believed due to simple orientational randomness is that of ice. To account for the experimental data the following assumptions are made.

(1) In ice each oxygen atom has two hydrogens attached to it at distances about 0.95 Å forming a molecule, the HOH angle being about 105° as in the gas molecule.
(2) Each HOH molecule is oriented so that its two H atoms are directed approximately towards two of the four O atoms which surround it tetrahedrally.
(3) The orientations of adjacent HOH molecules are such that only one H atom lies approximately along each O–O axis.
(4) Under ordinary conditions the interaction of non-adjacent molecules is not such as to stabilize appreciably any one of the many configurations satisfying the preceding conditions relative to the others.

On these assumptions Pauling* calculated that theoretically $o=\frac{3}{2}$. Experimentally this value is verified for both H_2O and D_2O.

* Pauling, J. Amer. Chem. Soc. 1935 57 2680.

§3.55 Isotopic mixing

The presence of isotopes can have three effects, which we shall now consider.
The first effect is that $\ln M$ must be replaced by the suitably weighted sum

$$\sum_i x_i \ln M_i \qquad 3.55.1$$

where x_i is the mole fraction of the particular isotope i having a proper mass M_i. Similarly $\ln \Theta_r$ must be replaced by

$$\sum_i x_i \ln \Theta_{ri} \qquad 3.55.2$$

where Θ_{ri} is the value of Θ_r for the particular isotope i. The terms in Θ_v must similarly be replaced by suitably weighted averages. It should not be necessary to give details, especially since in almost all cases it is sufficiently accurate to replace these averaging rules by the simpler rules of replacing

$$M \quad \text{by} \quad \sum_i x_i M_i \qquad 3.55.3$$

$$\Theta_r \quad \text{by} \quad \sum_i x_i \Theta_{ri} \qquad 3.55.4$$

$$\Theta_v \quad \text{by} \quad \sum_i x_i \Theta_{vi}. \qquad 3.55.5$$

It is only in the cases of H_2, D_2, and possibly other very light molecules containing H, D that these simpler averaging rules may not always be sufficiently accurate.

The term H^0 occurring in H_m and in μ, but not in S_m, must likewise be replaced by the weighted average

$$\sum_i x_i H_i^0. \qquad 3.55.6$$

The second effect is that any phase whether solid, liquid, or gaseous, consisting of a mixture of isotopic molecules in mole fractions x_i, has a proper entropy exceeding the proper entropy of similar phases of the pure isotopes at the same temperature and pressure, and this excess is

$$-R \sum_i x_i \ln x_i \qquad 3.55.7$$

which is always positive since $x_i < 1$. We shall meet formula (7) again in chapter 4. In the present context we need only note that as long as the solid and gas have the same isotopic composition, the terms of the form (7) cancel and so contribute nothing to o.

SYSTEMS OF A SINGLE COMPONENT   151

The third effect to be considered is that associated with difference in symmetry. Let us consider the particular example of Cl_2. There are three kinds of molecules $^{35}Cl^{35}Cl$, $^{37}Cl^{37}Cl$, and $^{35}Cl^{37}Cl$. For the molecules $^{35}Cl^{35}Cl$ and $^{37}Cl^{37}Cl$ the symmetry number s is 2, while for the molecule $^{35}Cl^{37}Cl$ it is 1. In the crystal, on the other hand, o for $^{35}Cl^{37}Cl$ will have the value 2 because each molecule can be reversed to give a physically distinct state of the crystal of effectively equal energy, whereas for $^{35}Cl^{35}Cl$ and $^{37}Cl^{37}Cl$ there are not two distinguishable orientations of effectively equal energy and so o is 1. Thus the product so has the same value 2 for all three types of molecules. Ignoring the isotopic composition means then assigning to $^{35}Cl^{37}Cl$ a fictitious value of $s=2$ instead of $s=1$ and to o a fictitious value $o=1$ instead of $o=2$. When we compare the entropies of the gas and the crystal, and it is only in such comparisons that the values assigned to S_m have any significance, the two errors cancel.

It is instructive to compare the behaviours of CO and N_2 with those of $^{35}Cl^{37}Cl$ and $^{35}Cl^{35}Cl$. We saw in §3.52 that for CO the value of o is 2 while of course $s=1$. We should however obtain correct results if we assumed as for N_2 that $o=1$ with $s=2$, using this effective symmetry number because CO is an effectively symmetrical molecule.

The same principle holds in more complicated cases. For example comparing the isotopic molecules CH_4, CH_3D, CH_2D_2 we see that for the first $s=12$, $o=1$, for the second $s=3$, $o=4$ and for the third $s=2$, $o=6$ so that in all three cases the product so is 12.

§3.56 *The exceptional case of hydrogen*

Hydrogen is exceptional in several respects. This is due partly to its molecule having such a small moment of inertia with a consequently high value of the rotational characteristic temperature $\Theta_r = 85.4$ K. It is also partly due to the molecules having an exceptionally small field of force so that even at very low temperatures they still rotate in the crystal. We shall not here go into the theory* of the behaviour of hydrogen as this would take us too far afield. We shall merely state the facts sufficiently to show how the various thermodynamic formulae must be used so as to obtain correct results.

For the sake of consistency we define the conventional zero of entropy precisely as for all other molecules, so the formulae of §3.26 to §3.28 are valid for the gas. As regards the physical meaning of this convention, instead of completely neglecting the intranuclear degrees of freedom we ignore the

---
* Fowler and Guggenheim, Statistical Thermodynamics 1939, Cambridge University Press § 531.

contributions due to any intranuclear degrees of freedom other than resultant nuclear spin, and the contribution due to the spin of the two nuclei in a hydrogen molecule is taken to be the same as if the nuclei were present in independent atoms. Any actual deviation from this will then appear in o.

We shall first consider the gas. The usual formulae for gases with diatomic molecules are applicable only at temperatures large compared with Θ_r and consequently for H_2 they are valid only above about 300 K.

At ordinary temperatures and a fortiori at lower temperatures the vibrational degree of freedom in H_2 may be ignored. As the temperature decreases from about 300 K to about 45 K the rotational contributions to the thermodynamic functions drop from the values for a pair of classical degrees of freedom to values for unexcited degrees of freedom. In particular at temperatures around 45 K or lower

$$-\mu_{\text{rot}}/RT = S_{\text{rot}}/R = \tfrac{3}{4} \ln 3 \qquad 3.56.1$$

$$H_{\text{rot}}/RT = 0. \qquad 3.56.2$$

The constant term $\tfrac{3}{4} \ln 3$ in (1) is due to the fact that hydrogen behaves as a mixture of $\tfrac{1}{4}$ para hydrogen with a proper rotational entropy zero at low temperatures and $\tfrac{3}{4}$ ortho hydrogen with a proper rotational entropy $R \ln 3$ at low temperatures.

Turning now to the crystal, let us first ignore any experimental data below 12 K and extrapolate smoothly the data between 20 K and 12 K in the usual way. We thus obtain well determined values of

$$S_m^G(T) - S_m^S(0) = S_m^G(T) - R \ln o. \qquad 3.56.3$$

We may use the usual formula for $S_m^G(T)$ with $T > 300$ K or alternatively formula (1) for $S_m^G(T)$ with $T < 45$ K; by either procedure we obtain a value for o agreeing within the experimental error with

$$\ln o = \tfrac{3}{4} \ln 3. \qquad 3.56.4$$

We notice that the conventional zero-temperature entropy of the crystal obtained by smooth extrapolation from 12 K is the same as the rotational entropy in the gas below 45 K.

This would complete the picture of ordinary hydrogen were it not for the existence of experimental data on the crystal between 12 K and 2 K. In this range the entropy decreases with anomalous rapidity. In fact the heat capacity not only is anomalously greater than corresponds to the form aT^3, but it actually increases as the temperature decreases below 6 K. On theoretical grounds it is clear that the ortho molecules are somehow beginning to 'line up' with a consequent decrease of entropy. There can be little doubt that if these experimental data extended to still lower temperatures the heat capacity would eventually become normal again after there had been a total loss of proper entropy $\tfrac{3}{4}R \ln 3$. If we then determined $S_m^S(0)$ from here instead of by extrapolation from 12 K, we should find

$$o = 1. \qquad 3.56.5$$

Up to this point we have assumed that the crystal, like the gas, consists of the ordinary metastable mixture of $\tfrac{1}{4}$ para hydrogen and $\tfrac{3}{4}$ ortho hydrogen. For a mixture of this composition the contributions of nuclear spin to the entropy are normal and so their conventional omission leads to no complications. If however the crystalline hydrogen were converted to stable pure para hydrogen there would be a decrease in the contributions to the proper entropy of $\tfrac{3}{4}R \ln 3$ from the nuclear spins and of $-R(\tfrac{3}{4} \ln \tfrac{3}{4} + \tfrac{1}{4} \ln \tfrac{1}{4})$ from the mixing of the para and ortho molecules. This would manifest itself as

$$o = \tfrac{1}{4}. \qquad 3.56.6$$

The conventional zero-temperature entropy of stable para hydrogen has the negative value

$$S_m^S(0) = -R \ln 4 \quad \text{(para hydrogen)}. \qquad 3.56.7$$

For deuterium D_2 the general picture is similar with several differences of detail. The gas behaves like other diatomic gases at temperatures exceeding 200 K. Between this temperature and about 25 K, the rotational contributions to the thermodynamic functions drop from their values for a pair of classical degrees of freedom to values for unexcited degrees of freedom. In particular below 25 K

$$-\mu_{\text{rot}}/RT = S_{\text{rot}}/R = \tfrac{1}{3} \ln 3 \qquad 3.56.8$$

$$H_{\text{rot}}/RT = 0. \qquad 3.56.9$$

The constant term $\tfrac{1}{3} \ln 3$ in (8) is due to the fact that D_2 behaves as a mixture of $\tfrac{2}{3}$ ortho deuterium with a proper rotational entropy zero at low temperatures and $\tfrac{1}{3}$ para deuterium with a proper rotational entropy $R \ln 3$ at low temperatures.

For the crystal similarly, if one extrapolates in the usual way from a temperature between 20 K and 10 K one obtains

$$S_m^S(0)/R = \ln o = \tfrac{1}{3} \ln 3. \qquad 3.56.10$$

For the ordinary metastable mixture of $\tfrac{2}{3}$ ortho deuterium and $\tfrac{1}{3}$ para deuterium the contributions of nuclear spin to the entropy are normal and so their conventional omission leads to no complications. When, however, the crystalline deuterium is converted to stable pure ortho deuterium there is a decrease of $R(\frac{1}{3} \ln 3 + \frac{2}{3} \ln 6)$ in the contribution of nuclear spins to the proper entropy and a decrease of $-R(\frac{1}{3} \ln \frac{1}{3} + \frac{2}{3} \ln \frac{2}{3})$ from the mixing of the ortho and para molecules. If we assume that at the lowest temperature the molecules line up, as we know to be the case with H_2, this will manifest itself as

$$o = \tfrac{6}{9} = \tfrac{2}{3}.$$

The conventional zero-temperature entropy of stable ortho deuterium is then

$$S_m^s(0) = R \ln \tfrac{2}{3}$$

which is negative.

§3.57 Third law of thermodynamics and the Nernst heat theorem

We recall our formulation of the third law in §1.66 which we now repeat.

By the standard methods of statistical thermodynamics it is possible to derive for certain entropy changes general formulae which cannot be derived from the zeroth, first, or second laws of classical thermodynamics. In the present chapter we have had three distinct examples of this type.

In the first place we have quoted in §§3.26–3.29 results of completely general validity for the entropy of gases at sufficiently high temperatures.

In the second place we have quoted a result of completely general validity for the increase of entropy when isotopes, or for that matter any other very similar molecules, are mixed at constant temperature and pressure.

In the third place we have quoted a result concerning the conventional zero-temperature entropy of a crystal, namely that its conventional value is usually but not always zero.

This last result, in the form quoted, is not altogether satisfactory because it admits exceptions without indicating how or when these occur. It is therefore desirable to try to replace this statement by a more definite statement not admitting exceptions. The following statement fulfils these requirements.

If ΔS denotes the increase in entropy in any isothermal change which we represent symbolically by

$$\alpha \to \beta \qquad\qquad 3.57.1$$

and we extrapolate ΔS to $T=0$ smoothly in the usual way, then if the states α and β are both internally stable, or if any kind of internal metastability present is not affected by the change $\alpha \to \beta$, then

$$\lim_{T \to 0} \Delta S = 0. \qquad\qquad 3.57.2$$

If on the other hand α is internally metastable, while β is stable, so that the change α→β removes the metastability, then

$$\lim_{T \to 0} \Delta S < 0. \qquad 3.57.3$$

The case where α is stable and β metastable does not arise, since the change α→β would then be impossible. The above statements constitute an amended form* of a theorem first stated by Nernst and usually known as the *Nernst heat theorem*.

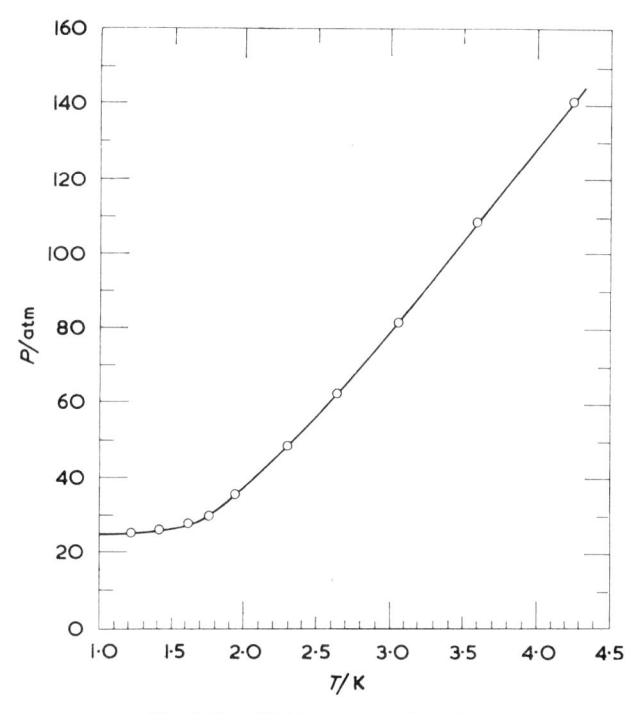

Fig. 3.13. Melting curve of helium

We shall now verify that the behaviour already described of crystals in the limit $T \to 0$ is in accord with the above general statement.

We observe that the several exceptional crystals for which o is not unity are in fact in internally metastable states with some form of randomness of arrangement of the molecules. If by any means it were possible to change such a crystal to the stable completely ordered modifications o would be reduced from a value greater than unity to the value unity and so (3) is satisfied.

* Simon, Ergeb. Exakt. Naturw. 1930 **9** 222.

Examples of changes satisfying (2) are allotropic changes such as

$$\text{white tin} \to \text{grey tin}$$
$$\text{monoclinic sulphur} \to \text{rhombic sulphur}.$$

In each of these examples, although at low temperatures and ordinary pressures the first form is metastable with respect to the second, both forms are completely stable with respect to internal changes. In each case for both phases $o=1$ and so the equality (2) is obeyed.

Another interesting example is that of helium, the only substance which remains liquid down to $T=0$. The liquid is changed to solid under pressure. The relation between the pressure and the freezing temperature is shown in figure 3.13 from which it is clear that

$$\lim_{T \to 0} dP/dT = 0. \qquad 3.57.4$$

But according to the Clapeyron relation this is equivalent to

$$\lim_{T \to 0} \Delta S/\Delta V = 0. \qquad 3.57.5$$

But ΔV is certainly finite and so (5) implies

$$\lim_{T \to 0} \Delta S = 0. \qquad 3.57.6$$

The most numerous and important examples of the relation (2) are those of chemical reactions between solid phases, for example

$$\text{Ag} + \text{I} \to \text{AgI}.$$

These will be discussed in §6.11.

§3.58 *Thermal expansion at low temperatures*

It is an experimental fact that the coefficient of thermal expansion of solids and of liquid helium tends towards zero as the temperature is decreased. But according to Maxwell's relation (1.47.4), this implies that

$$\lim_{T \to 0} (\partial S/\partial P)_T = 0. \qquad 3.58.1$$

If we integrate this from P_1 to P_2, we obtain

$$\lim_{T \to 0} \{S(T, P_2) - S(T, P_1)\} = 0 \qquad 3.58.2$$

which is in accordance with the general relation (3.57.2).

SYSTEMS OF A SINGLE COMPONENT 157

This is the only example of the application of (3.57.2) to a simple physical change which we can discuss at this stage. In chapter 11 we shall consider an interesting application to variation of the strength of an applied magnetic field.

§3.59 *Unattainability of zero temperature*

The general laws formulated in the preceding sections concerning the behaviour of matter extrapolated to $T=0$ are equivalent to the following theorem*.

It is impossible by any procedure, no matter how idealized, to reduce the temperature of any system to zero temperature in a finite number of finite operations.

We shall now prove this equivalence. Let us consider a process (e.g. change of volume, change of external field, allotropic change) denoted formally by

$$\alpha \to \beta. \qquad 3.59.1$$

We shall use the superscripts $^\alpha$ and $^\beta$ to denote properties of the system in the states α and β respectively. Then the proper entropies of the system in these two states depend on the temperature according to the formulae

$$S_m^\alpha = S_m^{0\alpha} + \int_0^T (C^\alpha/T) dT \qquad 3.59.2$$

$$S_m^\beta = S_m^{0\beta} + \int_0^T (C^\beta/T) dT \qquad 3.59.3$$

where $S_m^{0\alpha}$, $S_m^{0\beta}$ are the limiting values of S_m^α, S_m^β for $T \to 0$. It is known from quantum theory that both the integrals converge. Suppose now that we start with the system in the state α at the temperature T' and that we can make the process $\alpha \to \beta$ take place adiabatically. Let the final temperature after the system has reached the state β be T''. We shall now consider the possibility or impossibility of T'' being zero. From the second law of thermodynamics we know that for an adiabatic process defined by its initial and final states the entropy increases if there is any irreversible change and remains constant if the change is completely reversible. It is therefore clear that the chances of attaining as low a final T as possible are most favourable when the change is completely reversible. We need therefore consider only such changes.

* Simon, Science Museum Handbook 1937 3 p. 61. All earlier discussions are unnecessarily restricted. Cf. Fowler and Guggenheim, Statistical Thermodynamics, Cambridge University Press 1939 § 538.

For a reversible adiabatic change (1) we have then by (2) and (3)

$$S_m^{0\alpha} + \int_0^{T'} (C^\alpha/T) dT = S_m^{0\beta} + \int_0^{T''} (C^\beta/T) dT. \qquad 3.59.4$$

If T'' is to be zero, we must then have

$$S_m^{0\beta} - S_m^{0\alpha} = \int_0^{T'} (C^\alpha/T) dT. \qquad 3.59.5$$

Now if $S_m^{0\beta} - S_m^{0\alpha} > 0$ it will always be possible to choose an initial T' satisfying (5) and by making the process $\alpha \to \beta$ take place from this initial T' it will be possible to reach $T'' = 0$. From the premise of the unattainability of $T = 0$ we can therefore conclude that

$$S_m^{0\beta} \leq S_m^{0\alpha}. \qquad 3.59.6$$

Similarly we can show that if we can make the reverse process take place reversibly and adiabatically then we could reach $T'' = 0$ from an initial temperature T' satisfying

$$S_m^{0\alpha} - S_m^{0\beta} = \int_0^{T'} (C^\beta/T) dT. \qquad 3.59.7$$

Further if $S_m^{0\alpha} - S_m^{0\beta} > 0$, we can always choose an initial T' satisfying (7). From the unattainability of $T = 0$ we can therefore conclude that

$$S_m^{0\alpha} \leq S_m^{0\beta}. \qquad 3.59.8$$

From (6) and (8) we deduce

$$S_m^{0\alpha} = S_m^{0\beta} \qquad 3.59.9$$

which is precisely formula (3.57.2) of Nernst's heat theorem.

We can also show conversely that given (9), neither the process $\alpha \to \beta$ nor the reverse process $\beta \to \alpha$ can be used to reach $T = 0$. For, assuming (9) to be true, we now have for the adiabatic process the initial temperature T' and the final temperature T'' related by

$$\int_0^{T'} (C^\alpha/T) dT = \int_0^{T''} (C^\beta/T) dT. \qquad 3.59.10$$

To reach $T'' = 0$ we should require

$$\int_0^{T'} (C^\alpha/T) dT = 0. \qquad 3.59.11$$

But, since $C^\alpha > 0$ always, for any non-zero T' it is impossible to satisfy (11).

SYSTEMS OF A SINGLE COMPONENT

Hence the process cannot be used to reach $T=0$. The proof for the reverse process $\beta \to \alpha$ is exactly similar.

In the above argument we have assumed that the states α and β are connected by reversible paths. If all the phases concerned are phases in complete internal equilibrium the changes concerned must presumably be regarded as reversible. If any phase occurs naturally in metastable internal equilibrium, a process affecting it may or may not disturb the frozen metastability. If it does not disturb it, then the change may still be regarded as reversible, but otherwise it will be a natural irreversible change. We shall now verify that by using internally metastable phases we are still unable to reach $T=0$. In fact as foreshadowed above the irreversibility involved makes the task more difficult.

Suppose for example that α is internally metastable, while β is internally stable. Then according to the Nernst heat theorem

$$S_m^{0\alpha} > S_m^{0\beta}. \qquad 3.59.12$$

But the change $\alpha \to \beta$ is a natural irreversible process and the opposite change is impossible; hence the adiabatic change $\alpha \to \beta$ takes place with increase of entropy, so that

$$S_m^{0\alpha} + \int_0^{T'} (C^\alpha/T) \mathrm{d}T < S_m^{0\beta} + \int_0^{T''} (C^\beta/T) \mathrm{d}T. \qquad 3.59.13$$

Thus to attain $T''=0$ we must have

$$\int_0^{T'} (C^\alpha/T) \mathrm{d}T < S_m^{0\beta} - S_m^{0\alpha} < 0 \qquad 3.59.14$$

using (12). But since $C^\alpha > 0$ always, it is impossible to satisfy (14) and so we again find it impossible to reach $T=0$.

We shall revert to the subject of the unattainability of $T=0$ at the end of chapter 11 on magnetic systems.

§3.60 *Interfacial layers*

We complete this chapter by a consideration of interfacial layers. In a one-component system we cannot usually have more than one liquid phase and so we need consider only the interface between a liquid and its vapour. The interfacial tension of such an interface is called the *surface tension of the liquid*.

As we have seen, a one-component system with two bulk phases has one degree of freedom. We may accordingly treat the temperature as the independent variable; the pressure is then determined by the temperature. Thus the properties of the interfacial layer, in particular the surface tension, will be completely determined by the temperature. Our main task is therefore to consider how the surface tension depends on the temperature.

§3.61 Temperature dependence of surface tension

We begin with formula (1.57.3) which for a single component reduces to

$$-d\gamma = S_A^\sigma dT - \tau dP + \Gamma d\mu \qquad 3.61.1$$

where S_A^σ denotes S^σ/A.

From the equilibrium between the liquid phase, denoted by the superscript ^{L}, and the gas phase, denoted by the superscript ^{G}, we have as in §3.37

$$d\mu = -S_m^L dT + V_m^L dP = -S_m^G dT + V_m^G dP. \qquad 3.61.2$$

When we eliminate $d\mu$ and dP from (1) and (2) we obtain

$$-d\gamma/dT = (S_A^\sigma - \Gamma S_m^L) - (\tau - \Gamma V_m^L)(S_m^G - S_m^L)/(V_m^G - V_m^L). \qquad 3.61.3$$

This formula relates the temperature coefficient of the surface tension to certain entropy changes. Before we examine this formula in any detail, we shall show how it can be transformed to another relation involving energy changes instead of entropy changes.

For the two bulk phases we have as usual

$$\mu = G_m^L = U_m^L - TS_m^L + PV_m^L \qquad 3.61.4$$

$$\mu = G_m^G = U_m^G - TS_m^G + PV_m^G. \qquad 3.61.5$$

For the surface layer we have by applying to unit area the formulae of §1.56

$$\Gamma\mu = G_A^\sigma = U_A^\sigma - TS_A^\sigma + P\tau - \gamma \qquad 3.61.6$$

where G_A^σ denotes G^σ/A and U_A^σ denotes U^σ/A. We now use (4), (5), and (6) to eliminate S_m^L, S_m^G, and S^σ from (3). We obtain

$$-T d\gamma/dT = (U_A^\sigma - \Gamma U_m^L) + P(\tau - \Gamma V_m^L) - \gamma$$
$$\qquad - (\tau - \Gamma V_m^L)(U_m^G - U_m^L)/(V_m^G - V_m^L) - P(\tau - \Gamma V_m^L). \qquad 3.61.7$$

The terms containing P cancel and (7) reduces to

$$\gamma - T d\gamma/dT = (U_A^\sigma - \Gamma U_m^L) - (\tau - \Gamma V_m^L)(U_m^G - U_m^L)/(V_m^G - V_m^L). \qquad 3.61.8$$

§3.62 *Invariance of relations*

We recall that according to the definition in §1.53 of a surface phase the properties associated with it depend on the position of the boundaries AA' and BB' in figure 1.2. We shall henceforth refer to these as the Lσ and the Gσ boundaries respectively. Since the precise placing of these boundaries is arbitrary, the values assigned to such quantities as τ, Γ, S^σ, U^σ are also arbitrary. We can nevertheless verify that our formulae are invariant with respect to shifts of either or both of the boundaries. It is hardly necessary to mention that the intensive variables T, P, and μ are unaffected by shifts of either boundary. It is also clear from the definition of γ in §1.58 that its value is invariant.

Let us now consider a shift of the plane boundary through a distance $\delta\tau$ away from the gas phase. Then Γ becomes increased by the amount of liquid in a cylinder of height $\delta\tau$ of cross-section unity and consequently of volume $\delta\tau$. Thus Γ becomes increased by $\delta\tau/V_m^L$. It follows immediately that $\tau - \Gamma V_m^L$ remains invariant. Similarly S_A^σ becomes increased by the entropy in a cylinder of liquid of volume $\delta\tau$ that is to say by an amount $S_m^L \delta\tau/V_m^L$. It follows immediately that $S_A^\sigma - \Gamma S_m^L$ remains invariant. Precisely similar considerations show that $U_A^\sigma - \Gamma U_m^L$ remains invariant.

We have now to consider a similar shift of the Gσ boundary through a distance $\delta\tau$ away from the liquid phase. Then Γ is increased by $\delta\tau/V_m^G$ and so $\tau - \Gamma V_m^L$ is increased by $(V_m^G - V_m^L)\delta\tau$. Similarly $S_A^\sigma - \Gamma S_m^L$ is increased by $(S_m^G - S_m^L)\delta\tau$. When we insert these values into (3.61.3) we see that the resulting variation vanishes. The same holds for (3.61.8).

§3.63 *Simplifying approximation*

The formulae of §3.61 are strictly accurate and involve no assumptions or approximations concerning the structure of the interfacial layer. We shall see that they can be greatly simplified by making use of our knowledge concerning this layer.

In §3.39 we mentioned that, at temperatures well below the critical, PV_m^L is small compared with RT and may usually be ignored. In the interfacial layer the density is comparable to that in the liquid phase so that τ/Γ is comparable to V_m^L and negligible compared with V_m^G. Consequently the terms containing the factor $\tau - \Gamma V_m^L$ may be neglected. Formulae (3.61.3) and (3.61.8) then reduce to, respectively,

$$-d\gamma/dT = S_A^\sigma - \Gamma S_m^L \qquad 3.63.1$$

$$\gamma - T\,d\gamma/dT = U_A^\sigma - \Gamma U_m^L. \qquad 3.63.2$$

It is worth noticing that the right side of (1) is the entropy of unit area of surface less the entropy of the same material content of liquid and the right side of (2) is the energy of unit area of surface less the energy of the same material content of liquid. More pictorially we may say that when unit area of surface is created isothermally and reversibly, the work done on the system is γ, the heat absorbed by the system is the right side of (1) multiplied by T, and the increase of energy, the sum of these two quantities, is equal to the right side of (2).

If however we are making the above simplifying approximations, then by making them at an earlier stage we can considerably simplify the derivations. We accordingly replace (3.61.1) by the approximation

$$-\mathrm{d}\gamma = S_A^\sigma \mathrm{d}T + \Gamma \mathrm{d}\mu \qquad 3.63.3$$

and (3.61.2) by the approximation

$$\mathrm{d}\mu = -S_m^L \mathrm{d}T = S_m^G \mathrm{d}T + RT \mathrm{d}\ln P. \qquad 3.63.4$$

Eliminating $\mathrm{d}\mu$ from (3) and (4) we obtain immediately

$$-\mathrm{d}\gamma = (S_A^\sigma - \Gamma S_m^L)\mathrm{d}T \qquad 3.63.5$$

in agreement with (1).

Furthermore we replace (3.61.4) by the approximation

$$\mu = U_m^L - TS_m^L \qquad 3.63.6$$

and (3.61.6) by the approximation

$$\Gamma\mu = U_A^\sigma - TS_A^\sigma - \gamma. \qquad 3.63.7$$

Eliminating S_A^σ and S_m^L from (5), (6), and (7) we recover (2).

We conclude this discussion with a warning against indiscriminately using the simplified formulae of this section in the neighbourhood of the critical point. The necessary condition for their use is that

$$\tau/\Gamma - V_m^L \ll V_m^G - V_m^L. \qquad 3.63.8$$

In the neighbourhood of the critical temperature V_m^L becomes nearly as great as V_m^G and this condition may no longer be taken for granted.

§3.64 *Vapour pressure of small drops*

Figure 3.14 represents a small spherical drop and a portion of liquid in bulk both at the same temperature. We denote the interiors of these liquid phases by α, β respectively and the vapour immediately outside them by α′, β′ respectively. Let us assume that the external pressures $P^{\alpha'}$ and $P^{\beta'}$ are equal, that is

$$P^{\alpha'} = P^{\beta'}. \qquad 3.64.1$$

Then, according to (1.60.9) the pressure P^α at the interior α of the drop is greater than that P^β of the liquid in bulk by

$$P^\alpha - P^\beta = 2\gamma/a \qquad 3.64.2$$

where a is the radius of the drop. But according to (3.46.11) the fugacity p is related to the pressure P by

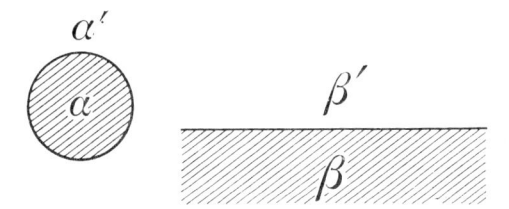

$$d \ln p/dP = V_m^L/RT. \qquad 3.64.3$$

Fig. 3.14. Vapour pressure of droplet

If then we neglect the compressibility of the liquid, the fugacity p^α of the liquid in the drop is related to the fugacity p^β of the liquid in bulk by

$$RT \ln(p^\alpha/p^\beta) = (P^\alpha - P^\beta)V_m^L. \qquad 3.64.4$$

Comparing (2) and (4) we find

$$RT \ln(p^\alpha/p^\beta) = 2\gamma V_m^L/a. \qquad 3.64.5$$

We see then that at the same external pressure the small drop always has a greater fugacity than the bulk liquid. Vapour will distil from the drop to the liquid and as the drop becomes smaller its fugacity increases still more. Thus small drops are essentially unstable relatively to the liquid in bulk.

§3.65 Empirical temperature dependence of surface tension

Since the surface tension of a liquid decreases with increasing temperature and vanishes at the critical point, the simplest possible form of empirical relation between γ and T is

$$\gamma = \gamma_0(1 - T/T_c)^{1+r} \qquad 3.65.1$$

where γ_0 and r are constants. For the substances having the simplest and most symmetrical molecules such as Ne, Ar, Xe, N_2, O_2 excellent agreement with the experimental data is obtained with $r=\frac{2}{9}$ as is shown in table 3.11. The data at the foot of this table will be discussed in §3.66.

The reason for the particular choice $r=\frac{2}{9}$ will be explained shortly. Ferguson* in a review of the experimental data for ten esters and four other organic compounds found $r=0.210\pm0.015$, which does not differ significantly from the value $\frac{2}{9}$ adopted above.

Another type of formula relates the surface tension to the coexisting proper volumes V^L of the liquid and V^G of the vapour. The simplest satisfactory formula of this type is the following

$$\gamma y^{-\frac{4}{3}} \propto (1-T/T_c) \qquad 3.65.2$$

where y is defined in terms of densities ϱ by

$$yV_c = (\varrho^L - \varrho^G)/\varrho_c. \qquad 3.65.3$$

This formula, due to Katayama†, is a striking improvement over the older and less accurate formula of Eötvos, which contained V^L instead of y^{-1}. This was shown by Katayama for various organic compounds and we shall now verify that this is also the case for the substances having the simplest molecules.

In §3.48 we verified that the substances having the simplest molecules follow with a high degree of accuracy formula (3.48.7), namely

$$(\varrho^L - \varrho^G)/\varrho_c = \tfrac{7}{2}(1-T/T_c)^{\frac{1}{3}}. \qquad 3.65.4$$

Using the definition (3) of y, this can be written

$$y \propto (1-T/T_c)^{\frac{1}{3}}. \qquad 3.65.5$$

If we now eliminate y between (2) and (5), we obtain

$$\gamma \propto (1-T/T_c)^{\frac{11}{9}} \qquad 3.65.6$$

of the form (1) with $r=\frac{2}{9}$. It follows that the verification of (6) in table 3.11 and the verification of (4) in figure 3.11 together constitute a verification of (2).

If instead of eliminating y between (2) and (5), we eliminate T/T_c between the same formulae we obtain

$$\gamma \propto y^{\frac{11}{3}}. \qquad 3.65.7$$

---

* Ferguson, Trans. Faraday Soc. 1923 **19** 407; Proc. Phys. Soc. London 1940 **52** 759.
† Katayama, Sci. Rep. Tôhoku Univ. 1916 **4** 373.

SYSTEMS OF A SINGLE COMPONENT 165

TABLE 3.11

$$\gamma=\gamma_0(1-T/T_c)^{\frac{11}{9}}$$

	Ne			Ar			Xe			N_2			O_2	
$T_c=44.8$ K			$T_c=150.7$ K			$T_c=289.8$ K			$T_c=126.0$ K			$T_c=154.3$ K		
$\gamma_0=15.1$ dyne cm^{-1}			$\gamma_0=36.31$ dyne cm^{-1}			$\gamma_0=50.70$ dyne cm^{-1}			$\gamma_0=28.4$ dyne cm^{-1}			$\gamma_0=38.4$ dyne cm^{-1}		
	γ/dyne cm^{-1}			γ/dyne cm^{-1}			γ/dyne cm^{-1}			γ/dyne cm^{-1}			γ/dyne cm^{-1}	
T/K	calc.	obs.	T/K	calc.	obs.	T/K	calc.	obs.	T/K	calc.	obs.	T/K	calc.	obs.
24.8	5.64	5.61	85.0	13.16	13.19	162	18.7	18.7	70.0	10.54	10.53	70.0	18.34	18.35
25.7	5.33	5.33	87.0	12.67	12.68	164	18.3	18.3	75.0	9.40	9.39	75.0	17.02	17.0
26.6	5.02	4.99	90.0	11.95	11.91				80.0	8.29	8.27	80.0	15.72	15.73
27.4	4.75	4.69							85.0	7.20	7.20	85.0	14.44	14.5
28.3	4.45	4.44							90.0	6.14	6.16	90.0	13.17	13.23
$V_c=41.7$ cm^3 mole^{-1}			$V_c=75.3$ cm^3 mole^{-1}			$V_c=118.8$ cm^3 mole^{-1}			$V_c=90.2$ cm^3 mole^{-1}			$V_c=74.5$ cm^3 mole^{-1}		
$\gamma_0 V_c^{\frac{2}{3}} T_c^{-1}=$			$\gamma_0 V_c^{\frac{2}{3}} T_c^{-1}=$			$\gamma_0 V_c^{\frac{2}{3}} T_c^{-1}=$			$\gamma_0 V_c^{\frac{2}{3}} T_c^{-1}=$			$\gamma_0 V_c^{\frac{2}{3}} T_c^{-1}=$		
4.05 erg K^{-1}mole$^{-\frac{2}{3}}$			4.3 erg K^{-1}mole$^{-\frac{2}{3}}$			4.2 erg K^{-1}mole$^{-\frac{2}{3}}$			4.5 erg K^{-1}mole$^{-\frac{2}{3}}$			4.4 erg K^{-1}mole$^{-\frac{2}{3}}$		

The better known relation of Macleod* with an index 4 instead of $\tfrac{11}{3}$ is less accurate, at least for the substances having the simplest molecules. Actually for the half dozen organic compounds considered by Macleod it is clear that γ in fact varies as some power of y less than 4.

§3.66 Corresponding states of surface tension

The principle of corresponding states, so far as it is applicable at all to surfaces, can on physical grounds be expected to hold only for substances having the simplest and most symmetrical molecules.

According to the principle it is clear from dimensional considerations that $\gamma V_c^{\frac{2}{3}} T_c^{-1}$ should be a common function of T/T_c for substances obeying the principle. In particular, if these substances obey (3.65.1) then $\gamma_0 V_c^{\frac{2}{3}} T_c^{-1}$ should have a common value. The data† at the bottom of table 3.11 show that this is in fact the case within about $\pm 2\%$ for Ar, Xe, N_2, O_2 while the value for Ne deviates by rather less than 10 %.

More recent measurements‡ are summarized in table 3.12.

TABLE 3.12

	Ar	N_2	CH_4
γ_0/dyne cm^{-1}	37.78	28.12	39.08
$\gamma_0 V_c^{\frac{2}{3}} T_c^{-1}$/erg K^{-1} mole$^{-\frac{2}{3}}$	4.25	4.48	4.39

§3.67 Sorption of a single gas

In our discussion of the interface between a liquid and a gas the surface area A was first introduced as an independent variable. The interfacial tension γ was then introduced through the relation

$$w = \gamma \, dA. \qquad 3.67.1$$

Both A and γ are well defined measurable quantities. The situation for a solid–gas interface is altogether different. The area of the interface may be difficult, if not impossible, to measure accurately especially if the solid is porous or a powder. Furthermore the surface area can not be varied reversibly and consequently there is no relation such as (1). There is no quantity

---
* Macleod, Trans. Faraday Soc. 1923 **19** 38.
† Guggenheim, J. Chem. Phys. 1945 **13** 259. Guggenheim, Proc. Phys. Soc. London 1965 **85** 811.
‡ Sprow and Prausnitz, Trans. Faraday Soc. 1966 **62** 1102.

analogous to interfacial tension. For discussing the equilibrium between a gas and a solid–gas interface a completely different approach is called for.

The name *sorption* was coined by McBain* to include as special cases absorption and adsorption which should be restricted to proven cases of solution and surface condensation respectively.

In the following discussion of sorption we shall assume that the sorbed gas is a single substance, but no restriction will be placed on the nature of the sorbent except that we assume absence of hysteresis. In other words we shall consider the equilibrium between a single gas and a sorbent which may be a piece of platinum gauze, a lump of impure charcoal, or some powdered glass, or in fact almost anything.

The first question to be considered is how to measure sorption. This may be done by a sorption balance. Essentially the sorbent is suspended on a spring balance so that its weight, and thence its mass, can be compared in vacuo and in equilibrium with a surrounding gas at a given temperature and pressure. If no correction is applied for the buoyancy due to the surrounding gas, the apparent increase in mass recorded by the balance is equal to that of the excess quantity of the sorbed substance due to sorption over and above the quantity which would be contained in the same volume, at the same temperature and pressure, in the absence of the sorbent. This mass divided by the proper mass of the gas is equal to the excess amount of the sorbed substance due to sorption over and above the amount of the gas which would be contained in the same volume, at the same temperature and pressure, in the absence of the sorbent. We denote this amount by n^a and we shall call n^a the sorbed excess. This quantity is the simplest and most convenient measure of sorption. At a first approach it might seem that a simpler quantity would be the amount of sorbed substance contained by the sorbent. Such a quantity would have to be calculated by adding to n^a the quantity $\varrho V^s/M$ where ϱ is the density of the gas, M is the proper mass of the gas, and V^s is the volume of the sorbent. On reflection it becomes clear that V^s is a rather vague quantity, difficult if not impossible to measure accurately especially if the sorbent is porous or a powder. This difficulty is completely avoided by using n^a, without any buoyancy correction as our measure of sorption. We shall adopt this approach and shall consider how n^a is related to the temperature and pressure.

§3.68 *Temperature dependence of sorption*

We consider a vessel of volume V containing the sorbent and a fixed amount

* McBain, Phil. Mag. 1909 **18** 916.

n of the sorbate at the temperature T and pressure P. If V_m denotes the proper volume of gaseous sorbate, then according to the definition of n^a in the previous section we have

$$V = (n - n^a) V_m. \quad 3.68.1$$

We now combine Maxwell's relation (1.47.4) with (1) and obtain

$$-(\partial S/\partial P)_T = (\partial V/\partial T)_P = (n - n^a)(\partial V_m/\partial T)_P - (\partial n^a/\partial T)_P V_m \quad 3.68.2$$

and consequently

$$\begin{aligned}-(\partial S/\partial n^a)_T &= -(\partial S/\partial P)_T (\partial P/\partial n^a)_T \\ &= (n-n^a)(\partial V_m/\partial T)_P (\partial P/\partial n^a)_T - (\partial n^a/\partial T)_P V_m (\partial P/\partial n^a)_T \\ &= (n-n^a)(\partial V_m/\partial T)_P (\partial P/\partial n^a)_T + (\partial P/\partial T)_{n^a} V_m. \end{aligned} \quad 3.68.3$$

We now compare our system with another system consisting of a vessel of the same volume V containing the same gas at the same temperature and pressure but without any sorbent. We shall use dashed symbols for quantities relating to this second system when they may differ from those relating to the first system. We have then

$$n' = n - n^a. \quad 3.68.4$$

Moreover since both systems are in equilibrium and the gas is in identical conditions we have also

$$\mu' = \mu. \quad 3.68.5$$

If now we denote by ΔS_m the entropy increase in the first system per unit decrease of n^a brought about by decreasing the pressure, we have according to (3)

$$\Delta S_m = (n - n^a)(\partial V_m/\partial T)_P (\partial P/\partial n^a)_T + (\partial P/\partial T)_{n^a} V_m. \quad 3.68.6$$

If further we denote by $\Delta' S$ the entropy increase in the second system corresponding to the same decrease in pressure, we have

$$\Delta' S_m = n'(\partial V_m/\partial T)_P (\partial P/\partial n^a)_T. \quad 3.68.7$$

Subtracting (7) from (6) and using (4) we obtain

$$\Delta S_m - \Delta' S_m = (\partial P/\partial T)_{n^a} V_m \quad 3.68.8$$

or

$$T \Delta S_m - T \Delta' S_m = T (\partial P/\partial T)_{n^a} V_m \quad 3.68.9$$

It follows immediately from (5) that

$$\Delta \mu = \Delta' \mu \quad 3.68.10$$

SYSTEMS OF A SINGLE COMPONENT

and so subtracting (10) from (9) we obtain the alternative formula

$$\Delta H_m - \Delta' H_m = T(\partial P/\partial T)_{n^a} V_m. \qquad 3.68.11$$

The left side of either (9) or (11) may be called* the *equilibrium proper enthalpy of desorption*. It is the heat that must be supplied per unit decrease of n^a resulting from decrease of pressure under isothermal equilibrium conditions less the heat that must be supplied to the second system when the pressure is isothermally decreased by the same amount. We emphasize that every quantity occurring in (9) and (11) is experimentally determinable without the use of any approximation or extraneous assumption. This contrasts with some other formulae which contain quantities such as the surface area of the sorbent, the spreading pressure, and the volume occupied by a sorbed molecule. Such quantities play their natural part as parameters in a model used for a kinetic or statistical mechanical theory of sorption, but, not being accurately measurable, they have no part in a treatment by classical thermodynamics.

In all normal applications the pressure will be sufficiently small to justify neglect of all virial coefficients higher than the second. We then have in accordance with (3.19.5)

$$V_m = RT/P + B. \qquad 3.68.12$$

Substituting (12) into (9) and (11) we obtain

$$T\Delta S_m - T\Delta' S_m = \Delta H_m - \Delta' H_m = RT^2 (\partial \ln P/\partial T)_{n^a}(1 + PB/RT). \qquad 3.68.13$$

When the term in B is negligible so that the gas is effectively perfect, (13) reduces to a form derivable by more elementary methods.

---

* The derivation of these formulae was presented to the Boston University Conference on Nucleation 1951.

CHAPTER 4

MIXTURES

§4.01 *Introduction*

In this chapter we shall discuss homogeneous mixtures of two or more substances which do not react chemically. Consideration of chemical reactions is postponed to chapter 6. The mixtures may be gaseous, liquid, or solid. So far as possible each of the several component substances of a mixture will be treated on the same basis. The opposite point of view is taken in chapter 5 where one substance is regarded as the solvent and the remaining substances as solutes.

As soon as we turn from single substances to mixtures we introduce the possibility of new degrees of freedom associated with differences of composition. For example we can have two or more liquid phases of different composition in mutual equilibrium.

There are no differences of principle between the treatments of systems of two components on the one hand and of systems of more than two components on the other. Many of the formulae for the former are however more compact than the corresponding formulae for the latter. For this reason we shall in some sections confine ourselves mainly if not entirely, to systems of two substances, i.e. *binary* systems.

§4.02 *Composition of mixture*

The most convenient quantities specifying the relative composition of a mixture are the mole fractions of the several components. These were defined in §1.29. We recall that in a mixture of c components only $c-1$ of the mole fractions are independent owing to the identity

$$x_1 + x_2 + \ldots + x_c = 1. \qquad 4.02.1$$

When we require independent variables it is convenient to use $x_2, x_3, \ldots x_c$ and regard x_1 as a dependent variable defined by

$$x_1 = 1 - x_2 - x_3 - \ldots - x_c. \qquad 4.02.2$$

In the simple case of a binary mixture the subscripts may be dropped so that we write x instead of x_2 and $1-x$ instead of x_1.

§4.03 Partial and proper quantities

We recall the definitions in §1.26 of partial quantities X_i and proper quantities X_m in terms of an extensive property X, namely

$$X_i = (\partial X/\partial n_i)_{T, P, n_j} \quad (j \neq i) \qquad 4.03.1$$

$$X_m = X/\sum_i n_i. \qquad 4.03.2$$

We also recall formula (1.26.3)

$$X = \sum_i n_i X_i. \qquad 4.03.3$$

If we differentiate this we obtain

$$dX = \sum_i X_i dn_i + \sum_i n_i dX_i \qquad 4.03.4$$

while (1) may be rewritten as

$$dX = \sum_i X_i dn_i \quad \text{(const. } T, P\text{).} \qquad 4.03.5$$

Subtracting (5) from (4) we obtain

$$\sum_i n_i dX_i = 0 \quad \text{(const. } T, P\text{)} \qquad 4.03.6$$

or dividing by $\Sigma_i n_i$

$$\sum_i x_i dX_i = 0 \quad \text{(const. } T, P\text{).} \qquad 4.03.7$$

In particular for a binary mixture

$$(1-x)(\partial X_1/\partial x) + x(\partial X_2/\partial x) = 0 \quad \text{(const. } T, P\text{).} \qquad 4.03.8$$

In the case of a binary mixture we can express X_1 and X_2 in terms of X_m and x. Formula (5) reduces to

$$dX = X_1 dn_1 + X_2 dn_2. \qquad 4.03.9$$

If we apply (9) to unit amount of varying composition, it becomes

$$dX_m = (X_2 - X_1) dx \qquad 4.03.10$$

or

$$(\partial X_m/\partial x)_{T, P} = X_2 - X_1. \qquad 4.03.11$$

If we apply (3) to unit amount it becomes

$$X_m = (1-x)X_1 + xX_2. \qquad 4.03.12$$

Solving (11) and (12) for X_1 and X_2 we obtain

$$X_1 = X_m - x(\partial X_m/\partial x)_{T,P} \qquad 4.03.13$$

$$X_2 = X_m + (1-x)(\partial X_m/\partial x)_{T,P}. \qquad 4.03.14$$

Formulae (13) and (14) have a simple geometrical interpretation shown in figure 4.1. The abscissa is x, increasing from zero at O representing the pure component 1 to unity at O' representing the pure component 2. Suppose the curve APB to be a plot of the proper quantity X_m as ordinate and P to be any point on it. Let the tangent QPR to this curve at P cut the O and O' ordinates at Q and R respectively. Then from (13) and (14) we see that the partial quantities X_1 and X_2 for the composition at P are represented by OQ and O'R respectively. It is clear from this construction or otherwise that for either pure component the partial quantity is equal to the proper quantity.

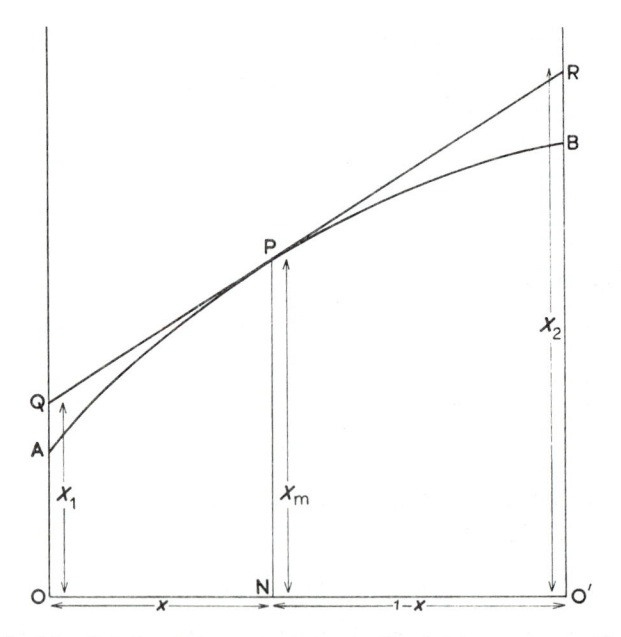

Fig. 4.1. Relation between partial quantities and proper quantity

An especially important example of the pair of formulae (13) and (14) is obtained by setting $X = G$. Taking account of (1.28.11) we obtain

$$\mu_1 = G_1 = G_m - x(\partial G_m/\partial x)_{T, P} \qquad 4.03.15$$

$$\mu_2 = G_2 = G_m + (1-x)(\partial G_m/\partial x)_{T, P}. \qquad 4.03.16$$

§4.04 Relations between partial quantities

As already mentioned in §1.26 from every homogeneous relation between extensive properties we obtain by differentiation with respect to n_i a corresponding relation between the partial quantities at a given composition. We now give important examples of such relations taking into account (1.28.11) that

$$\mu_i = G_i. \qquad 4.04.1$$

We have with T, P as independent variables

$$H_i = U_i + PV_i \qquad 4.04.2$$

$$\mu_i = G_i = U_i - TS_i + PV_i \qquad 4.04.3$$

$$S_i = -\partial G_i/\partial T = -\partial \mu_i/\partial T \qquad 4.04.4$$

$$H_i = G_i - T\partial G_i/\partial T = \mu_i - T\partial \mu_i/\partial T \qquad 4.04.5$$

$$V_i = \partial G_i/\partial P = \partial \mu_i/\partial P \qquad 4.04.6$$

$$C_i = T\partial S_i/\partial T = \partial H_i/\partial T. \qquad 4.04.7$$

Relations of precisely the same form hold between proper quantities.

From (5) and (3.15.2) we deduce

$$\partial \ln \lambda_i/\partial T = -H_i/RT^2. \qquad 4.04.8$$

§4.05 Partial quantities at high dilution

By rewriting (4.03.8) in the form

$$(\partial X_1/\partial x)/(\partial X_2/\partial x) = -x/(1-x) \qquad 4.05.1$$

we make the interesting observation that as $x \to 0$ either $(\partial X_1/\partial x) \to 0$ or $(\partial X_2/\partial x) \to \infty$. Both alternatives occur. We shall find that as $x \to 0$, the quantities $(\partial U_1/\partial x)$, $(\partial H_1/\partial x)$, $(\partial V_1/\partial x)$, $(\partial C_1/\partial x)$ all tend towards zero, while $(\partial S_2/\partial x)$ and $(\partial G_2/\partial x) = (\partial \mu_2/\partial x)$ tend to infinity.

In the limit $x \to 1$ we of course meet the converse behaviour.

§4.06 Perfect gaseous mixture

In discussing gaseous mixtures, or in discussing single gases, it is expedient

174   MIXTURES

to begin by taking T, V as independent variables and later to transform to T, P as independent variables.

We begin by rewriting some of the most important formulae for an amount n of a single perfect gas occupying a volume V at a temperature T. The Helmholtz function is

$$F = n\{\mu^\ominus - RT + RT \ln(nRT/P^\ominus V)\} \qquad 4.06.1$$

where $\mu^\ominus$ depends on T and on the standard pressure $P^\ominus$ but not on n or V. From (1) we derive by differentiation

$$S = -n\{d\mu^\ominus/dT + R \ln(nRT/P^\ominus V)\} \qquad 4.06.2$$

$$U = n\{\mu^\ominus - T d\mu^\ominus/dT - RT\} \qquad 4.06.3$$

$$P = nRT/V \qquad 4.06.4$$

$$H = n\{\mu^\ominus - T d\mu^\ominus/dT\} \qquad 4.06.5$$

$$G = n\{\mu^\ominus + RT \ln(nRT/P^\ominus V)\} = n\{\mu^\ominus + RT \ln(P/P^\ominus)\}. \qquad 4.06.6$$

Formula (6) may be regarded as defining $\mu^\ominus$.

We now define a perfect gaseous mixture by the property that at given T, V the Helmholtz function F of the mixture is equal to the sum of the Helmholtz functions of the unmixed gases each at the given T, V. We accordingly have for a perfect gaseous mixture

$$F = \sum_i n_i\{\mu_i^\ominus - RT + RT \ln(n_i RT/P^\ominus V)\}. \qquad 4.06.7$$

The molecular interpretation of formula (7) is that for a mixture of perfect gases in a container at given temperature and volume the molecules of each gaseous species behave as if the other species were absent. From the additivity of the Helmoltz function we deduce by differentiation the additivity of other functions. In particular we have

$$S = -\sum_i n_i\{d\mu_i^\ominus/dT + R \ln(n_i RT/P^\ominus V)\} \qquad 4.06.8$$

$$U = \sum_i n_i\{\mu_i^\ominus - T d\mu_i^\ominus/dT - RT\} \qquad 4.06.9$$

$$P = \sum_i n_i RT/V \qquad 4.06.10$$

$$H = \sum_i n_i\{\mu_i^\ominus - T d\mu_i^\ominus/dT\} \qquad 4.06.11$$

$$G = \sum_i n_i\{\mu_i^\ominus + RT \ln(n_i RT/P^\ominus V)\}$$

$$= \sum_i n_i\{\mu_i^\ominus + RT \ln(x_i P/P^\ominus)\}. \qquad 4.06.12$$

MIXTURES 175

We repeat that $\mu_i^\ominus$ is a function of T and $P^\ominus$ independent of V and the n_i's.

Formula (10) is called *Dalton's law of partial pressures*; a better name would be the *law of additivity of pressures*.

§4.07 Slightly imperfect gaseous mixture

A single gas or a gaseous mixture may be regarded as perfect if the interactions between molecules may be neglected. When these interactions between molecules are not negligible we use a *virial expansion* in powers of n/V. We recall the formulae for a single gas

$$F = n\{\mu^\ominus - RT + RT \ln(nRT/P^\ominus V)\} \\ + nRT\{nB_2/V + \tfrac{1}{2}n^2 B_3/V^2 + \tfrac{1}{3}n^3 B_4/V^3\} \quad 4.07.1$$

where B_2, B_3, and B_4 are the second, third, and fourth virial coefficients respectively. Higher terms may be added, but when three virial coefficients are insufficient the virial expansion ceases to be useful.

From (1) we obtain by differentiation with respect to V

$$P = (nRT/V)\{1 + nB_2/V + n^2 B_3/V^2 + n^3 B_4/V^3\}. \quad 4.07.2$$

Either formula (1) or formula (2) defines the virial coefficients B_2, B_3, B_4. For the sake of brevity and simplicity we shall omit the terms in B_3, B_4. There is in principle no difficulty in including them if required. We accordingly abbreviate (1) to

$$F = n\{\mu^\ominus - RT + RT \ln(nRT/P^\ominus V)\} + n^2 B/V \quad 4.07.3$$

where we have dropped the subscript 2 from B_2.

It is known from statistical mechanics that B_2 takes account of interactions between pairs of molecules, B_3 between triplets, B_4 between quadruplets. This tells us how to generalize formula (1) to mixtures. In particular for a binary mixture

$$F = n_1\{\mu_1^\ominus - RT + RT \ln(n_1 RT/P^\ominus V)\} \\ + n_2\{\mu_2^\ominus - RT + RT \ln(n_2 RT/P^\ominus V)\} \\ + \{n_1^2 B_{(11)} + 2n_1 n_2 B_{(12)} + n_2^2 B_{(22)}\}RT/V. \quad 4.07.4$$

Differentiating (4) with respect to V we obtain

$$PV = (n_1 + n_2)RT + \{n_1^2 B_{(11)} + 2n_1 n_2 B_{(12)} + n_2^2 B_{(22)}\}RT/V. \quad 4.07.5$$

From (4) and (5) we obtain by addition for the Gibbs function

$$G = n_1\{\mu_1^\ominus + RT \ln(n_1 RT/P^\ominus V)\} \\ + n_2\{\mu_2^\ominus + RT \ln(n_2 RT/P^\ominus V)\} \\ + 2\{n_1^2 B_{(11)} + 2n_1 n_2 B_{(12)} + n_2^2 B_{(22)}\}RT/V. \quad 4.07.6$$

If we may neglect terms in the squares of $B_{(11)}$, $B_{(12)}$, $B_{(22)}$ as well as terms in the third and higher virial coefficients, we call the mixture *slightly* imperfect. From (5) we have to this approximation

$$\ln(PV/RT) = \ln(n_1+n_2) + \{n_1^2 B_{(11)} + 2n_1 n_2 B_{(12)} + n_2^2 B_{(22)}\}/(n_1+n_2)V. \qquad 4.07.7$$

Substituting from (7) into (6) we obtain

$$\begin{aligned}G &= n_1\{\mu_1^\ominus + RT\ln(P/P^\ominus) + RT\ln[n_1/(n_1+n_2)]\} \\ &+ n_2\{\mu_2^\ominus + RT\ln(P/P^\ominus) + RT\ln[n_2/(n_1+n_2)]\} \\ &+ \{n_1^2 B_{(11)} + 2n_1 n_2 B_{(12)} + n_2^2 B_{(22)}\}P/(n_1+n_2)\end{aligned} \qquad 4.07.8$$

and consequently for the proper Gibbs function

$$\begin{aligned}G_m &= x_1\{\mu_1^\ominus + RT\ln(P/P^\ominus) + RT\ln x_1\} \\ &+ x_2\{\mu_2^\ominus + RT\ln(P/P^\ominus) + RT\ln x_2\} \\ &+ \{x_1^2 B_{(11)} + 2x_1 x_2 B_{(12)} + x_2^2 B_{(22)}\}P.\end{aligned} \qquad 4.07.9$$

It is convenient to define a quantity $\delta_{(12)}$ by

$$B_{12} = \tfrac{1}{2}(B_{(11)} + B_{(22)}) + \delta_{(12)}. \qquad 4.07.10$$

We can now rewrite (9) as

$$\begin{aligned}G_m &= x_1\{\mu_1^\ominus + RT\ln(P/P^\ominus) + RT\ln x_1 + B_{(11)}P\} \\ &+ x_2\{\mu_2^\ominus + RT\ln(P/P^\ominus) + RT\ln x_2 + B_{(22)}P\} \\ &+ 2x_1 x_2 \delta_{(12)}P.\end{aligned} \qquad 4.07.11$$

Experimental data on $B_{(12)}$ or $\delta_{(12)}$ are much scantier than data on $B_{(11)}$ and $B_{(22)}$. When the species 1 and 2 conform to the principle of corresponding states it is possible to estimate $B_{(12)}$ from $B_{(11)}$ and $B_{(22)}$ with useful accuracy*. When this is not the case it is usual to neglect $\delta_{(12)}$. This procedure is due to Lewis and Randall[†].

From (11) by use of (4.03.13) and (4.03.14) we deduce

$$RT\ln\lambda_1 = \mu_1 = G_1 = \mu_1^\ominus + RT\ln(P/P^\ominus) + RT\ln x_1 + (B_{11} + 2x_2^2 \delta_{(12)})P \qquad 4.07.12$$

$$RT\ln\lambda_2 = \mu_2 = G_2 = \mu_2^\ominus + RT\ln(P/P^\ominus) + RT\ln x_2 + (B_{22} + 2x_1^2 \delta_{(12)})P. \qquad 4.07.13$$

By further differentiations we obtain

$$-S_1 = d\mu_1^\ominus/dT + R\ln(P/P^\ominus) + R\ln x_1 + (d/dT)(B_{11} + 2x_2^2 \delta_{(12)})P \qquad 4.07.14$$

* Guggenheim and McGlashan, Proc. Roy. Soc. A 1951 **206** 448.
† Lewis and Randall, Thermodynamics and the Free Energy of Chemical Substances, McGraw-Hill 1923 p. 226.

$$H_1 = \mu_1^\ominus - T\,d\mu_1/dT + (1 - T\,d/dT)(B_{11} + 2x_2^2 \delta_{(12)})P \qquad 4.07.15$$

$$V_1 = RT/P + B_{11} + 2x_2^2 \delta_{(12)}. \qquad 4.07.16$$

There are analogous formulae for S_2, H_2, and V_2.

§4.08 *Fugacities of gases*

The fugacity p_i of each species i in a gas is defined by

$$p_i/\lambda_i = \text{const.} \qquad (T \text{ const.}) \qquad 4.08.1$$

$$p_i/x_i P \to 1 \quad \text{as} \quad P \to 0 \qquad (T \text{ const.}). \qquad 4.08.2$$

Using these definitions we obtain for a binary mixture from (4.07.12) and (4.07.13)

$$p_1 = x_1 P \exp\{(B_{11} + 2x_2^2 \delta_{(12)})P/RT\} \qquad 4.08.3$$

$$p_2 = x_2 P \exp\{(B_{22} + 2x_1^2 \delta_{(12)})P/RT\}. \qquad 4.08.4$$

The quantity $x_i P$ is called the *partial pressure* of i.

In a perfect gas these simplify to

$$p_1 = x_1 P \qquad 4.08.5$$

$$p_2 = x_2 P \qquad 4.08.6$$

so that the fugacity of each species is equal to its partial pressure.

§4.09 *Liquid mixtures*

We now turn to liquid mixtures and the equilibrium between such phases and other phases, especially a gas phase. We begin by certain general considerations applying to all such liquid mixtures. We shall next consider a special class of such mixtures, called *ideal*, which exhibit an especially simple behaviour. We shall then show how the behaviour of non-ideal mixtures can conveniently be compared and correlated with that of ideal mixtures. The procedure will be illustrated in greater detail for a class of mixtures called *simple*.

§4.10 *Liquid–vapour equilibrium*

Let us now consider from a general point of view the equilibrium conditions between a liquid mixture of c independent species or components and a vapour phase. Each phase by itself has evidently $c+1$ degrees of freedom, which we can take as given by the $c+1$ independent variables T, P, x_2, x_3, ... x_c. Alternatively if we use the $c+2$ variables T, P, μ_1, μ_2, ... μ_c these are not independent, being connected by the Gibbs–Duhem relation

$$S_m dT - V_m dP + x_1 d\mu_1 + x_2 d\mu_2 + \ldots + x_c d\mu_c = 0. \qquad 4.10.1$$

We now consider two phases, one liquid and the other vapour, in mutual equilibrium. We shall continue to use x_i to denote a mole fraction in the liquid phase but shall henceforth denote a mole fraction in the vapour phase by y_i. The variables T, P, μ_1, μ_2, ... μ_c are connected by two Gibbs–Duhem relations, one for each phase. Thus, using the superscripts ^{L} to denote liquid and ^{G} to denote gas, we have

$$S_m^L dT - V_m^L dP + x_1 d\mu_1 + x_2 d\mu_2 + \ldots + x_c d\mu_c = 0 \qquad 4.10.2$$

$$S_m^G dT - V_m^G dP + y_1 d\mu_1 + y_2 d\mu_2 + \ldots + y_c d\mu_c = 0. \qquad 4.10.3$$

It is hardly necessary to point out that we need not attach superscripts to the variables T, P, μ_1, μ_2, ... μ_c since at equilibrium each of these has the same value in both phases.

From (2) and (3) we could, if we wished, eliminate any one of the quantities dT, dP, $d\mu_1$, $d\mu_2$, ... $d\mu_c$ thus obtaining a single relation between the remaining $c+1$ quantities. Whether we do this or not, it is clear that only c of these quantities are independent. We conclude that a system of two phases and c component species in equilibrium has c degrees of freedom in agreement with Gibbs' phase rule.

§4.11 Azeotropy

For a binary liquid–vapour system the relations (4.10.2) and (4.10.3) reduce to

$$S_m^L dT - V_m^L dP + (1-x)d\mu_1 + x d\mu_2 = 0 \qquad 4.11.1$$

$$S_m^G dT - V_m^G dP + (1-y)d\mu_1 + y d\mu_2 = 0 \qquad 4.11.2$$

where as usual x, y denote mole fractions of the second species. If we subtract (1) from (2) we obtain

$$(S_m^G - S_m^L)dT - (V_m^G - V_m^L)dP + (x-y)(d\mu_1 - d\mu_2) = 0. \qquad 4.11.3$$

We shall show that this leads to particularly simple and interesting results when the compositions of the two phases are the same, that is to say when

$$x = y. \qquad 4.11.4$$

Such mixtures are called *azeotropic*, which means that their composition is not changed by boiling.

Let us first consider variations of pressure and composition at constant temperature. Then (3) becomes

$$(V_m^G - V_m^L)dP/dx = (x - y)(d\mu_1/dx - d\mu_2/dx) \quad \text{(const. } T\text{)}. \quad 4.11.5$$

Hence for an azeotropic mixture, according to (4)

$$(V_m^G - V_m^L)dP/dx = 0 \quad \text{(const. } T\text{)}. \quad 4.11.6$$

and since $V_m^G \neq V_m^L$ it follows that

$$dP/dx = 0 \quad \text{(const. } T\text{)}. \quad 4.11.7$$

This tells us that at a given temperature the total vapour pressure of a binary liquid mixture is a maximum or a minimum at the composition of the azeotropic mixture.

Similarly if we consider variations of temperature and composition at constant pressure, then (3) becomes

$$-(S_m^G - S_m^L)dT/dx = (x - y)(d\mu_1/dx - d\mu_2/dx) \quad \text{(const. } P\text{)}. \quad 4.11.8$$

and consequently for an azeotropic mixture

$$(S_m^G - S_m^L)dT/dx = 0 \quad \text{(const. } P\text{)}. \quad 4.11.9$$

Since $S_m^G \neq S_m^L$, it follows that

$$dT/dx = 0 \quad \text{(const. } P\text{)}. \quad 4.11.10$$

This tells us that at a given pressure the boiling temperature of a binary liquid mixture is a maximum or a minimum at the composition of the azeotropic mixture.

These conclusions are almost obvious when expressed by diagrams. For example figure 4.2 shows the boiling point T plotted against compositions of the two phases. For instance the points L and G represent the liquid and gas phases in equilibrium at one temperature; L', G' is another such pair and L'', G'' another. The point M represents liquid and gas of the same composition and in this example the equilibrium temperature or boiling point is a minimum.

Let us now consider simultaneous variation of temperature and pressure such that the mixture remains azeotropic. Returning to formula (3) and substituting the condition for azeotropy (4), we have

$$(S_m^G - S_m^L)dT - (V_m^G - V_m^L)dP = 0 \quad 4.11.11$$

which we can rewrite as

$$dP/dT = (S_m^G - S_m^L)/(V_m^G - V_m^L) = \Delta_e S/\Delta_e V \qquad 4.11.12$$

of the same form as Clapeyron's relation (3.37.3).

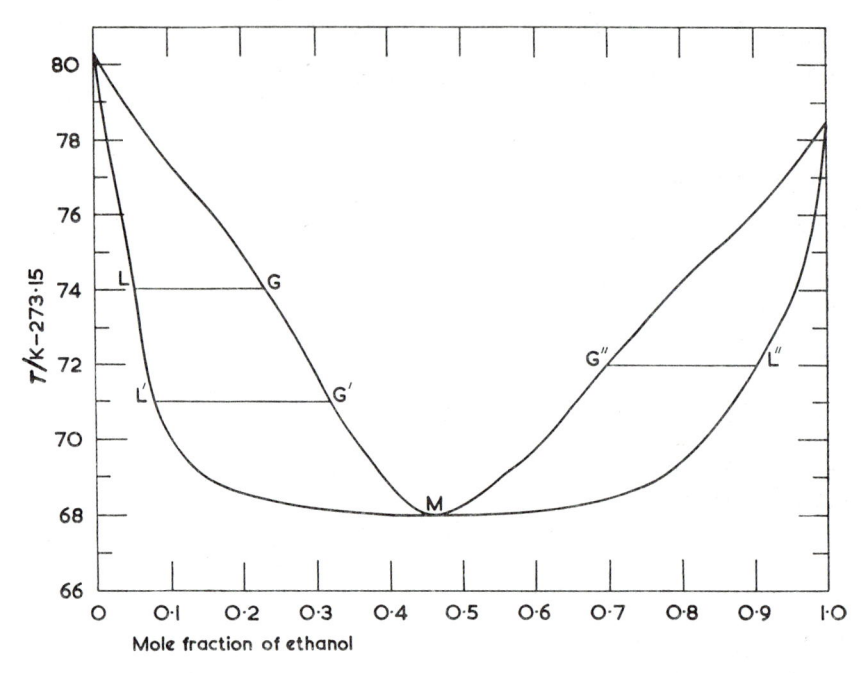

Fig. 4.2. Boiling point of mixtures of benzene and ethanol at one atmosphere

§4.12 *Relative activities and fugacities in liquids*

All the equilibrium properties of each species *i* are determined by its chemical potential μ_i or by its absolute activity λ_i related to μ_i by the definition

$$\mu_i = RT \ln \lambda_i. \qquad 4.12.1$$

Up to the present we have mentioned absolute activities from time to time and have given formulae for them with the object of familiarizing the reader with them. We have however hitherto made little use of absolute activities. Henceforth we shall make considerably increasing use of them, for in the treatment of mixtures they are often more convenient than chemical potentials.

In our consideration of liquid mixtures we shall be concerned particularly with a comparison of the equilibrium properties of the mixture with those of the pure components. Consequently we shall be concerned not so much with μ_i itself as with the difference $\mu_i - \mu_i^0$ where the superscript ^{0} denotes the value for the pure liquid at the same temperature and pressure. From (1) we have

$$\mu_i - \mu_i^0 = RT \ln(\lambda_i/\lambda_i^0) \qquad 4.12.2$$

where the superscript ^{0} is used again with the same meaning. We shall be particularly concerned with the ratios λ_i/λ_i^0. These ratios are called *relative activities* and will be denoted by a_i. This name and this symbol are due to G. N. Lewis*. We have then

$$a_i = \lambda_i/\lambda_i^0. \qquad 4.12.3$$

We must mention that quantities other than the relative activities defined here have sometimes also been called activities and denoted by the same symbol a_i. In order to avoid confusion we shall make no use or further mention of such other quantities.

For the equilibrium of the species i between any two phases α and β we have according to (3.15.6) the simple condition

$$\lambda_i^\alpha = \lambda_i^\beta \qquad 4.12.4$$

and in particular for the equilibrium between a liquid phase L and a gas phase G

$$\lambda_i^L = \lambda_i^G. \qquad 4.12.5$$

From (5) and (4.08.1) it follows that the ratio of the absolute activities of the species i in any two liquid phases α, β is equal to the ratio of the fugacities of the species in the gas phases in equilibrium with α, β respectively.

We now define the *fugacity* p_i of a species i in a liquid phase as follows. We begin by defining the fugacity p_i^0 of the pure liquid i at a given temperature as equal to the fugacity of its saturated vapour. We then define the fugacity of i in liquid mixtures at given temperatures but variable pressure and variable composition by

$$p_i/\lambda_i = \text{const.} \quad (\text{const. } T). \qquad 4.12.6$$

An important application of (6) is to the comparison between the absolute activity λ_i of i in a liquid mixture and its absolute activity λ_i^0 in the pure liquid at the same temperature and pressure. We then have

$$a_i = \lambda_i/\lambda_i^0 = p_i/p_i^0. \qquad 4.12.7$$

* Lewis, J. Amer. Chem. Soc. 1913 **35** 17.

§4.13 Pressure dependence

We must now describe how p_i is determined in a liquid mixture of given composition at a given temperature and at a given pressure. At the given temperature and composition we have to determine the total saturation vapour pressure P_{sat} and the composition of the vapour. If the vapour may be regarded as a perfect gas then the fugacity p_i is equal to the partial vapour pressure $y_i P_{sat}$. If the vapour is not a perfect gas we have to apply a correction for non-ideality by use of a formula such as (4.08.3). We then obtain the value of the fugacity at the given pressure P by means of

$$\partial \ln p_i / \partial P = \partial \ln \lambda_i / \partial P = V_i / RT \qquad 4.13.1$$

or in the integrated form

$$\ln p_i(P) - \ln p_i(P_{sat}) = (P - P_{sat}) V_i / RT \qquad 4.13.2$$

where we have neglected compressibility.

Although at ordinary pressures the quantity on either side of (2) may be negligible, nevertheless formula (2) is in principle important as showing that $p_i(P)$ is precisely defined and precisely determinable for any chosen value of P, not merely for $P = P_{sat}$.

Formula (2) is also important in the discussion of osmotic equilibrium in the following section.

§4.14 Osmotic equilibrium

Suppose we have two liquid mixtures α and β separated by a membrane permeable to the species 1 but impermeable to all other species present in either mixture. In this connection we shall follow the customary practice of calling the permeant species 1 the *solvent* and the nonpermeant species *solutes*. We assume that the two phases are at the same temperature, but not necessarily at the same pressure. The condition that the two phases should be in equilibrium with respect to the solvent species 1 is

$$\lambda_1^\alpha = \lambda_1^\beta \qquad 4.14.1$$

or if we use (4.12.6)

$$p_1^\alpha = p_1^\beta. \qquad 4.14.2$$

For the relations to be satisfied it will generally be necessary for the two phases to be at different pressures. There is then equilibrium with respect to the solvent species 1, but not with respect to the solute species; nor is there hydrostatic equilibrium between the two phases, the difference of pressure being balanced by a force exerted by the membrane. A partial equilibrium of this kind is called *osmotic equilibrium* of the solvent species 1.

By using the relation (4.13.1)

$$\partial \ln p_1/\partial P = \partial \ln \lambda_1/\partial P = V_1/RT \qquad 4.14.3$$

we can determine the pressure $P^\alpha - P^\beta$ required to preserve osmotic equilibrium.

We shall use the notation $p_1(P, x)$ to denote the value of the fugacity of the solvent in a liquid phase of composition x at a pressure P. We do not refer to the temperature as this is assumed constant throughout. The condition (2) for osmotic equilibrium becomes in this notation

$$p_1(P^\alpha, x^\alpha) = p_1(P^\beta, x^\beta). \qquad 4.14.4$$

Dividing both sides of (4) by $p_1(P^\beta, x^\alpha)$ and taking logarithms, we obtain

$$\ln\{p_1(P^\alpha, x^\alpha)/p_1(P^\beta, x^\alpha)\} = \ln\{p_1(P^\beta, x^\beta)/p_1(P^\beta, x^\alpha)\}. \qquad 4.14.5$$

If we integrate (3) from P^β to P^α and substitute the result on the left side of (5) we find

$$\frac{1}{RT}\int_{P^\beta}^{P^\alpha} V_1^\alpha dP = \ln\{p_1(P^\beta, x^\beta)/p_1(P^\beta, x^\alpha)\}. \qquad 4.14.6$$

In order to evaluate the integral in (6) it is for most purposes sufficient to ignore compressibility and treat V_1 as independent of P. In case greater refinement should be desired, we can obtain all the accuracy that can ever be required by assuming that V_1 varies linearly with P. We then obtain

$$\langle V_1^\alpha \rangle (P^\alpha - P^\beta)/RT = \ln\{p_1(P^\beta, x^\beta)/p_1(P^\beta, x^\alpha)\} \qquad 4.14.7$$

where the symbol $\langle V_1^\alpha \rangle$ denotes the value of V_1^α at a pressure equal to the mean of P^α and P^β.

Formula (7) is the general relation determining the pressure difference across the membrane at osmotic equilibrium. The case of greatest interest is when the phase β consists of the pure solvent. The pressure difference $P^\alpha - P^\beta$ is then called the *osmotic pressure* and is denoted by Π. We can in this case replace the superscript $^\beta$ by ^{0} and drop the superscript $^\alpha$. We thus have

$$\Pi \langle V_1 \rangle/RT = \ln\{p_1^0(P)/p_1(P)\} \qquad 4.14.8$$

or if, as is often the case, we may ignore compressibility

$$\Pi V_1/RT = \ln\{p_1^0(P)/p_1(P)\}. \qquad 4.14.9$$

If moreover the pressure P on the pure solvent is roughly atmospheric, then regardlesss of how great Π may be we may regard p_1^0/p_1 as essentially independent of P. Formula (9) can then be simplified to

$$\Pi V_1/RT = \ln(p_1^0/p_1) \qquad 4.14.10$$

from which we see that, provided the pressure P on the pure solvent is low, the osmotic pressure Π does not depend significantly on P.

If instead of dividing both sides of (4) by $p_1(P^\beta, x^\alpha)$, we divide both sides by $p_1(P^\alpha, x^\beta)$ and otherwise proceed in the same way, we obtain instead of (7)

$$\langle V_1^\beta \rangle (P^\alpha - P^\beta)/RT = \ln\{p_1(P^\alpha, x^\beta)/p_1(P^\alpha, x^\alpha)\} \qquad 4.14.11$$

and instead of (8)

$$\Pi \langle V_1^0 \rangle/RT = \ln\{p_1^0(P+\Pi)/p_1(P+\Pi)\}. \qquad 4.14.12$$

It can be shown that the alternative formulae (8) and (12) are equivalent. On the whole formula (8) is the more useful.

§4.15 Pressure on semi-permeable membrane

The osmotic pressure is by definition a pressure that must be applied to the solution to bring it into a certain equilibrium condition. It is not a pressure exerted by the solution or part of the solution at its normal low pressure. It is analogous to the freezing point of a solution, which has no relation to the actual temperature of the solution, but is the temperature to which it must be brought to reach a certain equilibrium state. The osmotic pressure is nevertheless sometimes defined as the pressure exerted on a membrane, permeable only to the solvent, separating the solution from pure solvent. This definition, unless carefully qualified, is incorrect. Another definition sometimes given is the pressure exerted by the solute molecules on a membrane permeable only to the solvent. This definition is still more incorrect than the last. The truth as regards the pressure on the membrane is as follows. When the solution is at the same pressure e.g. atmospheric, as the solvent, there will be a resultant flow of solvent through the membrane from the solvent to the solution, but the resultant pressure on the membrane itself is negligibly small, and may be in either direction. If, however, the solution is subjected to a certain high external pressure, the flow of solvent through the membrane is equal in either direction; there is then osmotic equilibrium and the excess pressure on the solution over the pressure of the solvent is by definition the osmotic pressure. Under conditions of osmotic equilibrium, but only under these conditions, is the external pressure difference required to prevent the membrane from moving equal to the osmotic pressure.

§4.16 Duhem–Margules relation

We recall the Gibbs–Duhem relation

$$\sum_i n_i \mathrm{d}\mu_i = 0 \quad \text{(const. } T, P\text{)} \qquad 4.16.1$$

or

$$\sum_i n_i \mathrm{d} \ln \lambda_i = 0 \quad \text{(const. } T, P\text{)} \qquad 4.16.2$$

From (4.12.6) and (2) we deduce

$$\sum_i n_i \mathrm{d} \ln p_i = 0 \quad \text{(const. } T, P\text{)} \qquad 4.16.3$$

or dividing by $\Sigma_i n_i$

$$\sum_i x_i \mathrm{d} \ln p_i = 0 \quad \text{(const. } T, P\text{)}. \qquad 4.16.4$$

This important relation is known as the *Duhem–Margules relation**.

In the simple case of a binary mixture (4) reduces to

$$(1-x)\partial \ln p_1/\partial x + x \partial \ln p_2/\partial x = 0 \quad \text{(const. } T, P\text{)} \qquad 4.16.5$$

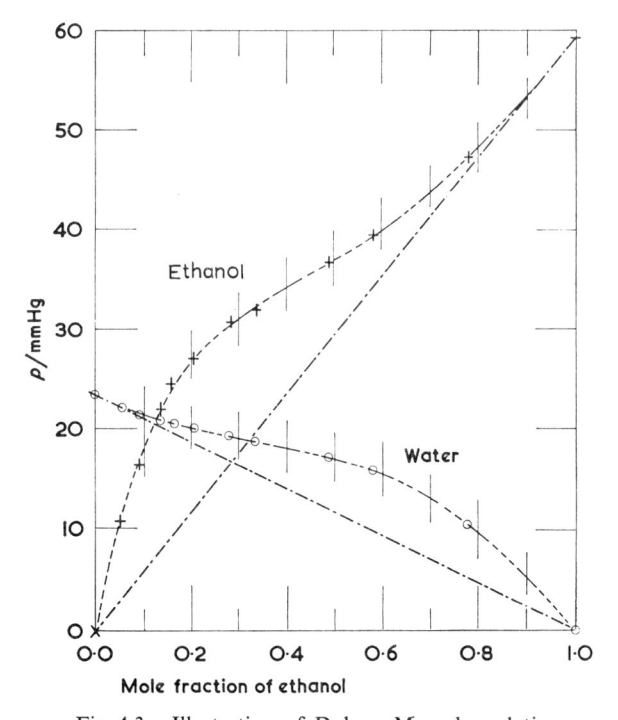

Fig. 4.3. Illustration of Duhem–Margules relation

* Margules, Sitz. ber. Akad. Wiss. Wien 1895 **104** 1258–1260.

where as usual x denotes the mole fraction of species 2. It follows from this relation that if p_1 and p_2 are plotted against x, the shape of either curve completely determines the shape of the other. An example of this interrelation* between the pair of curves is shown in figure 4.3 and table 4.1, where the subscript $_1$ denotes water and $_2$ denotes ethanol. In this illustration no correction has been applied for gas imperfection. In other words p_1 has been taken as equal to $(1-y)P_{sat}$ and p_2 as equal to yP_{sat}.

TABLE 4.1

Verification of Duhem–Margules relation for mixtures of water and ethanol at 25 °C

x	$p_1 \approx (1-y)P_{sat}$ mmHg	$p_2 \approx y P_{sat}$ mmHg	$-(1-x)\,\partial \ln p_1/\partial x$ $= x\,\partial \ln p_2/\partial x$
0	23.75	0.0	1.00
0.1	21.7	17.8	0.76
0.2	20.4	26.8	0.41
0.3	19.4	31.2	0.37
0.4	18.35	34.2	0.355
0.5	17.3	36.9	0.41
0.6	15.8	40.1	0.53
0.7	13.3	43.9	0.655
0.8	10.0	48.3	0.77
0.9	5.5	53.3	0.915
1.0	0.0	59.0	1.00

§ 4.17 *Temperature coefficients*

Formula (4.04.5) is equivalent to

$$\partial(\mu_i/T)/\partial T = -H_i/T^2 \qquad 4.17.1$$

or

$$\partial \ln p_i/\partial T = -H_i/RT^2. \qquad 4.17.2$$

Consequently from (4.12.7)

$$\partial \ln a_i/\partial T = \partial \ln(p_i/p_i^0)/\partial T = -(H_i - H_i^0)/RT^2. \qquad 4.17.3$$

§ 4.18 *Ideal mixtures*

In order to obtain more detailed information concerning the equilibrium properties of liquid mixtures it is necessary to know or assume something

* Adam and Guggenheim, Proc. Roy. Soc. A 1933 **139** 231.

about the dependence of the chemical potentials μ_i or the absolute activities λ_i on the composition of the mixture. Thermodynamic considerations alone cannot predict the form of this dependence, but only impose certain restrictions such as the Gibbs–Duhem relation.

We shall begin by considering liquid mixtures having the property that at constant temperature and pressure the dependence of the Gibbs function G, and consequently also of functions derived from G, on the composition is of the same form as for a perfect gaseous mixture. This means that the value of G in a mixture containing amount n_i of the species i exceeds the value of G for the unmixed species at the same temperature and pressure by the negative amount

$$\Delta G = RT \sum_i n_i \ln\{n_i / \sum_k n_k\}. \qquad 4.18.1$$

Such mixtures are called *ideal mixtures*. We shall devote considerable attention to such mixtures for several reasons.

In the first place the behaviour of ideal mixtures is the simplest conceivable either from a mathematical or from a physical aspect.

In the second place statistical theory predicts that mixtures of very similar species, in particular isotopes, will be ideal.

In the third place it is found experimentally that almost ideal mixtures exist, for example benzene + bromobenzene.

In the fourth place although real mixtures, other than isotopic mixtures, are not ideal, in many cases the resemblances between a real mixture and an ideal mixture are more striking than the differences.

§4.19 Thermodynamic functions of ideal mixtures

From formula (4.18.1) we can immediately derive formulae for all the thermodynamic functions of an ideal mixture. In particular we have, using the superscript ^{0} to refer to the pure liquids at the same pressure,

$$\mu_i = RT \ln \lambda_i = \partial G / \partial n_i = \mu_i^0 + RT \ln\{n_i / \sum_k n_k\} = \mu_i^0 + RT \ln x_i \qquad 4.19.1$$

$$a_i = \lambda_i / \lambda_i^0 = x_i \qquad 4.19.2$$

$$G_m = \sum_i x_i \mu_i = \sum_i x_i \mu_i^0 + RT \sum_i x_i \ln x_i \qquad 4.19.3$$

$$S_m = -\sum_i x_i \{\partial \mu_i^0 / \partial T\} - R \sum_i x_i \ln x_i \qquad 4.19.4$$

$$H_m = \sum_i x_i \{\mu_i^0 - T(\partial \mu_i^0 / \partial T)\} = \sum_i x_i H_i^0 \qquad 4.19.5$$

$$V_m = \sum_i x_i \{\partial \mu_i^0/\partial P\} = \sum_i x_i V_i^0 \qquad 4.19.6$$

$$S_i = -\partial \mu_i^0/\partial T - R \ln x_i \qquad 4.19.7$$

$$H_i = \mu_i^0 - T\{\partial \mu_i^0/\partial T\} = H_i^0 \qquad 4.19.8$$

$$V_i = \partial \mu_i^0/\partial P = V_i^0. \qquad 4.19.9$$

Furthermore if we use the symbol Δ_m to denote the increase in a function when unit amount of mixture is formed from the pure components at constant temperature and pressure we have

$$\Delta_m G = RT \sum_i x_i \ln x_i \qquad 4.19.10$$

$$\Delta_m S = -R \sum_i x_i \ln x_i \qquad 4.19.11$$

$$\Delta_m H = 0 \qquad 4.19.12$$

$$\Delta_m V = 0. \qquad 4.19.13$$

The relations (10), (11), (12), (13) have precisely the same form as for the formation of a perfect gaseous mixture from the pure component gases at constant temperature and pressure.

It must be emphasized that this similarity between ideal liquid mixtures and perfect gaseous mixtures as regards dependence of the thermodynamic properties on the composition holds only when the other independent variables are T and P. There are no correspondingly simple relations in terms of the variables T and V, which are moreover an inconvenient set of independent variables for any phase other than a gas.

There is, of course, no similarity between liquid and gaseous mixtures as regards dependence of properties on the pressure. For example, in a perfect gaseous mixture

$$\partial \mu_i/\partial P = V_i = RT/P \qquad 4.19.14$$

while in a liquid ideal mixture

$$\partial \mu_i/\partial P = \partial \mu_i^0/\partial P = V_i^0 = V_i^{0\ominus}(1 + \kappa P^\ominus - \kappa P) \qquad 4.19.15$$

where $V_i^{0\ominus}$ denotes the value of V_i^0 when $P = P^\ominus$ which varies only slightly with P.

§4.20 *Fugacities in ideal mixtures*

From formulae (4.08.1) and (4.19.2) we deduce immediately for each species in an ideal mixture

$$p_i = x_i p_i^0 \quad \text{(const. } T, P\text{)}. \qquad 4.20.1$$

Since at ordinary pressures the equilibrium properties of a liquid are insensitive to the pressure, we may often with negligible error replace (1) by

$$p_i = x_i p_i^0 \quad \text{(const. } T; \ P = P_{\text{sat}}\text{)} \qquad 4.20.2$$

where P_{sat} denotes the total pressure of the saturated vapour.

If moreover we may neglect deviations of the gas from ideality, we may replace each fugacity p_i by the partial pressure $y_i P$ so that (2) becomes

$$y_i P_{\text{sat}} = x_i P_{\text{sat}, i}^0 \quad \text{(const. } T; \ P = P_{\text{sat}}\text{)} \qquad 4.20.3$$

where $P_{\text{sat}, i}^0$ denotes the saturated vapour pressure of the pure liquid i. Formula (3) is called *Raoult's law*.

For a binary mixture (1) becomes

$$p_1 = (1-x) p_1^0 \qquad p_2 = x p_2^0 \quad \text{(const. } T, P\text{)} \qquad 4.20.4$$

and formula (3) becomes

$$(1-y) P_{\text{sat}} = (1-x) P_{\text{sat}, 1}^0 \qquad y P_{\text{sat}} = x P_{\text{sat}, 2}^0 \quad \text{(const. } T; \ P = P_{\text{sat}}\text{)}. \qquad 4.20.5$$

Thus if the fugacities, or less exactly the partial vapour pressures, of the two components of an ideal binary mixture are plotted against the mole fraction of one of them two straight lines are obtained. The experimental data* for the mixture ethylene bromide and propylene bromide at the temperature 85 °C are shown in figure 4.4 and we see that this mixture is

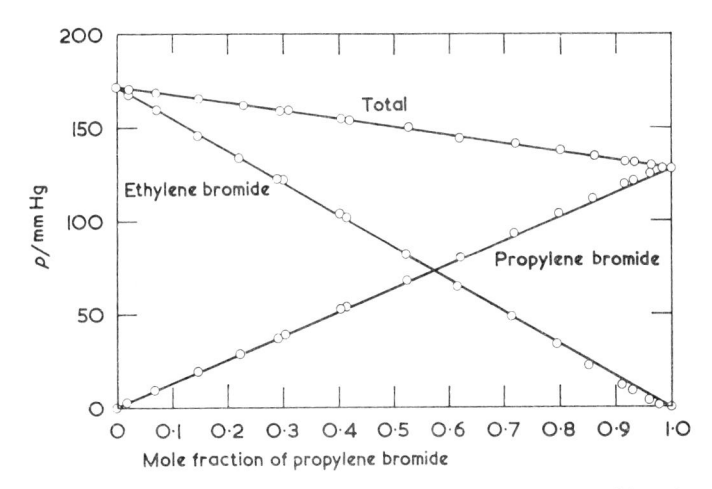

Fig. 4.4. Partial and total vapour pressures of mixtures of ethylene bromide and propylene bromide at 85 °C

* Von Zawidzki, Z. Phys. Chem. 1900 **35** 129.

nearly ideal. In figure 4.5 we see a similar plot* for the mixture benzene and bromobenzene at 80 °C which is also nearly ideal in spite of the considerable difference between the vapour pressures of the two pure components.

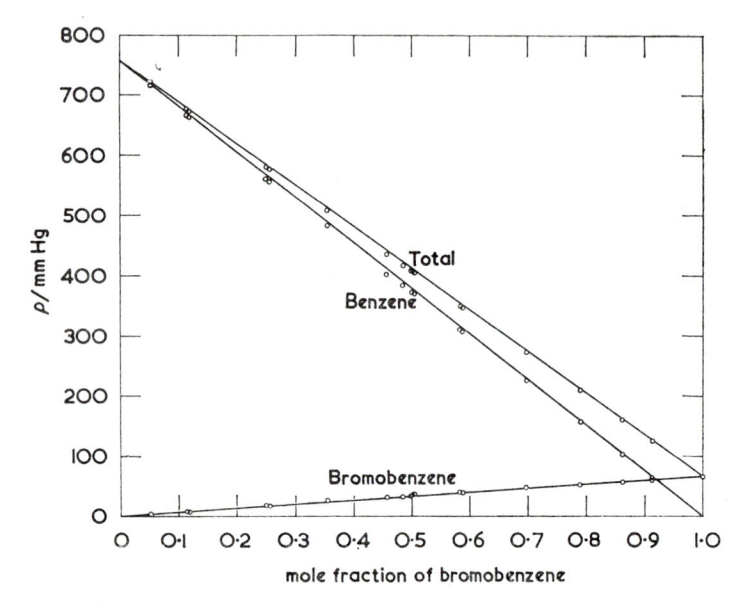

Fig. 4.5. Partial and total vapour pressures of mixtures of benzene and bromobenzene at 80 °C

§4.21 *Osmotic pressure of ideal solution*

To obtain the osmotic pressure of an ideal mixture or ideal solution, regarding the component 1 to which the membrane is permeable as solvent, we have merely to substitute (4.20.4) into (4.14.8). We thus obtain

$$\Pi \langle V_1 \rangle / RT = -\ln x_1 \qquad 4.21.1$$

wherein we recall that $\langle V_1 \rangle$ denotes the value of V_1 averaged between the pressures of the two phases in osmotic equilibrium. When we neglect compressibility (1) reduces to

$$\Pi V_1 / RT = -\ln x_1. \qquad 4.21.2$$

§4.22 *Non-ideal mixtures*

Few, if any, real mixtures are ideal, but it is convenient to correlate the thermodynamic properties of each real mixture with those of an ideal

* McGlashan and Wingrove, Trans. Faraday Soc. 1956 **52** 470.

mixture. This is achieved most conveniently by the use of *excess functions*. For the sake of brevity we shall confine ourselves almost entirely to binary mixtures.

§4.23 Functions of mixing and excess functions

Consider the process of mixing an amount $1-x$ of the liquid species 1 with an amount x of the liquid species 2, at constant temperature and pressure, so as to form unit amount of a liquid mixture. The increase of G in this process is called the *proper Gibbs function of mixing* and is denoted by $\Delta_m G$. If the mixture were ideal we should have according to (4.19.3)

$$\Delta_m G^I = RT\{(1-x)\ln(1-x) + x \ln x\} \qquad 4.23.1$$

where the superscript I denotes ideal. For a real mixture we denote the excess $\Delta_m G$ over its ideal value $\Delta_m G^I$ by G_m^E and call this the *excess proper Gibbs function*. We have then

$$\begin{aligned} G_m^E/RT &= (\Delta_m G - \Delta_m G^I)/RT \\ &= (1-x)\ln\{\lambda_1/\lambda_1^0(1-x)\} + x\ln\{\lambda_2/\lambda_2^0 x\} \\ &= (1-x)\ln\{a_1/(1-x)\} + x\ln\{a_2/x\} \\ &= (1-x)\ln\{p_1/p_1^0(1-x)\} + x\ln\{p_2/p_2^0 x\}. \end{aligned} \qquad 4.23.2$$

Since there has been confusion concerning the precise meaning of (4.23.2) and related formulae, we emphasize that all the quantities λ_1, λ_2, λ_1^0, λ_2^0, a_1, a_2, p_1, p_2, p_1^0, p_2^0 relate to the same pressure P as well as the same temperature. As explained in §4.13 we measure p_1, p_2 at the total saturation pressure P_{sat} of the mixture and then calculate the values of p_1, p_2 at the chosen pressure P by means of formula (4.13.2). The chosen pressure P is usually, but not necessarily, equal to the standard pressure $P^\ominus = 1$ atm.

Other excess functions are defined similarly. It is clear that the several excess functions are interrelated in the same way as the extensive functions from which they are derived. In particular we have

$$S_m^E = -\partial G_m^E/\partial T \qquad 4.23.3$$

$$H_m^E = G_m^E - T\partial G_m^E/\partial T \qquad 4.23.4$$

$$V_m^E = \partial G_m^E/\partial P \qquad 4.23.5$$

$$U_m^E = G_m^E - T\partial G_m^E/\partial T - P\partial G_m^E/\partial P \qquad 4.23.6$$

$$C_m^E = -T\partial^2 G_m^E/\partial T^2. \qquad 4.23.7$$

Using (4.03.15) and (4.03.16) we also have

$$RT \ln\{a_1/(1-x)\} = \mu_1^E = G_m^E - x \partial G_m^E/\partial x \qquad 4.23.8$$

$$RT \ln(a_2/x) = \mu_2^E = G_m^E + (1-x) \partial G_m^E/\partial x. \qquad 4.23.9$$

Differentiating (2) with respect to x we obtain

$$(RT)^{-1} \partial G_m^E/\partial x = (1-x)\partial \ln p_1/\partial x + x\partial \ln p_2/\partial x + \ln\{p_1^0 p_2(1-x)/p_2^0 p_1 x\}. \qquad 4.23.10$$

By the Duhem–Margules relation we have

$$(1-x)\partial \ln p_1/\partial x + x\partial \ln p_2/\partial x = 0. \qquad 4.23.11$$

Subtracting (11) from (10) we obtain

$$(RT)^{-1}\partial G_m^E/\partial x = \ln(p_1^0/p_2^0) + \ln\{(1-x)p_2/xp_1\}. \qquad 4.23.12$$

Integrating (12) from $x=0$ to x and observing that G_m^E vanishes when $x=0$, we obtain

$$(RT)^{-1} G_m^E(x) = x \ln(p_1^0/p_2^0) + \int_0^x \ln \alpha' \, dx \qquad 4.23.13$$

where α' is defined by

$$\alpha' = (1-x)p_2/xp_1. \qquad 4.23.14$$

Setting $x=1$ in (13) and remembering that G_m^E vanishes when $x=1$ we obtain

$$\int_0^1 \ln \alpha' \, dx = \ln(p_2^0/p_1^0). \qquad 4.23.15$$

If then $\ln \alpha'$ is plotted against x, the two domains separated by the straight line parallel to the x-axis and distant $\ln(p_2^0/p_1^0)$ from the x-axis have equal areas.

§4.24 Volatility ratio

The ratio $(1-x)y/x(1-y)$, where y denotes the mole fraction of component 2 in the saturated vapour, is called the *volatility ratio* or *relative volatility* and is denoted by α. For the purpose of a rough check on the reliability of dubious measurements it is often sufficient to use the approximation of a perfect gas. To this approximation we have

$$p_1/p_2 = (1-y)/y \qquad 4.24.1$$

and (4.23.14) becomes

$$\alpha' = (1-x)y/x(1-y) = \alpha. \qquad 4.24.2$$

Consequently (4.23.15) becomes

$$\int_0^1 \ln \alpha \, dx \approx \ln(P^0_{2,\text{sat}}/P^0_{1,\text{sat}}) \qquad 4.24.3$$

where $P^0_{1,\text{sat}}$ denotes the saturated vapour pressure of pure 1 and $P^0_{2,\text{sat}}$ denotes the saturated vapour pressure of pure 2. Formula (3) furnishes a rough check on the consistency of measurements of α as a function of x. We can illustrate this by using the experimental data on mixtures of water and ethanol at 25 °C given in table 4.2.

TABLE 4.2

Volatility ratio of mixtures of water and ethanol at 25 °C

x	y	α	$\ln \alpha$
0.0252	0.1790	8.421	2.131
0.0523	0.3163	8.387	2.127
0.0916	0.4334	7.582	2.026
0.1343	0.5127	6.782	1.914
0.1670	0.5448	5.969	1.787
0.2022	0.5684	5.197	1.648
0.2848	0.6104	3.935	1.370
0.3368	0.6287	3.334	1.204
0.4902	0.6791	2.201	0.789
0.5820	0.7096	1.755	0.562
0.7811	0.8161	1.244	0.218

In figure 4.6 $\ln \alpha$ is plotted against x and the straight line is at a distance $\ln(P^0_{2,\text{sat}}/P^0_{1,\text{sat}})$ from the x-axis. The two domains separated by this straight

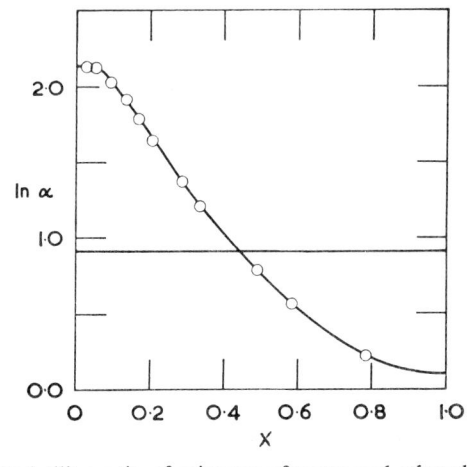

Fig. 4.6. Volatility ratio of mixtures of water and ethanol at 25 °C

line have equal areas in accordance with formula (3). This procedure for roughly checking experimental consistency was recommended independently and almost simultaneously by Redlich and Kister* and by Herington[†]. This kind of plot had previously been recommended by Scatchard and Raymond[‡].

§4.25 *Internal stability with respect to composition*

We turn now to a discussion of internal stability with respect to composition. We can conveniently study this problem for a binary system by reference to a plot of the proper Gibbs function G_m against mole fraction x at given T, P. Examples of such plots are shown in figure 4.7.

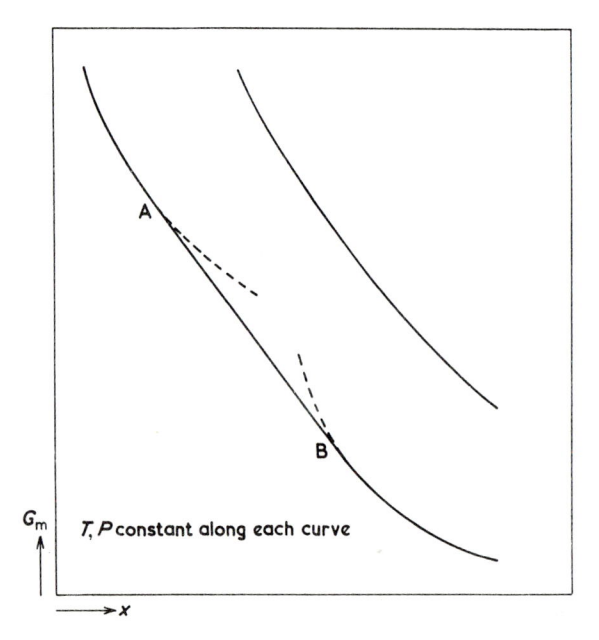

Fig. 4.7. Stable and metastable isotherms

If we now imagine a phase of composition x to split into two, one of slightly greater and the other of slightly smaller x, the new value of G_m is then given by a point on the straight line joining the two points representing the two new phases. If this point lies above the one representing the

---
* Redlich and Kister, Ind. Eng. Chem. 1948 **40** 345 (paper received 25 November 1946)
[†] Herington, Nature 1947 **160** 610 (letter dated 11 July 1947).
[‡] Scatchard and Raymond, J. Amer. Chem. Soc. 1938 **60** 1281.

original phase, the system will revert to its original state which is stable. In the contrary case the original phase is unstable. It is then clear from the diagram that while the upper curve represents phases all stable, the phases represented by the dotted portion of the lower curve between A and B are metastable with respect to a mixture of phases represented by A and B.

Since according to (4.03.11)

$$\partial G_m/\partial x = G_2 - G_1 = \mu_2 - \mu_1 \qquad 4.25.1$$

we see that the slope of the curve at any point is equal to $\mu_2 - \mu_1$. Since the two phases A and B are in mutual equilibrium they have equal values of μ_1, μ_2 and consequently of $\mu_2 - \mu_1$ in agreement with the fact that the straight line AB touches the curve at A and B.

§4.26 Critical mixing

It can happen that at some temperatures the behaviour corresponds to a curve such as the upper one in figure 4.7 while at other temperatures, lower or higher, the behaviour corresponds to a curve such as the lower one. There will then be some temperature at which the change in type of behaviour takes place. This state of affairs is called *critical mixing*. At temperatures on one side of the temperature of critical mixing the two liquids are miscible in all proportions; at temperatures on the other side the miscibility is limited, only phases to the left of A or to the right of B being stable.

We shall now determine the conditions of critical mixing. The lower curve in figure 4.7 is concave upwards in the stable regions and in the dotted metastable regions. In these parts of the curve

$$\partial^2 G_m/\partial x^2 > 0. \qquad 4.26.1$$

If we imagine the two dotted curves joined into a single curve then in the middle there must be a part of the curve convex upwards corresponding to completely unstable phases. Hence between A and B there are two points of inflexion where

$$\partial^2 G_m/\partial x^2 = 0. \qquad 4.26.2$$

At the temperature of critical mixing these two points merge into a single point at which as well as (2)

$$\partial^3 G_m/\partial x^3 = 0. \qquad 4.26.3$$

Formulae (2) and (3) together express the conditions of critical mixing.

It is convenient to express these conditions of critical mixing (2) and (3) in terms of the excess proper Gibbs function G_m^E. According to the definition of G_m^E we have

$$G_m = G_m^E + (1-x)\mu_1^0 + x\mu_2^0 + RT(1-x)\ln(1-x) + RTx\ln x. \quad 4.26.4$$

By successive differentiation with respect to x we obtain

$$\partial G_m/\partial x = \partial G_m^E/\partial x - \mu_1^0 + \mu_2^0 + RT\ln\{x/(1-x)\} \quad 4.26.5$$

$$\partial^2 G_m/\partial x^2 = \partial^2 G_m^E/\partial x^2 + RT/x(1-x) \quad 4.26.6$$

$$\partial^3 G_m/\partial x^3 = \partial^3 G_m^E/\partial x^3 + RT(2x-1)/x^2(1-x)^2. \quad 4.26.7$$

Substituting (6) into (2) and (7) into (3) we obtain the conditions of critical mixing in the form

$$\partial^2 G_m^E/\partial x^2 = -RT/x(1-x) \quad 4.26.8$$

$$\partial^3 G_m^E/\partial x^3 = -RT(2x-1)/x^2(1-x)^2. \quad 4.26.9$$

The use of these formulae will be illustrated in §4.30.

§4.27 Excess functions expressed as polynomials

It is convenient to express the excess proper Gibbs function G_m^E of a binary mixture as a polynomial in x. We might write such a polynomial as a succession of integral powers of x but such an expression would obscure any symmetry between the two component species. Bearing in mind that G_m^E must vanish identically both when $x_1 = 1 - x = 0$ and when $x_2 = x = 0$, we find it most convenient to write the polynomial in the form*

$$\begin{aligned} G_m^E &= x_1 x_2 \{A_0 + A_1(x_1 - x_2) + A_2(x_1 - x_2)^2\} \\ &= x(1-x)\{A_0 + A_1(1-2x) + A_2(1-2x)^2\}. \end{aligned} \quad 4.27.1$$

Higher powers of $x_1 - x_2$ can be included if required. The coefficients A_0, A_1, A_2 are by definition independent of x but will usually depend on T and on P.

§4.28 Symmetrical mixtures

We shall begin by considering those mixtures for which G_m^E, and consequently G_m also, is symmetrical with respect to x_1 and x_2. Such mixtures are of interest because they correspond to the model of molecular species of the

---

* Guggenheim, Trans. Faraday Soc. 1937 **33** 151 (formula 4.1); Redlich and Kister, Ind. Eng. Chem. 1948 **40** 345 (formula 8); Scatchard, Chem. Rev. 1949 **44** 9.

same size and shape. Moreover mixtures are known which conform within the experimental accuracy with formulae symmetrical in x_1 and x_2. We call such mixtures *symmetrical mixtures*.

For a symmetrical mixture $A_1 = 0$ and formula (4.27.1) reduces to

$$G_m^E = x(1-x)\{A_0 + A_2(1-2x)^2\}$$
$$= (A_0 + A_2)x(1-x) - 4A_2 x^2(1-x)^2. \qquad 4.28.1$$

From (1) we derive for other excess functions

$$-S_m^E = x(1-x)[\partial A_0/\partial T + (\partial A_2/\partial T)(1-2x)^2] \qquad 4.28.2$$

$$H_m^E = x(1-x)[A_0 - T(\partial A_0/\partial T) + \{A_2 - T(\partial A_2/\partial T)\}(1-2x)^2] \quad 4.28.3$$

$$V_m^E = x(1-x)[\partial A_0/\partial P + (\partial A_2/\partial P)(1-2x)^2] \qquad 4.28.4$$

$$\mu_1^E = RT \ln(a_1/x_1) = x_2^2\{A_0 + A_2(x_2 - x_1)(x_2 - 5x_1)\} \qquad 4.28.5$$

$$\mu_2^E = RT \ln(a_2/x_2) = x_1^2\{A_0 + A_2(x_1 - x_2)(x_1 - 5x_2)\}. \qquad 4.28.6$$

In (5) or (6) the term in A_2 vanishes when $x = \tfrac{5}{6}$ or $\tfrac{1}{6}$ as well as when $x = \tfrac{1}{2}$.

When the deviation of the vapour from a perfect gas is neglected, we obtain for the volatility ratio α using (4.23.12) and (4.24.2)

$$RT \ln(P_{1,\,\text{sat}}^0/P_{2,\,\text{sat}}^0) + RT \ln \alpha = -(x_2 - x_1)\{A_0 + (1 - 8x_1 x_2)A_2\}. \qquad 4.28.7$$

We observe that the term in A_2 vanishes when $x_1 x_2 = \tfrac{1}{8}$ as well as when $x = \tfrac{1}{2}$.

§4.29 Simple mixtures

We shall now consider those symmetrical mixtures for which the terms in A_2 and higher terms are negligible. Writing w in place of A_0 we have

$$G_m^E = x(1-x)w \qquad w = w(T, P) \qquad 4.29.1$$

where w or A_0 is independent of x but will in general depend on T and P. We call mixtures having properties defined by (1) *simple mixtures*. Such mixtures are important for several reasons.

In the first place the behaviour of these mixtures is the simplest conceivable after ideal mixtures either from a mathematical or from a physical aspect.

In the second place some binary mixtures show a behaviour which can be represented either accurately or approximately by the formulae of simple mixtures.

In the third place statistical theory predicts that a mixture of two kinds of non-polar molecules of similar simple shape and similar size should obey certain laws to which the formulae of simple mixtures are a useful approximation.

The formulae of simple mixtures, as here defined, were used by Porter[*] to express empirically partial vapour pressure measurements on mixtures of ethyl ether and acetone at 30 °C and a few measurements at 20 °C. The best value of w at 20 °C was found to be slightly greater than that at 30 °C. Later on Heitler[†] related these formulae to the model of liquids now usually called the 'quasi-crystalline' model and these formulae have been applied to experimental measurements on various mixtures especially by Hildebrand[‡]. It was assumed by Heitler and subsequently generally accepted that the value of w should be independent of temperature although this by no means follows from the quasi-crystalline model used in the derivation of the formulae.

From (1) we derive

$$-S_m^E = x(1-x)\partial w/\partial T \qquad 4.29.2$$

$$H_m^E = x(1-x)(w - T\partial w/\partial T) \qquad 4.29.3$$

$$V_m^E = x(1-x)\partial w/\partial P \qquad 4.29.4$$

$$\mu_1^E = RT \ln\{a_1/(1-x)\} = wx^2 \qquad 4.29.5$$

$$\mu_2^E = RT \ln\{a_2/x\} = w(1-x)^2. \qquad 4.29.6$$

We shall now compare the formulae of simple mixtures with the experimental data on mixtures of carbon tetrachloride and cyclohexane. For these mixtures G_m^E has been determined by vapour pressure measurements at 30, 40, 50, 60, and 70 °C. There are also measurements of the enthalpy of mixing $\Delta_m H$ at 10, 25, 40, and 55 °C. The experimental values[§] of G_m^E are shown plotted against $x(1-x)$ in figure 4.8. According to formula (1) the slopes of the straight lines are the values of w at each temperature. The experimental values[‖] of $H_m^E = \Delta_m H$ are shown similarly plotted in figure 4.9. According to formula (3) the slopes of the straight lines are the values of $w - T(\partial w/\partial T)$ at each temperature. The fact that the experimental points for $x < \frac{1}{2}$ shown to the left and those for $x > \frac{1}{2}$ shown to the right lie on the pair of straight lines confirms that the laws of simple mixtures are valid within the experimental accuracy. The thermodynamic consistency of the two sets of data requires that both should be fitted by the same values of w and

---

[*] Porter, Trans. Faraday Soc. 1920 **16** 336.
[†] Heitler, Ann. Phys., Lpz. 1926 **80** 629.
[‡] Hildebrand and Scott, Solubility of Nonelectrolytes, Reinhold 1950.
[§] Scatchard, Wood and Mochel, J. Amer. Chem. Soc. 1939 **61** 3206; Brown and Ewald, Australian J. Sci. Res. A 1950 **3** 306.
[‖] Adcock and McGlashan, Proc. Roy. Soc. A 1954 **226** 266.

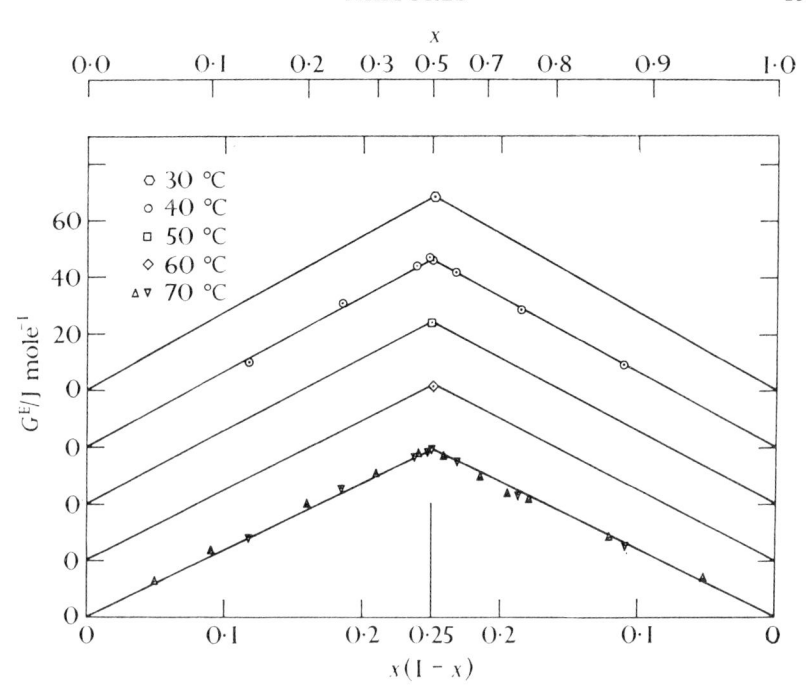

Fig. 4.8. Excess Gibbs function in mixtures of carbon tetrachloride and cyclohexane

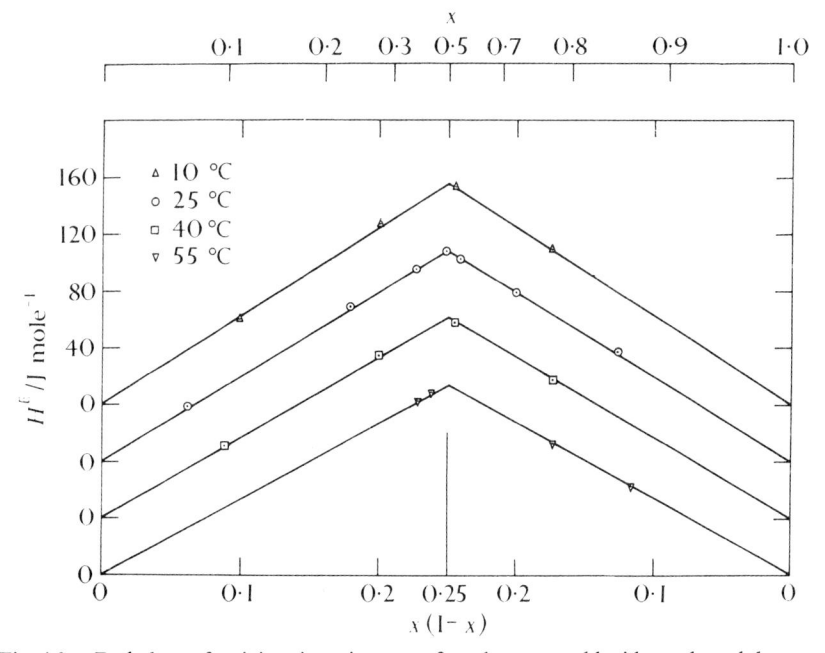

Fig. 4.9. Enthalpy of mixing in mixtures of carbon tetrachloride and cyclohexane

∂w/∂T. The straight lines in both figure 4.8 and figure 4.9 in fact correspond to the single relation*

$$w/\text{J mole}^{-1} = 1176 + 1.96T \ln T - 14.18T \qquad 4.29.7$$

so that

$$(w - T\partial w/\partial T)/\text{J mole}^{-1} = 1176 - 1.96T. \qquad 4.29.8$$

§4.30 Critical mixing in simple mixtures

In figure 4.10 the quantities $a_1 = p_1/p_1^0$ and $a_2 = p_2/p_2^0$ are plotted against x for simple mixtures with $w/RT = 1$ and $w/RT = -2$. When w is positive, the curves lie above the straight lines representing the behaviour of an ideal mixture; this situation is called a *positive deviation* from ideality. On the other hand when w is negative, both curves lie below the straight line of the ideal mixture and this situation is described as a *negative deviation* from ideality.

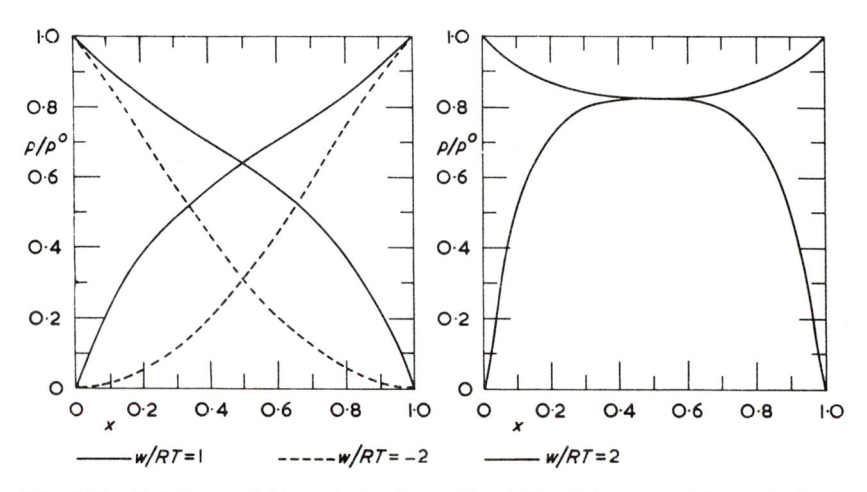

——— $w/RT = 1$ ----- $w/RT = -2$ ——— $w/RT = 2$

Fig. 4.10. Relative activities of simple mixtures: (complete mixing)

Fig. 4.11. Relative activities of simple mixture: (temperature of critical mixing)

Figure 4.11 gives similar plots for $w/RT = 2$. We shall now show that this determines the temperature of critical mixing. We begin by recalling the general conditions for critical mixing (4.26.8) and (4.26.9)

$$\partial^2 G_m^E/\partial x^2 = -RT/x(1-x) \qquad 4.30.1$$

$$\partial^3 G_m^E/\partial x^3 = -RT(2x-1)/x^2(1-x)^2. \qquad 4.30.2$$

* Adcock and McGlashan, Proc. Roy. Soc. A 1954 **226** 266.

We also recall formula (4.29.1) which defines simple mixtures

$$G_m^E = x(1-x)w \qquad w = w(T, P). \qquad 4.30.3$$

By successive differentiation of (3) with respect to x we obtain

$$\partial G_m^E/\partial x = (1-2x)w \qquad 4.30.4$$

$$\partial^2 G_m^E/\partial x^2 = -2w \qquad 4.30.5$$

$$\partial^3 G_m^E/\partial x^3 = 0. \qquad 4.30.6$$

By substituting (5) into (1) and (6) into (2) we obtain as the conditions for critical mixing in a simple mixture

$$-2w = -RT/x(1-x) \qquad 4.30.7$$

$$0 = RT(2x-1)/x^2(1-x)^2. \qquad 4.30.8$$

From (8) we deduce $x = \tfrac{1}{2}$, which is incidentally obvious from considerations of symmetry, and substituting this value of x into (7) we obtain for the temperature T_c of critical mixing

$$2RT_c = w \qquad 4.30.9$$

in agreement with figure 4.11.

When $w/RT > 2$ there is incomplete mixing. A typical example, namely $w/RT = 3$, is shown in figure 4.12. If x', x'' denote the compositions of the two phases in mutual equilibrium at a given temperature below that of critical mixing, then x', x'' are determined by the pair of simultaneous equations

$$p_1(x') = p_1(x'') \qquad 4.30.10$$

$$p_2(x') = p_2(x''). \qquad 4.30.11$$

Dividing (10) by p_1^0 and dividing (11) by p_2^0 we obtain the equivalent pair of simultaneous equations

$$a_1(x') = a_1(x'') \qquad 4.30.12$$

$$a_2(x') = a_2(x''). \qquad 4.30.13$$

The conditions (12) and (13) hold for the two-phase equilibrium of any binary mixture. In the particular case of simple mixtures there is complete symmetry between a_1 as a function of x and a_2 as a function of $1-x$. It follows from this symmetry that

$$x' + x'' = 1 \qquad 4.30.14$$

and consequently (12) and (13) lead to

$$a_1(x') = a_1(x'') = a_2(1-x'') = a_2(x'). \qquad 4.30.15$$

Hence x', x'' are determined by the intersections of the two curves. These are the points L, M in figure 4.12. The curves between L and M represent solutions either metastable near L, M or completely unstable towards the middle of the diagram.

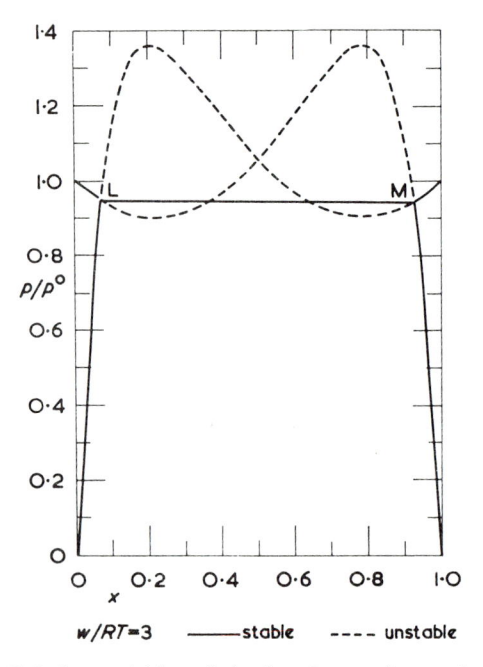

Fig. 4.12. Relative activities of simple mixture: (incomplete mixing)

When we substitute from (4.29.5) and (4.29.6) into (15) we obtain as the equation for either x' or x''

$$(1-x)/x = \exp\{(1-2x)w/RT\}. \qquad 4.30.16$$

If we use the abbreviation $s = 1 - 2x$ we can rewrite (16) as an equation for s in the form

$$s = \tanh(sw/2RT) \qquad 4.30.17$$

which can be solved numerically by inspection of tables of the tanh function. Incidentally we notice from (17) that $s \to 0$ as $w/2RT \to 1$, that is to say at critical mixing.

Pairs of liquids are known, for example water and nicotine, which are completely miscible above a certain critical temperature and below another critical temperature, but are incompletely miscible in the intermediate temperature range. It is interesting to note that even simple mixtures can behave in this way when w is a quadratic function of the temperature provided the three coefficients in the quadratic expression have suitable signs and magnitudes. To be precise if w has the quadratic form*

$$w/R = 2T + \{t^2 - (T - T_0)^2\}/\Theta \qquad 4.30.18$$

where Θ, T_0, and t are positive constants and $t < T_0$, then it is clear that $w/R = 2T$ when $T = T_0 - t$ or $T = T_0 + t$. It can also be verified that $w/R > 2T$ when $T_0 - t < T < T_0 + t$ and that $w/R < 2T$ when $T > T_0 + t$ or $T < T_0 - t$. Consequently the temperature range of incomplete miscibility extends from $T_0 - t$ to $T_0 + t$.

Incidentally the converse behaviour occurs, that is complete miscibility only between the two critical temperatures, if Θ is negative.

§4.31 Critical mixing in symmetrical mixtures

We shall now consider briefly the condition for critical mixing in a symmetrical mixture which is not a simple mixture. If we assume

$$G_m^E = x(1-x)\{A_0 + A_2(1-2x)^2 + A_4(1-2x)^4\} \qquad 4.31.1$$

then it is obvious from symmetry that at the critical point $x = \frac{1}{2}$. For this value of x

$$\partial^2 G_m^E/\partial x^2 = -2(A_0 - A_2). \qquad 4.31.2$$

Consequently by (4.26.8)

$$-2(A_0 - A_2) = -4RT \qquad 4.31.3$$

or

$$(A_0 - A_2)/RT = 2. \qquad 4.31.4$$

It is interesting to note that this condition is independent of A_4, and incidentally also of coefficients of higher powers of $(1-2x)^2$.

§4.32 Example of unsymmetrical mixture

By way of contrast with symmetrical mixtures we shall now briefly illustrate the opposite type of behaviour by a particular hypothetical example. We return to formula (4.27.1). Instead of setting $A_1 = 0$ and so obtaining

---

* Guggenheim, Faraday Soc. Discussion No. 15 1953 271.

the formulae of symmetrical mixtures, we now set $A_0+A_1=0$ and $A_2=0$. We then obtain

$$G_m^E = Ax^2(1-x) \qquad 4.32.1$$

where we have written A instead of $2A_0$.

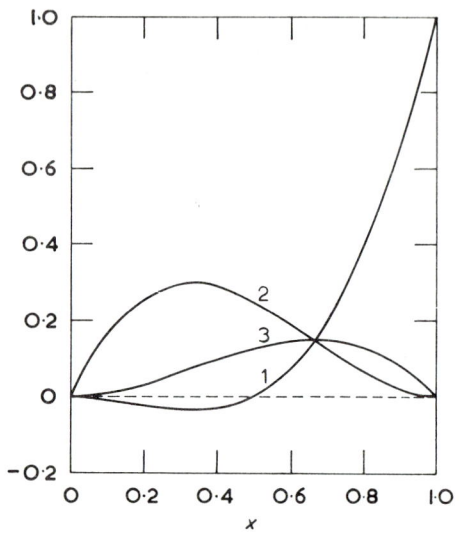

Fig. 4.13. Example of unsymmetrical excess functions
Curve 1 $(RT/A)\ln\{a_1/(1-x)\}$
Curve 2 $(RT/A)\ln\{a_2/x\}$
Curve 3 G_m^E/A

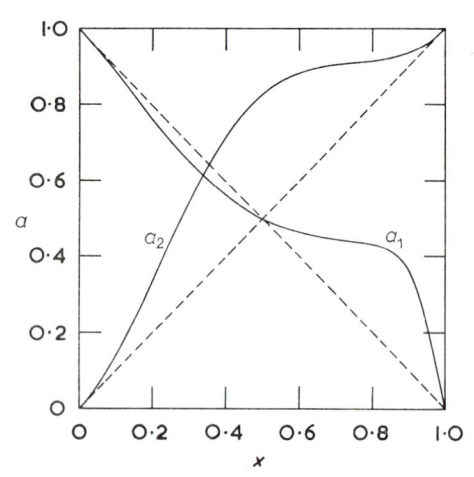

Fig. 4.14. Example of relative activities in a mixture having unsymmetrical excess functions

From (1) we obtain, using (4.23.8) and (4.23.9)

$$RT \ln\{a_1/(1-x)\} = Ax^2(2x-1) \qquad 4.32.2$$

$$RT \ln\{a_2/x\} = 2Ax(1-x)^2. \qquad 4.32.3$$

A remarkable feature* of this pair of formulae is that, whereas $\ln\{a_2/x\}$ has for all values of x the same sign as A, the sign of $\ln\{a_1/(1-x)\}$ changes at $x = \frac{1}{2}$. This behaviour with $A = \frac{1}{2}RT$ is illustrated in figure 4.14.

§4.33 Athermal mixtures of small and large molecules

According to (4.19.12) when two or more species form an ideal mixture, then they mix isothermally without increase or decrease of the enthalpy. Zero enthalpy of mixing is thus a necessary condition for two or more species to form an ideal mixture, but it is not a sufficient condition. Mixtures, not necessarily ideal, having zero enthalpy of mixing at all compositions are called *athermal mixtures*. Statistical mechanics indicates as a further necessary condition for a mixture to be ideal that the several kinds of molecules should not differ greatly in size. It is accordingly of interest to consider the properties of mixtures of two kinds of molecules sufficiently similar to mix in all proportions without any enthalpy of mixing, but differing widely in size. This is a complicated problem in statistical mechanics which has not been solved completely. It is probable that the shapes of the molecules matter as well as their sizes. However, when we ignore such complications, there are reasons for believing that the behaviour due to wide differences in size between the two species of molecule can be at least semi-quantitatively described by means of relatively simple formulae in which the only new parameter is the ratio of the molecular volumes.

If ϱ denotes the ratio of the volume of a molecule of type 2 to that of a molecule of type 1, then subject to various restrictions and approximations which we shall not here go into, we may write for the proper Gibbs function of mixing[†]

$$\Delta_m G = RT(1-x) \ln\{(1-x)/(1-x+\varrho x)\} + RTx \ln\{\varrho x/(1-x+\varrho x)\}. \qquad 4.33.1$$

Formula (1) is more easily memorized if written in a slightly different form involving the *volume fraction* ϕ of the second species defined by

$$\phi = \varrho x/(1-x+\varrho x). \qquad 4.33.2$$

---

* McGlashan, J. Chem. Ed. 1963 **40** 516.
† Guggenheim, Mixtures, Clarendon Press 1952.

Using (2) we can rewrite (1) in the shorter form

$$\Delta_m G = RT(1-x)\ln(1-\phi) + RTx\ln\phi. \qquad 4.33.3$$

This simple formula is due to Flory.*

From (3) we deduce using (4.03.15) and (4.03.16)

$$\ln a_1 = \ln(p_1/p_1^0) = \ln(\lambda_1/\lambda_1^0) = \ln\{(1-x)/(1-x+\varrho x)\} + (\varrho-1)x/(1-x+\varrho x)$$
$$= \ln(1-\phi) + (1-\varrho^{-1})\phi \qquad 4.33.4$$
$$\ln a_2 = \ln(p_2/p_2^0) = \ln(\lambda_2/\lambda_2^0) = \ln\{\varrho x/(1-x+\varrho x)\} - (\varrho-1)(1-x)/(1-x+\varrho x)$$
$$= \ln\phi - (\varrho-1)(1-\phi). \qquad 4.33.5$$

We notice that when $\varrho = 1$ we recover the formulae of ideal mixtures. Of especial interest is the opposite extreme when ϱ is so great that $1/\varrho$ may be neglected compared with unity. Formula (4) then reduces to

$$\ln a_1 = \ln(p_1/p_1^0) = \ln(\lambda_1/\lambda_1^0) = \ln(1-\phi) + \phi \qquad (\varrho \to \infty). \qquad 4.33.6$$

We then have the remarkable situation that the lowering of the vapour pressure of the 'solvent' species 1 is completely determined by the volume fraction of the 'solute' species 2. Under these conditions determinations of the vapour pressure of the solvent give no information concerning the size of the solute molecules, except that they are much larger than the solvent molecules. These formulae are relevant to solutions of rubber or polystyrene in certain non-polar solvents such as benzene and toluene.

§ 4.34 *Osmotic pressure in athermal mixtures*

By substituting (4.33.4) into (4.14.10) we obtain for the osmotic pressure with respect to a membrane permeable only to the species 1 with small molecules

$$\Pi\langle V_1\rangle/RT = -\ln(1-\phi) - (1-\varrho^{-1})\phi. \qquad 4.34.1$$

If we expand $\ln(1-\phi)$ in powers of ϕ we obtain

$$\Pi\langle V_1\rangle/RT = \phi/\varrho + \tfrac{1}{2}\phi^2 + \tfrac{1}{3}\phi^3 + \ldots \qquad 4.34.2$$

From (2) and (4.33.2) we see that in the limit of infinite dilution $\Pi \propto x$ as usual, but for this state of affairs it is not sufficient that $\phi \ll 1$, the much more stringent condition $\varrho\phi \ll 1$ being required. If we merely assume that $\phi \ll 1$ formula (2) reduces to

$$\Pi\langle V_1\rangle/RT = \phi/\varrho + \tfrac{1}{2}\phi^2 \qquad (\phi \ll 1) \qquad 4.34.3$$

* Flory, J. Chem. Phys. 1941 **9** 660; 1942 **10** 51.

and the term $\frac{1}{2}\phi^2$ will swamp the term ϕ/ϱ unless $\phi \ll \varrho^{-1}$. It follows from this that in a solution of macromolecules measurements of osmotic pressure cannot yield simple or reliable information concerning the size of the solute macromolecules unless the solutions are so dilute that $\phi \ll \varrho^{-1}$ which implies that $x \ll \varrho^{-2}$.

§4.35 Interfacial layers

We shall now consider the thermodynamics of interfacial layers between two bulk phases each containing the same two components. There are two cases to distinguish: first an interface between a liquid mixture and its vapour, when the interfacial tension is called the *surface tension*; second an interface between two liquid layers containing in different proportions two incompletely miscible components.

We shall first discuss the liquid–vapour interface using an approximation sufficient for most if not all practical applications. We shall next give a similar approximate treatment of a liquid–liquid interface. Finally we shall give an accurate treatment applicable in principle to either type of interface, but of small practical use.

§4.36 Liquid–vapour interface

We begin with formula (1.57.3) applied to a system of two components 1 and 2. Thus
$$-\mathrm{d}\gamma = S_A^\sigma \mathrm{d}T - \tau \mathrm{d}P + \Gamma_1 \mathrm{d}\mu_1 + \Gamma_2 \mathrm{d}\mu_2. \qquad 4.36.1$$
In the liquid phase we have according to (4.04.4) and (4.04.6)
$$\mathrm{d}\mu_1 = -S_1 \mathrm{d}T + V_1 \mathrm{d}P + (\partial \mu_1/\partial x)\mathrm{d}x \qquad 4.36.2$$
$$\mathrm{d}\mu_2 = -S_2 \mathrm{d}T + V_2 \mathrm{d}P + (\partial \mu_2/\partial x)\mathrm{d}x \qquad 4.36.3$$
where we have omitted superscripts from quantities relating to the liquid phase.

In our initial treatment of a liquid–vapour interface we shall make approximations similar to those used in §3.63 for a single-component interface.

In the first place we assume that in the liquid phase PV_1 and PV_2 are so small compared with RT that they may be neglected.

In the second place we assume that the two geometrical surfaces separating the interfacial phase from the two bulk phases are placed so near to each other that terms in $P\tau$ may also be neglected. We accordingly replace (1), (2), (3) by

$$-\mathrm{d}\gamma = S_A^\sigma \mathrm{d}T + \Gamma_1 \mathrm{d}\mu_1 + \Gamma_2 \mathrm{d}\mu_2 \qquad 4.36.4$$

$$\mathrm{d}\mu_1 = -S_1 \mathrm{d}T + (\partial\mu_1/\partial x)\mathrm{d}x \qquad 4.36.5$$

$$\mathrm{d}\mu_2 = -S_2 \mathrm{d}T + (\partial\mu_2/\partial x)\mathrm{d}x. \qquad 4.36.6$$

Substituting (5) and (6) into (4) we obtain

$$-\mathrm{d}\gamma = (S_A^\sigma - \Gamma_1 S_1 - \Gamma_2 S_2)\mathrm{d}T + \{\Gamma_1(\partial\mu_1/\partial x) + \Gamma_2(\partial\mu_2/\partial x)\}\mathrm{d}x. \qquad 4.36.7$$

The system of two components in liquid and vapour has two degrees of freedom. There are consequently two independent variables, for which we choose T and x. Formula (7) thus expresses variations of the surface tension γ in terms of variations $\mathrm{d}T$ and $\mathrm{d}x$ of the two independent variables.

Since the quantities $\partial\mu_1/\partial x$ and $\partial\mu_2/\partial x$ on the right of (7) are related by the Gibbs–Duhem relation

$$(1-x)\partial\mu_1/\partial x + x(\partial\mu_2/\partial x) = 0 \qquad (T, P \text{ const.}) \qquad 4.36.8$$

we can use this relation to eliminate either of them. If for example we eliminate $\partial\mu_1/\partial x$ we obtain

$$-\mathrm{d}\gamma = (S_A^\sigma - \Gamma_1 S_1 - \Gamma_2 S_2)\mathrm{d}T + \{\Gamma_2 - x\Gamma_1/(1-x)\}(\partial\mu_2/\partial x)\mathrm{d}x. \qquad 4.36.9$$

By this elimination we have unavoidably destroyed the symmetry between the components 1 and 2.

§4.37 *Invariance of relations*

We recall that according to the definition in §1.53 of a surface phase the properties associated with it depend on the positions of the boundaries AA' and BB' in figure 1.2. As in §3.62 we shall henceforth refer to the boundary between surface layer and liquid as Lσ and that between surface layer and gas as Gσ. Since the precise positions assigned to these geometrical boundaries are partly arbitrary, the values assigned to such quantities as Γ_1, Γ_2, S^σ are also arbitrary. We can nevertheless verify that our formulae are invariant with respect to shifts of either or both of these boundaries. It is hardly necessary to mention that the intensive variables T, μ_1, μ_2 are unaffected by shifts of either boundary. It is also clear from the definition of γ in §1.54 that its value is invariant.

Let us now consider a shift of the plane boundary Lσ a distance $\delta\tau$ away from the gas phase. Then Γ_1 becomes increased by the amount of the species 1 in a cylinder of liquid of height $\delta\tau$, of cross-section unity and so of volume $\delta\tau$. But the total amount of substance in the volume $\delta\tau$ is $\delta\tau/V_m$ of which the amount of species 1 is $(1-x)\delta\tau/V_m$. Similarly Γ_2 becomes

MIXTURES

increased by $x\delta\tau/V_m$. Consequently although shifting the boundary Lσ alters the values of Γ_1 and Γ_2, the quantity

$$\Gamma_2 - x\Gamma_1/(1-x) \qquad 4.37.1$$

remains unchanged. The invariant quantity (1) is essentially the same as a quantity defined by Gibbs in a more abstract manner which he denoted by the symbol $\Gamma_{2(1)}$.

Similarly when the boundary Lσ is shifted a distance $\delta\tau$ away from the gas phase, S^σ becomes increased by the entropy contained in a cylinder of liquid of volume $\delta\tau$, that is to say by an amount

$$S_m \delta\tau/V_m = \{(1-x)S_1 + xS_2\}\delta\tau/V_m. \qquad 4.37.2$$

At the same time $\Gamma_1 S_1$ is increased by $S_1(1-x)\delta\tau/V_m$ and $\Gamma_2 S_2$ by $S_2 x\delta\tau/V_m$. Consequently the quantity

$$S_A^\sigma - \Gamma_1 S_1 - \Gamma_2 S_2 \qquad 4.37.3$$

occurring in (4.36.9) remains invariant.

With regard to a shift of the geometrical surface Gσ, little need be said in the present connection. For our approximation of neglecting terms in $P\tau$, as we are doing, is equivalent to assuming that the amount of substance per unit volume in the gas phase is negligible compared with that in unit volume of the surface layer. Consequently if we shifted the geometrical surface Gσ away from the liquid even to the extent of doubling the value of τ, the change in the amount of substance contained in the surface layer would be negligible and consequently the values of Γ_1, Γ_2, S_A^σ would not be appreciably affected.

§4.38 *Temperature coefficient of surface tension*

If we apply formula (4.36.9) to variations of temperature at constant composition x of the liquid we obtain

$$-d\gamma/dT = S_A^\sigma - \Gamma_1 S_1 - \Gamma_2 S_2 \qquad \text{(const. } x\text{).} \qquad 4.38.1$$

This relation involving entropies can be transformed to one involving energies as follows. Since we are neglecting terms in PV_1 and PV_2 we may replace (4.04.3) by the approximation

$$\mu_1 = \mathscr{F}_1 = U_1 - TS_1 \qquad 4.38.2$$

$$\mu_2 = \mathscr{F}_2 = U_2 - TS_2. \qquad 4.38.3$$

Applying formula (1.56.1) to unit area and neglecting the term containing

$V^\sigma = \tau A$, we have

$$\Gamma_1 \mu_1 + \Gamma_2 \mu_2 = U_A^\sigma - TS_A^\sigma - \gamma. \qquad 4.38.4$$

We now use (2), (3), (4) to eliminate S_A^σ, S_1, S_2 from (1). We thus obtain

$$\gamma - T\,d\gamma/dT = U_A^\sigma - \Gamma_1 U_1 - \Gamma_2 U_2. \qquad 4.38.5$$

It is worth noticing that the right side of (1) is the entropy of unit area of interface less the entropy of the same material content in the liquid phase. Likewise the right side of (5) is the energy of unit area of interface less the energy of the same material content in the liquid phase. More pictorially we may say that it is the energy which must be supplied to prevent any change of temperature when unit area of surface is formed from the liquid. It may appropriately be called the *surface energy* of formation of unit area.

§4.39 Variations of composition

If we apply (4.36.7) to a variation of composition at constant temperature we obtain, using (4.12.7),

$$\begin{aligned} -\partial\gamma/\partial x &= \Gamma_1 \partial\mu_1/\partial x + \Gamma_2 \partial\mu_2/\partial x \\ &= RT(\Gamma_1 \partial \ln \lambda_1/\partial x + \Gamma_2 \partial \ln \lambda_2/\partial x) \\ &= RT(\Gamma_1 \partial \ln a_1/\partial x + \Gamma_2 \partial \ln a_2/\partial x) \\ &= RT(\Gamma_1 \partial \ln p_1/\partial x + \Gamma_2 \partial \ln p_2/\partial x) \quad (T\text{ const.}). \end{aligned} \qquad 4.39.1$$

When we combine (1) with the Gibbs–Duhem relation (4.16.1) or with the Duhem–Margules relation (4.16.4) we obtain

$$\begin{aligned} -\partial\gamma/\partial x &= -RT\{(1-x)\Gamma_2 - x\Gamma_1\}x^{-1} \partial \ln \lambda_1/\partial x \\ &= RT\{(1-x)\Gamma_2 - x\Gamma_1\}(1-x)^{-1} \partial \ln \lambda_2/\partial x \end{aligned} \qquad 4.39.2$$

or alternatively

$$\begin{aligned} -\partial\gamma/\partial x &= -RT\{(1-x)\Gamma_2 - x\Gamma_1\}x^{-1} \partial \ln p_1/\partial x \\ &= RT\{(1-x)\Gamma_2 - x\Gamma_1\}(1-x)^{-1} \partial \ln p_2/\partial x. \end{aligned} \qquad 4.39.3$$

From (3) we see that from measurements of γ and p_1 or p_2 over a range of compositions we can for each composition compute the value of the quantity I defined by

$$I = (1-x)\Gamma_2 - x\Gamma_1. \qquad 4.39.4$$

We have already verified in §4.37 that $I/(1-x)$ is invariant with respect to shift in position of the boundary Lσ between the liquid and the interface.

Obviously the same holds for I itself. Values can be assigned to Γ_1 and Γ_2 individually only by adopting some more or less arbitrary convention.* We shall illustrate this by a numerical example in the next section. Since Γ_1, Γ_2 must be finite, it follows from (4) that

$$\Gamma_2/I \to 1 \quad \text{as} \quad x \to 0 \qquad 4.39.5$$

$$-\Gamma_1/I \to 1 \quad \text{as} \quad x \to 1. \qquad 4.39.6$$

This implies that for small values of x the value of Γ_2 is unaffected by the position assigned to the boundary Lσ and that for small values of $1-x$ the value of Γ_1 is unaffected. Consequently when $x \ll 1$ we may regard I as a measure of the positive adsorption Γ_2 of species 2 at the surface and when $1-x \ll 1$ we may regard $-I$ as a measure of the adsorption Γ_1 of the species 1. At intermediate values of x no such simple physical meaning can be attached to I. We may however regard I as a measure of relative adsorption of the two species.

In the special case of an ideal mixture formula (3) becomes

$$-(RT)^{-1} \partial \gamma / \partial x = \Gamma_2/x - \Gamma_1/(1-x) = I/x(1-x). \qquad 4.39.7$$

§4.40 Example of water+ethanol

We shall now consider the experimental data for mixtures of water and ethanol in order to illustrate the use of the formulae of the preceding section. We neglect the difference between fugacity and partial pressure. The experimental data for the partial vapour pressures have already been given in figure 4.3 and table 4.1 where we verified that they are consistent with the Duhem–Margules relation. In table 4.3 the first three columns repeat those of table 4.1, the subscripts $_1$ denoting water and $_2$ ethanol. The fourth column gives experimental values of the surface tension γ. The fifth column gives values of $-\partial \gamma / \partial \ln p_2$ obtained by plotting γ against $\ln p_2$ and measuring slopes. The sixth column gives values of

$$I = (1-x)\Gamma_2 - x\Gamma_1 \qquad 4.40.1$$

calculated from (4.39.3) which can be rewritten in the form

$$-\partial \gamma / \partial \ln p_2 = RTI/(1-x). \qquad 4.40.2$$

The values of I are given in the sixth column. In the seventh column the corresponding molecular quantity I/L is given.

* Guggenheim and Adam, Proc. Roy. Soc. A 1933 **139** 231.

TABLE 4.3

Mixtures of water and ethanol at 25 °C
Determination of $I=(1-x)\Gamma_2-x\Gamma_1$

x	p_1 mmHg	p_2 mmHg	γ erg cm^{-2}	$-\partial\gamma/\partial \ln p_2$ erg cm^{-2}	$10^{10} I$ mole cm^{-2}	$10^2 I/L$ Å^{-2}
0.0	23.75	0.0	72.2	0.0	0.0	0.0
0.1	21.7	17.8	36.4	15.6	5.6	9.3
0.2	20.4	26.8	29.7	16.0	5.1	8.5
0.3	19.4	31.2	27.6	14.6	4.1	6.8
0.4	18.35	34.2	26.35	12.6	3.0	5.0
0.5	17.3	36.9	25.4	10.5	2.1	3.5
0.6	15.8	40.1	24.6	8.45	1.4	2.3
0.7	13.3	43.9	23.85	7.15	0.8	1.3
0.8	10.0	48.3	23.2	6.2	0.5	0.8
0.9	5.5	53.3	22.6	5.45	0.2	0.3
1.0	0.0	59.0	22.0	5.2	0.0	0.0

As we have repeatedly emphasized, this is as far as one can go without using some non-thermodynamic convention. We shall now give an example of such a convention*. Let us assume that the interfacial layer is unimolecular and that each molecule of water occupies a constant area of interface and likewise each molecule of ethanol. This assumption may be expressed by

$$A_1\Gamma_1+A_2\Gamma_2=1 \qquad (A_1, A_2 \text{ const.}). \qquad 4.40.3$$

We may call A_1, A_2 the *partial areas* of the two species in the interface. The essence of our assumption is not the definition of these quantities, but the assignment to them of definite constant values which can neither be determined nor be verified by thermodynamic means.

As an example we might assume

$$\begin{aligned} A_1 &= 0.04 \times 10^{10} \text{ cm}^2 \text{ mole}^{-1} \\ A_2 &= 0.12 \times 10^{10} \text{ cm}^2 \text{ mole}^{-1} \end{aligned} \qquad 4.40.4$$

corresponding to molecular cross-sections

$$\begin{aligned} A_1/L &= 7 \text{ Å}^2 \\ A_2/L &= 20 \text{ Å}^2. \end{aligned} \qquad 4.40.5$$

The relation (3) with the values of A_1, A_2 given by (4) is sufficient to determine values of Γ_1, Γ_2 from the values of the expression (1) already given in table 4.3. The results of the calculation are given in table 4.4.

* Guggenheim and Adam, Proc. Roy. Soc. A 1933 **139** 231.

TABLE 4.4

Mixtures of water and ethanol at 25 °C

Values of Γ_1 and Γ_2 calculated from $A_1\Gamma_1+A_2\Gamma_2=1$
with $A_1=0.04\times 10^{10}$ cm^2 mole^{-1} of water
$A_2=0.12\times 10^{10}$ cm^2 mole^{-1} of ethanol

x	$10^{10}I$ mole cm^{-2}	$10^{10}\Gamma_2$ mole cm^{-2}	$10^{10}\Gamma_1$ mole cm^{-2}	$\dfrac{\Gamma_2}{\Gamma_1+\Gamma_2}$
0.0	0.0	0.0	25.0	0.00
0.1	5.6	6.8	4.6	0.60
0.2	5.1	7.25	3.25	0.69
0.3	4.1	7.25	3.25	0.69
0.4	3.0	7.25	3.25	0.69
0.5	2.1	7.3	3.1	0.70
0.6	1.4	7.45	2.65	0.74
0.7	0.8	7.65	2.0	0.79
0.8	0.5	7.9	1.3	0.86
0.9	0.2	8.1	0.7	0.94
1.0	0.0	8.35	0.0	1.00

The first column gives the mole fraction x of ethanol, the second the values of I taken from the previous table, the third and fourth columns the values of Γ_1, Γ_2 calculated by means of (3). The fifth column gives the values of $\Gamma_2/(\Gamma_1+\Gamma_2)$ which we may call the *mole fraction* of ethanol *in the surface layer*. As the mole fraction, thus calculated, in the surface layer increases steadily with the mole fraction in the liquid, we may conclude that although the model on which the assumptions (3), (4), (5) were based is admittedly arbitrary, at least it does not lead to unreasonable or surprising results.

§4.41   *Interface between two binary liquids*

We turn now to consider the interface between two liquid phases of two components 1 and 2. Two such phases may or may not be simple, but they obviously cannot be ideal. In our initial treatment we shall make approximations similar to those in §4.36.

We assume that in a liquid phase PV_1 and PV_2 are so small compared with RT that they may be neglected. This assumption now applies to both liquid phases. Just as in §4.36 we also neglect terms in $P\tau$.

There is an important physical difference between the significance of our approximate treatment of a liquid–vapour interface in the previous sections and the approximate treatment we are now about to give of a liquid–liquid interface. In the case of the liquid–vapour system we took as independent variables the temperature and composition of the liquid phase. Since the system has two degrees of freedom, these determine the composition and pressure of the vapour phase. Moreover the consequent variations of pressure are significant in determining the thermodynamic properties of the vapour phase. In our present discussion of a liquid–liquid system we are assuming that the thermodynamic properties of all phases, that is both liquids and interface, are independent of the pressure. We are thus effectively suppressing variability of pressure as a possible degree of freedom. But when we do this, a single binary liquid phase has only two remaining degrees of freedom, so that we might take as variables either T, x which are independent or T, μ_1, μ_2 subject to the Gibbs–Duhem relation. Correspondingly in a system of two binary liquid phases the variables T, μ_1, μ_2 are subject to two Gibbs–Duhem relations, one in each phase. Thus the system has effectively only one degree of freedom instead of two. Hence the temperature completely determines the composition of both liquid phases and so also the properties of the interface.

We may alternatively describe the situation as follows. A binary liquid–liquid system, like a binary liquid–vapour system has two degrees of freedom. We may therefore take as independent variables T, P and these will then determine the composition of both phases and so also the properties of the interface. When however we use the approximation of treating the properties of every phase as effectively independent of P, then clearly all the equilibrium properties are completely determined by T.

We accordingly proceed to determine how the interfacial tension depends on the temperature.

§4.42 Temperature dependence of interfacial tension

We begin with formula (4.36.4)

$$-\mathrm{d}\gamma = S_A^\sigma \mathrm{d}T + \Gamma_1 \mathrm{d}\mu_1 + \Gamma_2 \mathrm{d}\mu_2 \qquad 4.42.1$$

which applies as well to a liquid–liquid as to a liquid–vapour interface. We also have a Gibbs–Duhem relation in each of the liquid phases. With the term $V_m \mathrm{d}P$ neglected, we have, denoting the two liquid phases by the superscripts $^\alpha$ and $^\beta$,

$$S_m^\alpha \mathrm{d}T + (1-x^\alpha)\mathrm{d}\mu_1 + x^\alpha \mathrm{d}\mu_2 = 0 \qquad 4.42.2$$

$$S_m^\beta \mathrm{d}T + (1-x^\beta)\mathrm{d}\mu_1 + x^\beta \mathrm{d}\mu_2 = 0 \qquad 4.42.3$$

wherein we have omitted the superscripts on T, μ_1, μ_2 since these have the

MIXTURES 215

same values throughout the system. We recall that S_m^α, S_m^β denote the proper entropies in the two phases.

To obtain the dependence of γ on the temperature, we have merely to eliminate $d\mu_1$, $d\mu_2$ from (1), (2), (3). We thus obtain

$$-\frac{d\gamma}{dT} = \frac{\begin{vmatrix} S_A^\sigma & S_m^\alpha & S_m^\beta \\ \Gamma_1 & 1-x^\alpha & 1-x^\beta \\ \Gamma_2 & x^\alpha & x^\beta \end{vmatrix}}{x^\beta - x^\alpha} = \frac{\begin{vmatrix} S_A^\sigma & S_m^\alpha & S_m^\beta \\ \Gamma_1+\Gamma_2 & 1 & 1 \\ \Gamma_2 & x^\alpha & x^\beta \end{vmatrix}}{x^\beta - x^\alpha}. \qquad 4.42.4$$

There seems to be no alternative simpler formula having as high an accuracy.

§ 4.43 *Accurate formulae*

For the sake of completeness we shall now derive formulae, in principle applicable to any interface in a system of two components, in which we do not neglect the terms in VdP or τdP. We however warn the reader that these formulae are too complicated to be of any practical use.

We accordingly revert to formula (4.36.1), namely

$$-d\gamma = S_A^\sigma dT - \tau dP + \Gamma_1 d\mu_1 + \Gamma_2 d\mu_2 \qquad 4.43.1$$

and formulae (4.36.2) and (4.36.3) applied to each of the two phases α, β

$$d\mu_1^\alpha = -S_1^\alpha dT + V_1^\alpha dP + (\partial\mu_1^\alpha/\partial x^\alpha)dx^\alpha \qquad 4.43.2$$

$$d\mu_2^\alpha = -S_2^\alpha dT + V_2^\alpha dP + (\partial\mu_2^\alpha/\partial x^\alpha)dx^\alpha \qquad 4.43.3$$

$$d\mu_1^\beta = -S_1^\beta dT + V_1^\beta dP + (\partial\mu_1^\beta/\partial x^\beta)dx^\beta \qquad 4.43.4$$

$$d\mu_2^\beta = -S_2^\beta dT + V_2^\beta dP + (\partial\mu_2^\beta/\partial x^\beta)dx^\beta. \qquad 4.43.5$$

We shall also use the Gibbs–Duhem relations for both the phases α, β

$$(1-x^\alpha)\partial\mu_1^\alpha/\partial x^\alpha + x^\alpha \partial\mu_2^\alpha/\partial x^\alpha = 0 \qquad 4.43.6$$

$$(1-x^\beta)\partial\mu_1^\beta/\partial x^\beta + x^\beta \partial\mu_2^\beta/\partial x^\beta = 0. \qquad 4.43.7$$

For any variations maintaining equilibrium, we have as usual

$$d\mu_1^\alpha = d\mu_1^\beta = d\mu_1 \qquad 4.43.8$$

$$d\mu_2^\alpha = d\mu_2^\beta = d\mu_2. \qquad 4.43.9$$

If we multiply (8) by $1-x^\beta$, (9) by x^β, substitute from (2), (3), (4), (5), and add, we obtain using (7)

$$0 = -\{(1-x^\beta)(S_1^\alpha - S_1^\beta) + x^\beta(S_2^\alpha - S_2^\beta)\}dT$$
$$+ \{(1-x^\beta)(V_1^\alpha - V_1^\beta) + x^\beta(V_2^\alpha - V_2^\beta)\}dP$$
$$+ \{(1-x^\beta)\partial\mu_1/\partial x^\alpha + x^\beta\partial\mu_2/\partial x^\alpha\}dx^\alpha. \qquad 4.43.10$$

If further we substitute (2), (3) into (1) we obtain

$$-d\gamma = (S_A^\sigma - \Gamma_1 S_1^\alpha - \Gamma_2 S_2^\alpha)dT - (\tau - \Gamma_1 V_1^\alpha - \Gamma_2 V_2^\alpha)dP$$
$$+ (\Gamma_1\partial\mu_1/\partial x^\alpha + \Gamma_2\partial\mu_2/\partial x^\alpha)dx^\alpha. \qquad 4.43.11$$

If we now eliminate dP between (10) and (11) we obtain

$$-d\gamma = (\Delta_{\alpha\sigma}S - \Delta_{\alpha\sigma}V\Delta_{\alpha\beta}S/\Delta_{\alpha\beta}V)dT$$
$$+ \{\Gamma_1 + \Delta_{\alpha\sigma}V(1-x^\beta)/\Delta_{\alpha\beta}V\}(\partial\mu_1/\partial x^\alpha)dx^\alpha$$
$$+ \{\Gamma_2 + \Delta_{\alpha\sigma}Vx^\beta/\Delta_{\alpha\beta}V\}(\partial\mu_2/\partial x^\alpha)dx^\alpha \qquad 4.43.12$$

where we have used the following abbreviations

$$\Delta_{\alpha\beta}S = (1-x^\beta)(S_1^\beta - S_1^\alpha) + x^\beta(S_2^\beta - S_2^\alpha) \qquad 4.43.13$$
$$\Delta_{\alpha\beta}V = (1-x^\beta)(V_1^\beta - V_1^\alpha) + x^\beta(V_2^\beta - V_2^\alpha) \qquad 4.43.14$$
$$\Delta_{\alpha\sigma}S = S_A^\sigma - \Gamma_1 S_1^\alpha - \Gamma_2 S_2^\alpha \qquad 4.43.15$$
$$\Delta_{\alpha\sigma}V = \tau - \Gamma_1 V_1^\alpha - \Gamma_2 V_2^\alpha. \qquad 4.43.16$$

From these definitions we observe that $\Delta_{\alpha\beta}S$ is the entropy increase and $\Delta_{\alpha\beta}V$ the volume increase when unit quantity of the phase β is formed at constant temperature and constant pressure by taking the required amounts of the two components from the phase α. Likewise $\Delta_{\alpha\sigma}S$ is the entropy increase and $\Delta_{\alpha\sigma}V$ the volume increase when unit area of the surface layer σ is formed at constant temperature and constant pressure by taking the required amounts of the two components from the phase α.

Finally we can eliminate $\partial\mu_1/\partial x^\alpha$ (or $\partial\mu_2/\partial x^\alpha$) between (6) and (12). Thus

$$-d\gamma = (\Delta_{\alpha\sigma}S - \Delta_{\alpha\sigma}V\Delta_{\alpha\beta}S/\Delta_{\alpha\beta}V)dT$$
$$+ \{\Gamma_2 - x^\alpha\Gamma_1/(1-x^\alpha) + \Delta_{\alpha\sigma}V(x^\beta - x^\alpha)/\Delta_{\alpha\beta}V(1-x^\alpha)\}(\partial\mu_2/\partial x^\alpha)dx^\alpha. \qquad 4.43.17$$

If we vary the temperature and the pressure so as to maintain x^α constant, (17) becomes

$$-d\gamma/dT = \Delta_{\alpha\sigma}S - \Delta_{\alpha\sigma}V\Delta_{\alpha\beta}S/\Delta_{\alpha\beta}V \qquad (x^\alpha \text{ const.}). \qquad 4.43.18$$

This formula applies in principle to any interface. For a liquid–vapour interface we may assume that $\Delta_{\alpha\sigma}V/\Delta_{\alpha\beta}V$ is negligibly small and then (18) reduces to

$$-d\gamma/dT = \Delta_{\alpha\sigma}S \qquad (x^\alpha \text{ const.}) \qquad 4.43.19$$

which is the same as (4.38.1). For a liquid–liquid interface formula (18), though strictly correct is of little use since the ratio $\Delta_{\alpha\sigma} V/\Delta_{\alpha\beta} V$ of two very small quantities is difficult, if not impossible, to estimate or measure.

§ 4.44 *Solid mixtures*

We turn now to a brief consideration of solid mixtures, especially binary solid mixtures. Much of the treatment of liquid mixtures is directly applicable mutatis mutandis to solid mixtures. Other parts of the treatment are obviously not applicable, in particular osmotic equilibrium and interfacial tensions.

There is a further difference between the treatments of liquid and of solid mixtures, a difference of degree or of emphasis rather than of kind. Most liquids are sufficiently volatile to have conveniently measurable vapour pressures and fugacities. Hence the partial vapour pressures and fugacities of a liquid mixture are familiar experimental quantities. There is consequently a natural and reasonable tendency so far as possible to express most other equilibrium properties in terms of the fugacities. Whereas a few solids also have readily measurable vapour pressures, many are effectively involatile. This being so, there is no particular merit in expressing other equilibrium properties in terms of the fugacities rather than in terms of the absolute activities. If then we compare, for example, the Gibbs–Duhem formula for a binary mixture

$$(1-x)\partial \mu_1/\partial x + x\partial \mu_2/\partial x = 0 \qquad 4.44.1$$

or its corollary

$$(1-x)\partial \ln \lambda_1/\partial x + x\partial \ln \lambda_2/\partial x = 0 \qquad 4.44.2$$

with the Duhem–Margules relation

$$(1-x)\partial \ln p_1/\partial x + x\partial \ln p_2/\partial x = 0 \qquad 4.44.3$$

whereas these three relations are all equivalent, it is natural to place the emphasis on (3) in the case of liquids, but on (1) or (2) in the case of solids.

One of the great similarities between solids and liquids, in contrast to gases, is their insensitivity to pressure. For most purposes we may ignore the pressure. When we do this, a single phase of two components has two degrees of freedom, so that we may use as independent variables T, x. A pair of such phases in equilibrium has then only one degree of freedom, the composition of both phases being determined by the temperature.

We shall deal extremely briefly with the aspects of solid mixtures which are parallel to those of liquid mixtures. We shall quote some formulae without repeating derivations previously given for liquids.

§4.45 Stationary melting points

In §4.11 we proved that whenever the relative compositions of a liquid and vapour in mutual equilibrium at a given pressure are identical, the equilibrium temperature is a minimum or maximum at the given pressure. By precisely the same proof the same result can be derived for a solid and vapour in equilibrium.

Of greater practical interest is the equilibrium between solid and liquid phases. Using the superscripts ^{S} and ^{L} to refer to these two phases respectively, we can derive a formula analogous to (4.11.8), namely

$$(S_m^L - S_m^S)dT/dx = (x^L - x^S)(d\mu_1/dx - d\mu_2/dx) \qquad 4.45.1$$

where x denotes the mole fraction of the component 2 in either phase. Whereas formula (4.11.8) was deduced for constant pressure conditions, as far as (1) is concerned the pressure is practically irrelevant. If the liquid and solid phases have identical compositions then

$$x^S = x^L \qquad 4.45.2$$

and so (1) reduces to

$$(S_m^L - S_m^S)dT/dx = 0. \qquad 4.45.3$$

Since $S_m^L \neq S_m^S$ it follows that

$$dT/dx = 0. \qquad 4.45.4$$

Thus when the compositions of the solid and liquid in mutual equilibrium are identical, the equilibrium temperature is stationary.

§4.46 Solid ideal mixtures

A solid *ideal mixture* is defined in the same manner as in the case of liquids, namely by

$$\Delta_m G = RT \sum_i x_i \ln x_i \qquad 4.46.1$$

and in particular for a binary ideal mixture

$$\Delta_m G/RT = (1-x)\ln(1-x) + x \ln x. \qquad 4.46.2$$

From this definition it follows immediately for a binary mixture that

$$\lambda_1 = \lambda_1^0(1-x) \qquad 4.46.3$$

$$\lambda_2 = \lambda_2^0 x \qquad 4.46.4$$

where the superscript ^{0} denotes the pure solid phase. Actual examples of

MIXTURES

ideal mixtures are as few among solids as among liquids, but the ideal mixture remains the convenient standard with which to compare a real mixture.

The thermodynamic functions and properties of ideal mixtures follow directly from (1) or (2) as in the case of liquids. In particular the enthalpies are additive; that is to say the enthalpy of mixing is zero. On the other hand the *proper entropy of mixing* is given by

$$\Delta_m S/R = -(1-x)\ln(1-x) - x \ln x. \qquad 4.46.5$$

Probably the most important application of this and related formulae is to isotopes, as in §3.55.

§ 4.47 *Excess functions*

Any real solid mixture, like any real liquid mixture, is conveniently described by the use of excess functions. For a binary mixture these are defined by

$$G_m^E = \Delta_m G - \Delta_m G^I = \Delta_m G - RT\{(1-x)\ln(1-x) + x \ln x\} \qquad 4.47.1$$

$$S_m^E = \Delta_m S - \Delta_m S^I = \Delta_m S + R\{(1-x)\ln(1-x) + x \ln x\} \qquad 4.47.2$$

$$H_m^E = \Delta_m H - \Delta_m H^I = \Delta_m H. \qquad 4.47.3$$

Solid mixtures may be classed, like liquid mixtures, into symmetrical mixtures, including simple mixtures, and unsymmetrical mixtures.

CHAPTER 5

SOLUTIONS, ESPECIALLY DILUTE SOLUTIONS

§5.01 Introduction

There is no fundamental difference between a liquid mixture and a solution. The difference is in the manner of description. In the description of mixtures in the previous chapter all the constituent species were treated in a like manner. In the description of solutions in the present chapter we shall on the contrary single out one species which we call the solvent. All the remaining species are called solutes. There is no rigid rule to determine which species shall be regarded as solvent, but it is usually the species present in the highest proportion, at least among those species which are liquid in the pure state at the given temperature and pressure. For example we should at room temperature speak of water as solvent and urea as solute, even if the urea were in excess, because pure urea at room temperature is a solid.

We shall always denote the solvent by the subscript $_1$ and the solutes by the subscripts $_2$, $_3$, ... in particular or by the general subscript $_s$.

§5.02 Mole ratios and molalities

We consider a phase containing an amount n_1 of the species 1, an amount n_2 of the species 2, and so on. When considering such a phase as a mixture we described the composition by the fractions $n_1/\Sigma_i n_i$, $n_2/\Sigma_i n_i$ and so on. When considering this phase as a solution we shall on the contrary describe its composition by the ratios n_2/n_1, n_3/n_1 and so on.

While the fractions $n_1/\Sigma_i n_i$, $n_2/\Sigma_i n_i$ were denoted by x_1, x_2,... and were called *mole fractions*, the fractions n_2/n_1, n_3/n_1, ... will be denoted by r_2, r_3, ... and will be called *solute–solvent mole ratios*. In a phase of c component species there are c different mole fractions, which we recall are related by $\Sigma_i x_i = 1$ so that only c−1 are independent. There are only c−1 mole ratios r_2, r_3, ... r_c and these are all independent. We shall use these as the independent variables together with T and P.

SOLUTIONS, ESPECIALLY DILUTE SOLUTIONS 221

For all purposes of general theory we shall use the variables $r_2, r_3, \ldots r_c$ together with T, P. For practical purposes, it is customary to use instead of r_s a quantity m_s directly proportional to r_s defined by

$$m_s = r_s/r^\ominus \qquad 5.02.1$$

where $r^\ominus$ is a standard value of r_s customarily and here defined by

$$r^\ominus = M_1/\text{kg mole}^{-1} \qquad 5.02.2$$

where M_1 is the proper mass of the solvent. It follows that $r_s = r^\ominus$ when there is one mole of the solute s for each kilogramme of solvent. Thus defined, r_s, $r^\ominus$, and m_s are all dimensionless. This quantity m_s is called *molality*. We shall derive most of our formulae in terms of the mole ratios r and shall transcribe only a few important ones into terms of molalities m by use of (1). The mole fractions x and mole ratios r are interrelated by

$$r_2 = x_2/x_1 = x_2/(1 - \sum_s x_s) \qquad 5.02.3$$

$$x_2 = r_2/(1 + \sum_s r_s) \qquad 5.02.4$$

and similar relations for the other solute species. We also note the relation

$$(1 + \sum_s r_s)(1 - \sum_s x_s) = 1. \qquad 5.02.5$$

§5.03 *Partial and apparent quantities*

If X denotes any extensive property such as V, U, S, F, G then the corresponding partial quantities are defined in §4.03 by

$$X_1 = (\partial X/\partial n_1)_{T, P, n_2, n_3}, \ldots \qquad 5.03.1$$

$$X_s = (\partial X/\partial n_s)_{T, P, n_1, n_2}, \ldots \qquad 5.03.2$$

According to (4.03.6) these are interrelated by

$$n_1 dX_1 + \sum_s n_s dX_s = 0 \qquad (T, P \text{ const.}). \qquad 5.03.3$$

Dividing (3) by n_1 we have

$$dX_1 + \sum_s r_s dX_s = 0 \qquad (T, P \text{ const.}). \qquad 5.03.4$$

In particular, if there is only one solute species 2 formula (4) reduces to

$$\partial X_1/\partial r + r \partial X_2/\partial r = 0 \qquad (T, P \text{ const.}). \qquad 5.03.5$$

When there is only one solute species 2 the quantity X_ϕ defined by

$$n_2 X_\phi = X - n_1 X_1^0 \qquad 5.03.6$$

where X_1^0 denotes the proper X of the pure solvent, is called the *apparent proper X of the solute**. We can obtain the relation between X_ϕ and X_2 by differentiating (6) with respect to n_2 keeping n_1 constant. We find

$$X_2 = \partial X/\partial n_2 = X_\phi + n_2(\partial X_\phi/\partial n_2)_{n_1} = X_\phi + r\, dX_\phi/dr. \qquad 5.03.7$$

There is no such quantity as X_ϕ when there is more than one solute species.

We recall the important relations between the several partial quantities in §4.04. These apply both to the partial quantities of the solvent and to those of the solute species.

We also recall the important equality (1.28.11) which holds both for the solvent and for the solute species

$$\mu_1 = G_1 \qquad \mu_s = G_s \qquad 5.03.8$$

with the consequent relations

$$\partial \mu_1/\partial T = -S_1 \qquad \partial \mu_s/\partial T = -S_s \qquad 5.03.9$$

$$\partial \ln \lambda_1/\partial T = -H_1/RT^2 \qquad \partial \ln \lambda_s/\partial T = -H_s/RT^2 \qquad 5.03.10$$

$$\partial \mu_1/\partial P = V_1 \qquad \partial \mu_s/\partial P = V_s. \qquad 5.03.11$$

§5.04 Gibbs–Duhem relation

We recall the Gibbs–Duhem relation

$$n_1 d\mu_1 + \sum_s n_s d\mu_s = 0 \qquad (T, P \text{ const.}) \qquad 5.04.1$$

or

$$n_1 d \ln \lambda_1 + \sum_s n_s d \ln \lambda_s = 0 \qquad (T, P \text{ const.}). \qquad 5.04.2$$

Dividing through by n_1 we obtain the alternative form

$$d\mu_1 + \sum_s r_s d\mu_s = 0 \qquad (T, P \text{ const.}) \qquad 5.04.3$$

or

$$d \ln \lambda_1 + \sum_s r_s d \ln \lambda_s = 0 \qquad (T, P \text{ const.}). \qquad 5.04.4$$

In the case of a single solute (3) reduces to

$$d\mu_1/dr + r\, d\mu_2/dr = 0 \qquad (T, P \text{ const.}) \qquad 5.04.5$$

or

$$d \ln \lambda_1/dr + r\, d \ln \lambda_2/dr = 0 \qquad (T, P \text{ const.}). \qquad 5.04.6$$

* The notation ϕ_v used by some authors instead of V_ϕ is deplorable.

§5.05 Partial quantities at high dilution

If X denotes any extensive property, so that X_1, X_2 are the corresponding partial quantities in a solution of a single solute, we have according to (5.03.5)

$$\partial X_1/\partial r + r\partial X_2/\partial r = 0 \qquad (T, P \text{ const.}). \qquad 5.05.1$$

It follows that when $r \to 0$ either $\partial X_1/\partial r \to 0$ or $\partial X_2/\partial r \to \infty$. The former case occurs when X denotes V or U or H; the latter occurs when X denotes S or F or G.

§5.06 Ideal dilute solutions

Let us consider a solution of a volatile solute 2 in the solvent 1 so dilute that $n_2 \ll n_1$. It is then physically obvious that, if the vapour may be regarded as a perfect gas, the partial pressure $y_2 P_{\text{sat}}$ of the solute will be directly proportional to n_2. More generally, whether or not the vapour may be regarded as a perfect gas, the fugacity p_2 of 2 will be directly proportional to n_2, that is to say

$$p_2 \propto n_2. \qquad 5.06.1$$

This however raises the question whether the proportionality (1) holds at constant $n_1 + n_2$, that is to say

$$p_2 \propto x_2 \qquad 5.06.2$$

or at constant n_1, that is to say

$$p_2 \propto r_2. \qquad 5.06.3$$

The answer is that in the limit as x_2 and r_2 tend to zero (2) and (3) become equivalent and it is only in this limit that either is obviously true. At finite values of x_2 and r_2 we must not expect either (2) or (3) to be accurate, but we may use either as a basis for comparison with the actual behaviour of solutions. It is true that (2) can under favourable conditions hold for all values of x from 0 to 1 in which case we have an ideal mixture as described in §4.18. Formula (3) on the contrary becomes untenable as we approach the state of the pure liquid 2, when $r_2 \to 1$, since it would lead to the absurd conclusion that p_2^0 is infinite. We must however remember that in this chapter our convention that the species 1 is the solvent implies that this species is present in excess and we are consequently not concerned with conditions approaching that of the pure liquid 2. In fact we are concerned mainly with the condition $r_2 \ll 1$. Bearing in mind this implied restriction we are free to choose either (2) or (3) as a basis of comparison with actual behaviour.

In practice it has been found that (3) is more convenient than (2) because the value of r_2 is unaffected by the addition to a given solution of other solute species. This practical advantage will become clearer in the next chapter when we consider chemical reactions between solute species.

We shall accordingly choose as a basis of comparison with actual behaviour formula (3) after we have generalized it for several solute species when it becomes

$$p_s \propto r_s \quad (T, P \text{ const.}). \qquad 5.06.4$$

It is clear from the relations in §4.12 that (4) is equivalent to

$$\lambda_s \propto r_s \quad (T, P \text{ const.}) \qquad 5.06.5$$

and (5) is applicable to solute species of immeasurably low volatility. We shall now adopt (5) as our basis of comparison with actual behaviour and we define a solution as being *ideal dilute* when the proportionality (5) is obeyed for all values of r_s less than or equal to that of the solution under consideration.

We can write (5) in the alternative form

$$\lambda_s = \lambda_s^\infty r_s \qquad 5.06.6$$

where λ_s^∞ depends on the nature of the solute, the nature of the solvent, the temperature, and the pressure, but not on the mole ratio r_s of the solute considered nor on the mole ratio of any other solute species. In numerical applications, as opposed to general theory, it is customary to use molalities instead of mole ratios. We then replace (6) by

$$\lambda_s = \lambda_s^\ominus m_s \qquad 5.06.7$$

$$\lambda_s^\ominus = \lambda_s^\infty / r^\ominus \qquad 5.06.8$$

where $r^\ominus$ is defined by (5.02.1).

Finally we may, if we prefer, use chemical potentials instead of absolute activities. We then have in analogous notation

$$\mu_s = \mu_s^\infty + RT \ln r_s \qquad 5.06.9$$

$$\mu_s = \mu_s^\ominus + RT \ln m_s \qquad 5.06.10$$

$$\mu_s^\ominus = \mu_s^\infty - RT \ln r^\ominus. \qquad 5.06.11$$

§5.07 Thermodynamic functions of ideal dilute solutions

Having defined an ideal dilute solution in terms of the absolute activities or chemical potentials of the solute species, we can deduce the relations for the properties of the solvent by means of the Gibbs–Duhem relation (5.04.4)

$$d \ln \lambda_1 + \sum_s r_s d \ln \lambda_s = 0 \qquad (T, P \text{ const.}). \qquad 5.07.1$$

From (5.06.5) we have

$$r_s d \ln \lambda_s = r_s d \ln r_s = dr_s \qquad (T, P \text{ const.}). \qquad 5.07.2$$

Substituting (2) into (1) we obtain

$$d \ln \lambda_1 = -\sum_s dr_s = -d(\sum_s r_s) \qquad (T, P \text{ const.}) \qquad 5.07.3$$

and so by integration

$$-\ln(\lambda_1/\lambda_1^0) = \sum_s r_s \qquad (T, P \text{ const.}) \qquad 5.07.4$$

where as usual the superscript ^{0} denotes the value for the pure liquid solvent. We can rewrite (4) in terms of chemical potentials

$$\mu_1 = \mu_1^0 - RT \sum_s r_s \qquad (T, P \text{ const.}) \qquad 5.07.5$$

and in terms of absolute activities

$$\ln \lambda_1 = \ln \lambda_1^0 - \sum_s r_s. \qquad 5.07.6$$

By use of (5.03.11) we deduce

$$V_1 = V_1^0 \qquad (T, P \text{ const.}) \qquad 5.07.7$$

$$V_s = V_s^\infty \qquad (T, P \text{ const.}) \qquad 5.07.8$$

where V_s^∞ denotes the limiting value of V_s at infinite dilution of all solute species. We see then that V_1 and V_s are in an ideal dilute solution in a given solvent independent of the composition.

By use of (5.03.9) we deduce

$$S_1 = S_1^0 + R \sum_s r_s \qquad (T, P \text{ const.}) \qquad 5.07.9$$

$$S_s = S_s^\infty - R \ln r_s \qquad (T, P \text{ const.}) \qquad 5.07.10$$

where S_s^∞ is defined by

$$S_s^\infty = -\partial \mu_s^\infty / \partial T \qquad 5.07.11$$

and is thus in an ideal dilute solution in a given solvent independent of the composition.

By use of (5.03.10) we deduce

$$H_1 = H_1^0 \qquad 5.07.12$$

$$H_s = H_s^\infty \qquad 5.07.13$$

where H_s^∞ denotes the limiting value of H_s at infinite dilution of all solute species.

It follows from (7), (8) and from (12), (13) that any two ideal dilute solutions in the same solvent mix at constant temperature and pressure without change of volume and without change of enthalpy.

§5.08 Real solutions

As already mentioned, we do not expect a real solution to be ideal dilute except in the limit of infinite dilution but it is convenient to compare the behaviour of any real solution with its hypothetical behaviour if it remained ideal dilute at all compositions extending from infinite dilution to its actual composition. We then express the deviations between the real behaviour and this hypothetical behaviour by means of certain coefficients as will be described in the succeeding sections.

§5.09 Activity coefficients of solute species

We define the activity coefficient γ_2 of the solute species 2 by the relations

$$\lambda_2 = \lambda_2^\infty r_2 \gamma_2 = \lambda_2^\ominus m_2 \gamma_2 \qquad (T, P \text{ const.}) \qquad 5.09.1$$

$$\gamma_2 \to 1 \quad \text{as} \quad \sum_s m_s \to 0. \qquad 5.09.2$$

Alternatively we may express (1) in terms of chemical potentials

$$\mu_2 = \mu_2^\infty + RT \ln(r_2 \gamma_2) = \mu_2^\ominus + RT \ln(m_2 \gamma_2) \qquad (T, P \text{ const.}) \quad 5.09.3$$

in conjunction with (2). It is clear that the deviation of γ_2 from unity or of $\ln \gamma_2$ from zero is a measure of deviation from an ideal dilute solution.

It need hardly be mentioned that similar relations hold for every solute species. Thus (1) may be generalized to

$$\lambda_s = \lambda_s^\infty r_s \gamma_s = \lambda_s^\ominus m_s \gamma_s \qquad (T, P \text{ const.}) \qquad 5.09.4$$

and (3) to

$$\mu_s = \mu_s^\infty + RT \ln(r_s \gamma_s) = \mu_s^\ominus + RT \ln(m_s \gamma_s) \qquad (T, P \text{ const.}). \quad 5.09.5$$

It is unfortunate that the same name *activity coefficient* is sometimes used for the quantity a/x of the previous chapter as well as for γ_s.

§5.10 Osmotic coefficient of solvent

Following Bjerrum* we define the osmotic coefficient ϕ of the solvent by

$$-\ln(\lambda_1/\lambda_1^0) = -\ln(p_1/p_1^0) = \phi \sum_s r_s \qquad (T, P \text{ const.}). \qquad 5.10.1$$

By comparing (1) with (5.07.4) we see that $\phi - 1$ is a measure of deviation of behaviour from that of an ideal dilute solution.

We can also write (1) in terms of chemical potentials as

$$\mu_1 = \mu_1^0 - RT\phi \sum_s r_s \qquad (T, P \text{ const.}). \qquad 5.10.2$$

§5.11 Relation between activity coefficients and osmotic coefficient

When we substitute (5.10.1) and (5.09.4) into the Gibbs–Duhem relation (5.04.4) we obtain

$$d(\phi \sum_s r_s) = \sum_s r_s d \ln(r_s \gamma_s) \qquad (T, P \text{ const.}) \qquad 5.11.1$$

which can be rewritten as

$$d\{(\phi-1) \sum_s r_s\} = \sum_s r_s d \ln \gamma_s \qquad (T, P \text{ const.}) \qquad 5.11.2$$

a relation due to Bjerrum.† In the case of a single solute species (2) reduces to

$$d\{(\phi-1)r\}/dr = r \, d \ln \gamma / dr \qquad (T, P \text{ const.}) \qquad 5.11.3$$

or

$$(\phi-1)/r + d\phi/dr = d \ln \gamma/dr \qquad (T, P \text{ const.}). \qquad 5.11.4$$

If for example ϕ is related to r by

$$\phi - 1 = Ar^n \qquad (A, n \text{ const.}) \qquad 5.11.5$$

then by substituting from (5) into (4) we obtain after integration

$$\ln \gamma = (1+n^{-1})Ar^n = (1+n^{-1})(\phi-1). \qquad 5.11.6$$

§5.12 Temperature dependence

By substitution of (5.09.4) into the second of equations (5.03.10) we obtain

$$\partial \ln \lambda_s^\infty/\partial T + \partial \ln \gamma_s/\partial T = -H_s/RT^2. \qquad 5.12.1$$

---

* Bjerrum, Fysisk Tidskr. 1916 **15** 66; Z. Electrochem. 1918 **24** 325.
† Bjerrum, Z. Phys. Chem. 1923 **104** 406.

In the limit of infinite dilution (1) reduces to

$$\partial \ln \lambda_s^\infty / \partial T = -H_s^\infty / RT^2. \qquad 5.12.2$$

Subtracting (2) from (1) we obtain for the temperature dependence of the activity coefficient

$$\partial \ln \gamma_s / \partial T = -(H_s - H_s^\infty)/RT^2. \qquad 5.12.3$$

In particular we observe that γ_s will be independent of temperature if H_s is independent of composition.

By substitution of (5.10.1) into the first of equations (5.03.10) we obtain

$$-\sum_s r_s \partial \phi / \partial T + \partial \ln \lambda_1^0 / \partial T = -H_1/RT^2. \qquad 5.12.4$$

For the pure solvent (4) reduces to

$$\partial \ln \lambda_1^0 / \partial T = -H_1^0 / RT^2. \qquad 5.12.5$$

By subtraction of (5) from (4) we find for the temperature dependence of the osmotic coefficient

$$\sum_s r_s \partial \phi / \partial T = (H_1 - H_1^0)/RT^2. \qquad 5.12.6$$

In particular we observe that ϕ will be independent of temperature if H_1 is independent of composition.

§5.13 Pressure dependence

By substitution of (5.09.5) into the second of equations (5.03.11) we obtain

$$\partial \mu_s^\infty / \partial P + RT \partial \ln \gamma_s / \partial P = V_s. \qquad 5.13.1$$

In the limit of infinite dilution (1) reduces to

$$\partial \mu_s^\infty / \partial P = V_s^\infty. \qquad 5.13.2$$

By subtraction of (2) from (1) we find for the pressure dependence of the activity coefficient

$$\partial \ln \gamma_s / \partial P = (V_s - V_s^\infty)/RT. \qquad 5.13.3$$

By substitution of (5.10.2) into the first of equations (5.03.11) we obtain

$$\partial \mu_1^0 / \partial P - RT \sum_s r_s \partial \phi / \partial P = V_1. \qquad 5.13.4$$

For the pure solvent (4) reduces to

$$\partial \mu_1^0 / \partial P = V_1^0. \qquad 5.13.5$$

By subtraction of (5) from (4) we obtain for the pressure dependence of the osmotic coefficient

$$\sum_s r_s \partial\phi/\partial P = -(V_1 - V_1^0)/RT. \qquad 5.13.6$$

All these pressure dependences are usually negligible at ordinary pressures.

§5.14 Temperature dependence of fugacity of solvent

From (5.10.1) we have

$$-\ln(p_1/p_1^0) = \phi \sum_s r_s. \qquad 5.14.1$$

Differentiating (1) with respect to T and using (5.12.6) we obtain for a solution of given composition

$$-\partial \ln(p_1/p_1^0)/\partial T = \sum_s r_s \partial\phi/\partial T = (H_1 - H_1^0)/RT^2. \qquad 5.14.2$$

§5.15 Temperature dependence of fugacity of solute

For a volatile solute species we may replace (5.09.4) by

$$p_s = p_s^\infty r_s \gamma_s = p_s^\ominus m_s \gamma_s \qquad 5.15.1$$

where p_s^∞, $p_s^\ominus$ are independent of the composition but depend on the nature of the solute s and the solvent. Differentiating (1) with respect to T and using (5.12.3) we obtain

$$-\partial \ln(p_s/p_s^\infty)/\partial T = -\partial \ln(p_s/p_s^\ominus)/\partial T = (H_s - H_s^\infty)/RT^2. \qquad 5.15.2$$

§5.16 Osmotic pressure

We recall formula (4.14.8)

$$\Pi \langle V_1 \rangle / RT = \ln(p_1^0/p_1) \qquad 5.16.1$$

where $\langle V_1 \rangle$ denotes the value of V_1 at a pressure equal to the mean of the pressure P on the pure solvent and the pressure $P+\Pi$ on the solution at osmotic equilibrium while both p_1 and p_1^0 are values at an external pressure P. Since formula (1) does not contain mole fractions it is equally applicable to solutions described in terms of mole ratios.

Substituting (5.10.1) into (1) we obtain

$$\Pi \langle V_1 \rangle / RT = \phi \sum_s r_s. \qquad 5.16.2$$

If we use the superscript ^{id} to denote a hypothetical ideal dilute solution with the same composition as the actual solution we have

$$\Pi^{\mathrm{id}}\langle V_1\rangle/RT=\sum_s r_s. \qquad 5.16.3$$

Dividing (2) by (3) we find

$$\Pi=\phi\Pi^{\mathrm{id}} \qquad 5.16.4$$

and this relation explains the origin of the name *osmotic coefficient*.

§5.17 Freezing point

Let us now consider the equilibrium between the liquid solution and the pure solid solvent 1. We assume that the pressure is either constant or irrelevant. We use the superscript ^{S} to denote the solid phase, the superscript ^{0} for the pure liquid, and no superscript for the liquid mixture. Then for equilibrium between the pure solid and the liquid mixture at its freezing point temperature T

$$\lambda_1(T)=\lambda_1^S(T). \qquad 5.17.1$$

If T^0 denotes the corresponding equilibrium temperature for the pure liquid, that is to say the freezing point of the pure liquid, we have correspondingly

$$\lambda_1^0(T^0)=\lambda_1^S(T^0). \qquad 5.17.2$$

Dividing (2) by (1) we obtain

$$\lambda_1^0(T^0)/\lambda_1(T)=\lambda_1^S(T^0)/\lambda_1^S(T) \qquad 5.17.3$$

which can be rewritten in the form

$$\lambda_1^0(T)/\lambda_1(T)=\{\lambda_1^0(T)/\lambda_1^0(T^0)\}/\{\lambda_1^S(T)/\lambda_1^S(T^0)\}. \qquad 5.17.4$$

Taking logarithms we have

$$\ln\{\lambda_1^0(T)/\lambda_1(T)\}=\ln\{\lambda_1^0(T)/\lambda_1^0(T^0)\}-\ln\{\lambda_1^S(T)/\lambda_1^S(T^0)\}. \qquad 5.17.5$$

Now applying the first relation (5.03.10)

$$\partial \ln \lambda_1/\partial T=-H_1/RT^2 \qquad 5.17.6$$

to the pure solid and pure liquid in turn and integrating we obtain

$$\ln\{\lambda_1^S(T)/\lambda_1^S(T^0)\}=-\int_{T^0}^T (H_1^S/RT^2)\mathrm{d}T \qquad 5.17.7$$

$$\ln\{\lambda_1^0(T)/\lambda_1^0(T^0)\}=-\int_{T^0}^T (H_1^0/RT^2)\mathrm{d}T. \qquad 5.17.8$$

Substituting (7) and (8) into (5) we obtain

$$\ln\{\lambda_1^0(T)/\lambda_1(T)\}=-\int_{T^0}^T \{(H_1^0-H_1^S)/RT^2\}\mathrm{d}T=-\int_{T^0}^T (\Delta_f H_1^0/RT^2)\mathrm{d}T \qquad 5.17.9$$

where $\Delta_f H_1^0$ is the proper enthalpy of fusion of the pure solvent. We now substitute (5.10.1) into (9) and obtain

$$\phi \sum_s r_s = -\int_{T^0}^{T} (\Delta_f H_1^0/RT^2) dT \qquad 5.17.10$$

where ϕ denotes the osmotic coefficient of the solution at its freezing point. We can rewrite (10) more simply as

$$\phi \sum_s r_s = \langle \Delta_f H_1^0 \rangle (1/RT - 1/RT^0) \qquad 5.17.11$$

where $\langle \Delta_f H_1^0 \rangle$ denotes the average value of $\Delta_f H_1^0$ over the reciprocal temperature interval $1/T^0$ to $1/T$. Since $\Delta_f H_1^0$ is always positive it follows that $T < T^0$. Thus the freezing point of the solution is always below that of the pure solvent if the solid phase is pure solvent.

For dilute solutions when $T^0 - T \ll T^0$ we may replace (11) by the approximation

$$\phi \sum_s r_s = \Delta_f H_1^0 (T^0 - T)/RT^{0\,2} \qquad 5.17.12$$

or

$$T^0 - T = \phi \sum_s r_s (RT^{0\,2}/\Delta_f H_1^0). \qquad 5.17.13$$

In numerical calculations it is customary to use the molalities m_s instead of the solute–solvent mole ratios r_s. We recall the definition (5.02.1) of molality

$$m_s = r_s/r^\ominus \qquad 5.17.14$$

where $r^\ominus$ is a standard value of r_s customarily chosen so that $r_s = r_s^\ominus$ when there is one mole of the solute s for each kilogramme of solvent. We accordingly rewrite (13) as

$$T^0 - T = \phi \sum_s m_s (r^\ominus RT^{0\,2}/\Delta_f H_1^0). \qquad 5.17.15$$

The factor $r^\ominus RT^{0\,2}/\Delta_f H_1^0$ which is a property of the solvent but common to all solute species, is called the *cryoscopic constant*. We note that when $r^\ominus$ is given the value $M_1/\text{kg mole}^{-1}$ then $\Delta_f H_1^0/r^\ominus$ is numerically equal to the enthalpy of fusion in joules per kilogramme of the solvent. For water we have

$$RT^0 = 2.2712 \times 10^3 \text{ J mole}^{-1}$$
$$T^0 = 273.15 \text{ K}$$
$$\Delta_f H_1^0/r^\ominus = 3.335 \times 10^5 \text{ J mole}^{-1}$$

so that the cryoscopic constant is

$$2.2712 \times 10^3 \text{ J mole}^{-1} \times 273.15 \text{ K}/3.335 \times 10^5 \text{ J mole}^{-1} = 1.860 \text{ K}. \qquad 5.17.16$$

§5.18 Boiling point

We shall now consider the equilibrium between the liquid solution and the gas phase in the case that all the solute species have negligible vapour pressures. We accordingly regard the gas phase as consisting entirely of the component 1 and we use the superscript ^{G} to denote this phase.

We then proceed to consider the equilibrium between the two phases at a given pressure precisely as in the case of equilibrium with a pure solid phase studied in the previous section. The steps of the argument are precisely analogous and we obtain eventually the relation

$$\phi \sum_s r_s = \langle \Delta_e H_1^0 \rangle (1/RT^0 - 1/RT) \qquad 5.18.1$$

where $\langle \Delta_e H_1^0 \rangle$ denotes the value of the proper enthalpy of evaporation $\Delta_e H_1^0$ for the pure liquid averaged over the reciprocal temperature interval $1/T$ to $1/T^0$ and ϕ is the osmotic coefficient at the boiling point of the solution. Since $\Delta_e H_1^0$ is always positive it follows that $T > T^0$. Thus the boiling point of any solution of non-volatile solutes is above that of the pure solvent.

For dilute solutions when $T - T^0 \ll T^0$ we may replace (1) by the approximation

$$\phi \sum_s r_s = \Delta_e H_1^0 (T - T^0)/RT^{02} \qquad 5.18.2$$

or

$$T - T^0 = \phi \sum_s r_s (RT^{02}/\Delta_e H_1^0). \qquad 5.18.3$$

For purposes of numerical calculation it is customary to use the molalities m_s instead of the solute–solvent mole ratios r_s. We accordingly rewrite (3) as

$$T - T^0 = \phi \sum_s m_s (r^\ominus RT^{02}/\Delta_e H_1^0). \qquad 5.18.4$$

The factor $r^\ominus RT^{02}/\Delta_e H_1^0$ is called the *ebullioscopic constant* of the solvent. We note that when $r^\ominus$ is as usual chosen to be M_1 mole kg^{-1} then $\Delta_e H_1^0/r^\ominus$ is numerically equal to the enthalpy of evaporation in joules per kilogramme of solvent. For water we have

$$RT^0 = 3.1026 \times 10^3 \text{ J mole}^{-1}$$
$$T^0 = 373.15 \text{ K}$$
$$\Delta_e H_1^0/r^\ominus = 2.2567 \times 10^6 \text{ J mole}^{-1}$$

so that the ebullioscopic constant is

$$3.1026 \times 10^3 \text{ J mole}^{-1} \times 373.15 \text{ K}/2.2567 \times 10^6 \text{ J mole}^{-1} = 0.513 \text{ K}. \qquad 5.18.5$$

§5.19 Distribution between two solvents

For the equilibrium of a solute species s between two solutions in different solvents we have

$$\lambda_s^\alpha = \lambda_s^\beta \qquad 5.19.1$$

where the superscripts $^\alpha$ and $^\beta$ relate to the two phases. Substituting from (5.09.4) into (1) we obtain

$$\lambda_s^{\ominus\alpha} m_s^\alpha \gamma_s^\alpha = \lambda_s^{\ominus\beta} m_s^\beta \gamma_s^\beta \qquad 5.19.2$$

or by rearrangement

$$m_s^\beta \gamma_s^\beta / m_s^\alpha \gamma_s^\alpha = l_s^{\alpha\beta} \qquad 5.19.3$$

where $l_s^{\alpha\beta}$ is independent of the composition of the two phases and is defined by

$$l_s^{\alpha\beta} = \lambda_s^{\ominus\alpha} / \lambda_s^{\ominus\beta}. \qquad 5.19.4$$

In the special case that both solutions are ideal dilute (3) reduces to

$$m_s^\beta / m_s^\alpha = l_s^{\alpha\beta} \qquad 5.19.5$$

which is known as *Nernst's distribution law*.

§5.20 Solubility of pure solid

For the equilibrium with respect to the species s between a solution and the pure solid phase we have the condition

$$\lambda_s = \lambda_s^S \qquad 5.20.1$$

where the superscript ^{S} denotes the pure solid phase. Substituting from (5.09.4) into (1) we obtain

$$m_s \gamma_s = \lambda_s^S / \lambda_s^\ominus \qquad \text{(saturated solution)}. \qquad 5.20.2$$

From (2) we see that if several solutions in the same solvent at the same temperature are all saturated with the same solid phase of the species s, then in all these solutions $m_s \gamma_s$ has the same value.

Taking logarithms of (2), differentiating with respect to T and using (5.12.3) we obtain

$$\partial \ln(m_s \gamma_s)/\partial T = (H_s^\infty - H_s^S)/RT^2 \qquad \text{(saturated solution)} \qquad 5.20.3$$

and we observe that the quantity $H_s^\infty - H_s^S$ occurring as the numerator on the right is the *proper enthalpy of dissolution* of s at infinite dilution.

§5.21 Experimental determination of ϕ

The most accurate direct method of determining ϕ experimentally is by measurements of freezing point and use of formula (5.17.11) which for a single solute species reduces to

$$\phi m_s = \phi r_s/r^\ominus = (T^0 - T)(\langle \Delta_f H_1^0 \rangle / r^\ominus RT^0 T). \quad 5.21.1$$

All the quantities m_s, $\Delta_f H_1^0$, T^0, T can be measured and substitution of their values into (1) leads to experimental values of ϕ at the freezing point.

Let us suppose that freezing-point measurements have been made so as to determine ϕ over a range of steadily decreasing values of m and let us consider what results are to be expected.

Since we know that as $m \to 0$ so $\phi \to 1$ we may reasonably expect that $\phi - 1$ can be expressed as a series of integral powers of m say

$$\phi - 1 = A_1 m + A_2 m^2 + \ldots. \quad 5.21.2$$

This is in fact the case for non-electrolytes and we may then hope to determine by a series of accurate freezing-point measurements the coefficients in such a formula as (2) so as to obtain a good fit. Formula (2) is not applicable to solutions of electrolytes; these will be discussed in chapter 7.

Let us now consider what will happen if the measurements are extended down to gradually decreasing values of m. If the measurements are performed with sufficient care, we may expect to reach a range where all terms of (2) are negligible except the first. In this range $(\phi - 1)/m$ has a constant value A_1 and we may confidently and reasonably assume that this behaviour persists down to $m = 0$. Suppose however we tried to confirm this experimentally, let us examine what would happen.

We may reasonably assume that the experimental error in measuring $T^0 - T$ is at least roughly independent of m. Since at low values of m the value of $T^0 - T$ is itself roughly proportional to m it follows that the fractional experimental error in ϕ is inversely proportional to m. Hence according to (2) the fractional error in $\phi - 1$ will be inversely proportional to m^2. It is therefore clear that by proceeding to experiment at smaller values of m we eventually reach a stage where the experiments tell us nothing.

The most reasonable procedure is then to carry the experiments down to values of m where one finds experimentally

$$(\phi - 1)/m = A_1 \quad (A_1 \text{ const.}) \quad 5.21.3$$

and then assume that this simple law persists down to $m = 0$.

We may mention that for solutions of non-electrolytes the limiting law

SOLUTIONS, ESPECIALLY DILUTE SOLUTIONS 235

(3) has not merely an empirical basis, but also a theoretical one based on statistical mechanics.

§5.22 Determination of γ from ϕ

We recall Bjerrum's relation for a single solute species (5.11.3)

$$d\{(\phi-1)r\}/dr = r\, d\ln\gamma/dr \qquad (T, P \text{ const.}) \qquad 5.22.1$$

which we may also write as

$$d\{(\phi-1)m\}/dm = m\, d\ln\gamma/dm \qquad (T, P \text{ const.}) \qquad 5.22.2$$

or as

$$d\ln\gamma = [d\{(\phi-1)m\}/dm]\, d\ln m \qquad (T, P \text{ const.}). \qquad 5.22.3$$

Integrating (3) from 0 to m and observing that $\phi-1$ and $\ln\gamma$ tend to zero as m tends to zero, we obtain

$$\ln\gamma = \int_0^m [d\{(\phi-1)m\}/dm]\, d\ln m = \phi - 1 + \int_0^m (\phi-1)\, d\ln m. \qquad 5.22.4$$

If ϕ has been determined at all molalities from 0 to m we see that by using (4) we can in principle calculate γ at a molality m but caution is required so as to avoid spurious results. We saw in the previous section that with regard to the experimental determination of ϕ there are three ranges of m arranged in order of decreasing m with the following characteristics.

1. Large molalities, where ϕ can be measured and fitted to a more or less complicated formula.
2. Intermediate molalities, where ϕ can be fitted to the formula

$$\phi - 1 = A_1 m \qquad (A_1 \text{ const.}). \qquad 5.22.5$$

3. Lowest molalities, where no useful information about ϕ can be obtained by experiment and we assume that (5) continues to hold.

In using (4) it is expedient to break the range of integration at some value m' of m in the range where (5) is found to hold. We accordingly rewrite (4) as

$$\ln\gamma = \phi - 1 + \int_0^{m'} (\phi-1)\, d\ln m + \int_{m'}^m (\phi-1)\, d\ln m. \qquad 5.22.6$$

We evaluate the first integral as follows

$$\int_0^{m'} (\phi-1)\, d\ln m = \int_0^{m'} A_1\, dm = A_1 m' = \phi' - 1 \qquad 5.22.7$$

where ϕ' denotes the value of ϕ at $m=m'$. Using (7) in (6) we obtain finally

$$\ln \gamma = \phi - 1 + \phi' - 1 + \int_{m'}^{m} (\phi - 1) \, d \ln m \qquad 5.22.8$$

and the integral in (8) can be evaluated from the experimental values of ϕ, either by fitting these to a formula or graphically.

The important point emerging from this discussion is that we cannot calculate γ from experimental determinations of ϕ for example by freezing-point measurements, without making an assumption concerning ϕ at low values of m. Since such an assumption has to be made anyway, it is just as well to make it explicitly and so obtain a closed formula for γ as well as for ϕ in the range of small m. For solutions of non-electrolytes, with which we are here concerned, the usual and most reasonable assumption is formula (5). In chapter 7, when we study solutions of electrolytes, we shall meet a different situation.

§5.23 Fugacity of saturated solution

Throughout this chapter and the previous one we have never yet considered any equilibrium involving more than two bulk phases, nor shall we do so in any detail. No new principles are involved and the methods already described are applicable. We shall confine ourselves to a single interesting example.

We consider the following problem. How does the fugacity of the solvent vary with the temperature in a solution kept saturated with a single non-volatile solid? Using the subscripts $_1$ for the solvent, $_2$ for the solute, and the superscripts ^{G} for the gas phase, ^{S} for the solid, and none for the solution, we have for variations maintaining equilibrium

$$d \ln \lambda_1 = d \ln \lambda_1^G \qquad 5.23.1$$

$$d \ln \lambda_2 = d \ln \lambda_2^S. \qquad 5.23.2$$

Expanding these, and neglecting the effect of pressure on each of the condensed phases we have

$$-(H_1/RT^2)dT + (\partial \ln \lambda_1/\partial m_2)dm_2 = -(H_1^G/RT^2)dT + d \ln p_1 \qquad 5.23.3$$

$$-(H_2/RT^2)dT + (\partial \ln \lambda_2/\partial m_2)dm_2 = -(H_2^S/RT^2)dT. \qquad 5.23.4$$

Using $\Delta_e H$ to denote the proper enthalpy of evaporation from the solution and $\Delta_f H$ to denote the proper enthalpy of fusion into the solution, we can write (3) and (4) as

$$d \ln p_1 = (\partial \ln \lambda_1/\partial m_2)dm_2 + (\Delta_e H_1/RT^2)dT \qquad 5.23.5$$

$$(\partial \ln \lambda_2/\partial m_2)dm_2 = (\Delta_f H_2/RT^2)dT. \qquad 5.23.6$$

SOLUTIONS, ESPECIALLY DILUTE SOLUTIONS 237

We now use the Gibbs–Duhem relation in the form

$$\partial \ln \lambda_1/\partial m_2 + r(\partial \ln \lambda_2/\partial m_2) = 0 \qquad 5.23.7$$

to eliminate λ_1, λ_2 from (5), (6). We thus obtain

$$d \ln p_1/dT = (\Delta_e H_1 - r\Delta_f H_2)/RT^2. \qquad 5.23.8$$

It is interesting to observe that the expression inside the brackets is equal and opposite to the *enthalpy of formation* of the quantity of solution containing unit amount of solvent from the gaseous solvent and from the solid solute.

§5.24 *Surface tension*

We conclude this chapter with a brief discussion of interfacial layers, particularly those between a liquid and its vapour. As described in §4.36 we shall neglect effects of pressure on the liquid phase and on the surface layer.

For the sake of brevity we use the symbol D to denote the operator $\Sigma_s dr_s \partial/\partial r_s$. We have then by analogy with (4.36.4), (4.36.5), and (4.36.6)

$$-d\gamma = S_A^\sigma dT + \Gamma_1 d\mu_1 + \sum_s \Gamma_s d\mu_s \qquad 5.24.1$$

$$d\mu_1 = -S_1 dT + D\mu_1 \qquad 5.24.2$$

$$d\mu_s = -S_s dT + D\mu_s \qquad 5.24.3$$

where (2) and (3) relate to the liquid phase*. We also have in the liquid phase the Gibbs–Duhem relation (5.04.3)

$$D\mu_1 + \sum_s r_s D\mu_s = 0. \qquad 5.24.4$$

Substituting (2) and (3) into (1) we obtain

$$-d\gamma = (S_A^\sigma - \Gamma_1 S_1 - \sum_s \Gamma_s S_s) dT + \Gamma_1 D\mu_1 + \sum_s \Gamma_s D\mu_s. \qquad 5.24.5$$

Now eliminating $D\mu_1$ between (4) and (5) we obtain finally

$$-d\gamma = (S_A^\sigma - \Gamma_1 S_1 - \sum_s \Gamma_s S_s) dT + \sum_s (\Gamma_s - r_s \Gamma_1) D\mu_s. \qquad 5.24.6$$

By reasoning similar to that of §4.37 we can verify the invariance of the coefficients of dT and $D\mu_s$ with respect to shifts of the geometrical surfaces bounding the surface layer.

---

* There should be no confusion between γ denoting surface tension and the activity coefficients γ_s.

§5.25 Temperature dependence

For the temperature dependence of the surface tension at constant composition of the liquid we obtain immediately from (5.24.6)

$$-d\gamma/dT = S_A^\sigma - \Gamma_1 S_1 - \sum_s \Gamma_s S_s \qquad 5.25.1$$

where the right side is the entropy of unit area of the surface layer less the entropy of the same material content in the liquid phase.

By proceeding as in §4.38 we can transform (1) to the equivalent relation

$$\gamma - T\,\partial\gamma/\partial T = U_A^\sigma - \Gamma_1 U_1 - \sum_s \Gamma_s U_s. \qquad 5.25.2$$

The right side is the energy which must be supplied to prevent any change of temperature when unit area of surface is formed from the liquid.

§5.26 Variations of composition

For variations of composition at constant temperature (5.24.6) reduces to

$$-d\gamma = \sum_s (\Gamma_s - r_s \Gamma_1) D\mu_s \qquad 5.26.1$$

or using

$$D\mu_s = RT\,D\ln \lambda_s = RT\,D\ln p_s \qquad 5.26.2$$

$$-d\gamma = RT \sum_s (\Gamma_s - r_s \Gamma_1) D\ln p_s. \qquad 5.26.3$$

Each of the quantities

$$\Gamma_s - r_s \Gamma_1 \qquad 5.26.4$$

occurring on the right side of (3) is called the *surface excess* per unit area of the solute species s. The corresponding quantity for the solvent species 1 vanishes by definition. As we have repeatedly stressed, each quantity (4), in contrast to the individual Γ's is invariant with respect to shift of the boundary between the liquid phase and the surface phase and is therefore physically significant. The quantities (4) are the same as the quantities which Gibbs* denoted by $\Gamma_{s(1)}$ but his definition of these quantities was more abstract and more difficult to visualize.

§5.27 Interfacial tension between two solutions

For the interface between two liquid phases α, β neglecting dependence on

* Gibbs, Collected Works, Longmans, vol. **1** pp. 234–235.

pressure, we have

$$-d\gamma = S_A^\sigma dT + \Gamma_1 d\mu_1 + \sum_s \Gamma_s d\mu_s. \qquad 5.27.1$$

The Gibbs–Duhem relation for phase α can be written

$$(1 + \sum_s r_s^\alpha) S_m^\alpha dT + d\mu_1 + \sum_s r_s^\alpha d\mu_s = 0 \qquad 5.27.2$$

and that for phase β

$$(1 + \sum_s r_s^\beta) S_m^\beta dT + d\mu_1 + \sum_s r_s^\beta d\mu_s = 0. \qquad 5.27.3$$

If there are $c-1$ solute species, there are $c+1$ quantities dT, $d\mu_1$, $d\mu_s$ in (1) of which any two can be eliminated by using (2) and (3). The results obtainable are complicated and we shall not pursue them here.

§5.28 Volume concentrations

In analytical work it has long been the usual practice to describe the composition of a solution by the *volume concentration* c_s of each solute species defined as

$$c_s = n_s/V = r_s n_1/V. \qquad 5.28.1$$

As long as we are concerned with the properties of the solution at only one temperature this practice is unobjectionable. But in thermodynamics we are much interested in the temperature dependence of properties, and volume concentrations are then inconvenient. For whatever quantities be used to describe the composition of a liquid solution, it is expedient to use as the other two independent variables temperature and pressure, so that differentiation with respect to temperature implies constant pressure. We therefore have

$$\partial c_s/\partial T = -\alpha c_s \qquad 5.28.2$$

where α is the thermal expansivity. It is evident from (2) that, if c_s is chosen as a variable, it will not be an independent variable. On these grounds volume concentrations are not convenient in liquid solutions and we shall not use them.

Volume concentrations of course play an important part in the theory of gas kinetics. The implication, sometimes met, that they must therefore play a parallel part in the theory of solution kinetics shows a lack of appreciation of the utterly different and much more complex meaning of the word *collision* applied to a solution in contrast to a gas. We have not yet a complete theoretical treatment of collisions in solution, but the author believes that a successful theory would be based on molecular ratios rather than on volume concentrations.

CHAPTER 6

SYSTEMS OF CHEMICALLY REACTING SPECIES

§6.01 *Notation and terminology*

Any chemical process may be written in the form

$$\sum_A \nu_A A \rightarrow \sum_B \nu_B B \qquad 6.01.1$$

where A, B denote chemical species and ν_A, ν_B are integers or simple rational fractions. Since the meaning of formula (1) has sometimes been misunderstood it is desirable to state unambiguously what it means and what it does not. It means that in a system containing a large amount of A, ..., B, ... the amounts reacting are ν_A, ..., ν_B, ... It does not mean that a system composed of an amount ν_A of A, ..., is changed completely into an amount ν_B of B, ...

We can measure the extent to which the process (1) takes place by the *extent of reaction* defined in §1.44 such that a change of ξ to $\xi + d\xi$ means that an amount $\nu_A d\xi$ of A and the like react to given an amount $\nu_B d\xi$ of B and the like. We also recall that the *affinity* of the reaction is defined as

$$-(\partial F/\partial \xi)_{T,V} = -(\partial G/\partial \xi)_{T,P^\alpha} \qquad 6.01.2$$

where P^α is the pressure of each phase α.

If ξ increases by $d\xi$ in a time dt we then have the concise universal law that in any natural process

$$-(\partial G/\partial \xi)_{T,P^\alpha} d\xi/dt > 0 \quad \text{(natural)} \qquad 6.01.3$$

and consequently for equilibrium

$$-(\partial G/\partial \xi)_{T,P^\alpha} = 0 \quad \text{(equilibrium)}. \qquad 6.01.4$$

Formula (4) is equivalent to

$$\sum_A \nu_A \mu_A = \sum_B \nu_B \mu_B \quad \text{(equilibrium)}. \qquad 6.01.5$$

We now recall the abbreviated notation described in §1.44 according to which we replace (1) by

$$0 = \sum_B \nu_B B \qquad 6.01.6$$

where now each ν_B is negative for a reactant and positive for a reaction product. In this abbreviated notation (5) becomes

$$\sum_B \nu_B \mu_B = 0. \qquad 6.01.7$$

Since the absolute activity is related to the chemical potential by

$$\mu = RT \ln \lambda \qquad 6.01.8$$

formula (7) is equivalent to

$$\prod_B (\lambda_B)^{\nu_B} = 1 \qquad \text{(equilibrium)}. \qquad 6.01.9$$

We shall now further abbreviate our notation. Let I_B denote any intensive property relating to the species B such as λ_B, p_B, x_B, r_B, m_B, γ_B. Then we shall use the contracted notation $\Pi(I)$ defined by

$$\prod(I) = \prod_B (I_B)^{\nu_B}. \qquad 6.01.10$$

When the I_B's have values corresponding to a state of chemical equilibrium we shall call $\Pi(I)$ the *equilibrium product* of the I_B's.

Our first application of this notation is to (9) which we contract to

$$\prod(\lambda) = 1 \qquad \text{(equilibrium)} \qquad 6.01.11$$

and the general condition for chemical equilibrium may be stated as: *the equilibrium product of the absolute activities* is unity.

§6.02 *Enthalpy of reaction*

Consider the constant temperature process

$$0 = \sum_B \nu_B B \qquad (T \text{ const.}) \qquad 6.02.1$$

and let the operator Δ denote the excess of a final over an initial value corresponding to unit increase in the extent of reaction. If the process occurs at constant pressure then the heat absorbed is equal to ΔH. For this reason ΔH is called either the *heat of reaction at constant pressure* or better the *enthalpy of reaction*.

If on the other hand the process occurs at constant volume the heat absorbed is equal to ΔU, which is therefore called the *heat of reaction at constant volume.* This quantity is of little importance except for reactions involving a gas phase, for which it is related to ΔH by

$$\Delta U = \Delta H - RT \sum_{B}{}' v_B \qquad 6.02.2$$

where Σ' denotes summation over gaseous species only, the second virial coefficients and proper volumes of condensed phases being neglected.

We recall that for a perfect gas H is independent of the pressure and for a condensed phase the effect of variations of pressure is negligible. It is therefore often unnecessary to specify the pressure when speaking of enthalpies of reaction.

§6.03 *Hess' law*

Since H is a function of the state of a system, ΔH is for successive processes at the same temperature an additive function. This property of ΔH, known as *Hess' law*, is useful in enabling us to calculate ΔH for a reaction, difficult to produce quantitatively, from other reactions which give less difficulty. The following simple example illustrates the point

$$C(\text{graphite}) + O_2(g) \rightarrow CO_2(g) \qquad -\Delta H = 393.5 \text{ kJ}$$
$$CO(g) + \tfrac{1}{2}O_2(g) \rightarrow CO_2(g) \qquad -\Delta H = 283.0 \text{ kJ}.$$

In both the above cases ΔH is readily measurable. By subtraction we obtain

$$C(\text{graphite}) + \tfrac{1}{2}O_2(g) \rightarrow CO(g) \qquad -\Delta H = 110.5 \text{ kJ}$$

a reaction difficult, if not impossible, to study quantitatively.

Other numerous examples are the calculations of the *enthalpies of formation* of organic compounds from the *enthalpies of combustion*. A simple example is

$$CH_4(g) + 2O_2(g) \rightarrow CO_2(g) + 2H_2O(l) \qquad -\Delta H = 890.3 \text{ kJ}$$
$$C(\text{graphite}) + O_2(g) \rightarrow CO_2(g) \qquad -\Delta H = 393.5 \text{ kJ}$$
$$2H_2(g) + O_2(g) \rightarrow 2H_2O(l) \qquad -\Delta H = 571.6 \text{ kJ}$$

from which we immediately deduce

$$C(\text{graphite}) + 2H_2(g) \rightarrow CH_4(g) \qquad -\Delta H = 74.8 \text{ kJ}.$$

Unfortunately in calculating an enthalpy of formation as the difference between much greater enthalpies of combustion there is considerable loss in percentage accuracy since the experimental errors add up. Nevertheless this is the standard method for determining enthalpies of formation of organic compounds from their elements.

Unfortunately some authors have used the name *heat of reaction* for $-\Delta H$ instead of for ΔH. This practice is deplorable. In particular the name 'heat of combustion' is commonly used for $-\Delta H$. It might be pleaded in excuse that in this case $-\Delta H$ is always positive and that there is no other convenient name. There is on the other hand no excuse whatever for the habit of calling $-\Delta H$ for adsorption a 'heat of adsorption' when there exists the perfectly good name *heat of desorption*. The simplest and safest way to avoid any possible ambiguity is to write explicitly $\Delta H = \ldots$ or $-\Delta H = \ldots$ as in the above examples.

§6.04 *Kirchhoff's relations*

We often need the value of ΔH at one temperature when it has been measured at a different temperature. This causes no difficulty provided the dependence of H on the temperature has been measured or is known theoretically for the initial and final states.

Let T denote the temperature at which we want the value of ΔH and T' the temperature at which it has been measured. Then

$$\Delta H(T) - \Delta H(T') = \sum_B v_B H_B(T) - \sum_B v_B H_B(T') = \sum_B v_B \{H_B(T) - H_B(T')\}. \quad 6.04.1$$

Although (1) is the form in which the experimental data are available and should be used, it is customary to express it in the differential form

$$d\Delta H/dT = \sum_B v_B dH_B/dT = \sum_B v_B C_B. \quad 6.04.2$$

Formula (2) is known as *Kirchhoff's relation*. Since values of the heat capacities C are usually obtained by differentiating experimental measurements of $H(T) - H(T')$ and formula (2), if used, has to be integrated, it is difficult to see any advantage of (2) over (1). As already mentioned in §3.03 the main function of a heat capacity is to serve as the connecting link between the enthalpy and the entropy.

There is a second formula also associated with Kirchhoff, similar to (2), but relating the energy change ΔU with the heat capacities at constant volume, but this formula is not needed.

§6.05 *Prescription of standards*

As already explained in §3.25 the formulae for chemical equilibrium require a consistent choice of standards $P^\ominus$, $\lambda^\ominus$, $H^\ominus$, and $S^\ominus$. We now prescribe the choice used almost universally and used henceforth in this text.

The standard pressure $P^\ominus$ is prescribed as

$$P^\ominus = 1 \text{ atm.} \qquad 6.05.1$$

The standard enthalpy is prescribed by stating that

$$H^\ominus = 0 \qquad (T = 298.15 \text{ K}) \qquad 6.05.2$$

for every *element in its stable state*. The standard entropy is prescribed by stating that

$$S^\ominus = 0 \qquad (T \to 0) \qquad 6.05.3$$

for every *element in its stable state*.

Since $H^\ominus(T) - H^\ominus(298.15 \text{ K})$ is determinable by purely calorimetric measurements, the convention expressed by (2) determines unambiguously the value of $H^\ominus(T)$ for any T. Similarly since $S^\ominus(T) - S^\ominus(0)$ is determinable by purely calorimetric measurements, the convention expressed by (3) determines unambiguously the value of $S^\ominus(T)$ for any T. Then $\lambda^\ominus$ is unambiguously defined by

$$\ln \lambda^\ominus = H^\ominus/RT - S^\ominus/R. \qquad 6.05.4$$

Extensive tables exist of values of $H^\ominus(298.15 \text{ K})$ and of $S^\ominus(298.15 \text{ K})$. Less extensive tables exist of $H^\ominus(T)$ and of $S^\ominus(T)$ for other values of T.

Most of these tables give values of $H^\ominus$ in kcal mole^{-1} and of $S^\ominus$ in cal K^{-1} mole^{-1} although all precise calorimetric measurements are made in terms of joules. It would save considerable unnecessary calculation if the tabulated quantities were $H^\ominus/R$ and $S^\ominus/R$.

§6.06 Construction of tables

As already mentioned extensive tables exist of values of $H^\ominus$ and of $S^\ominus$ for $T = 298.15$ K. We now summarize briefly how these are constructed. We begin with $H^\ominus$.

The first step in determining $H^\ominus$ for a given substance is to choose a set of reactions for which the enthalpy of reaction can be measured directly and which add up to the process of formation of the given substance from its elements. Two simple examples have already been mentioned in §6.03. The set of reactions may include isothermal changes of pressure so as to convert each measured ΔH to the required $\Delta H^\ominus$. The values of the enthalpies of reaction at the several experimental temperatures are reduced to values at 298.15 K by use of Kirchhoff's relation. The values at 298.15 K for the several reactions are combined according to Hess' law to obtain the heat of formation $\Delta_f H^\ominus$. Finally from the chosen convention that $H^\ominus$ at 298.15 K is zero for every element in its stable state it follows that for each compound $H^\ominus = \Delta_f H^\ominus$ and this is the quantity tabulated. In the most extensive tables* $H^\ominus(298.15\text{ K})$ is denoted by $\Delta H f^0$.

We now consider the tabulation of $S^\ominus$. Purely calorimetric measurements, as illustrated in §3.52, lead directly to $S^\ominus(298.15\text{ K}) - S^\ominus(\text{crystal}, T \to 0)$. When we choose the convention that $S^\ominus(T \to 0)$ is zero for every element in its stable crystalline form we have for any substance

$$S^\ominus(\text{crystal}, T \to 0) = R \ln o \qquad 6.06.1$$

and consequently

$$S^\ominus(298.15\text{ K}) = \{S^\ominus(298.15\text{ K}) - S^\ominus(\text{crystal}, T \to 0)\} + R \ln o. \qquad 6.06.2$$

The expression { } is often called the *calorimetric entropy*. In order to determine the tabulated quantity $S^\ominus(298.15\text{ K})$ we need to know or assume the value of o. There are three different possibilities. For about thirty substances with simple molecules $S^\ominus(298.15\text{ K})$ for the gas has been determined from purely spectroscopic data and this value, often called the *spectroscopic entropy*, is found to be equal to the *calorimetric entropy*. It follows for all these substances that $o = 1$ or $\ln o = 0$. For a few substances with simple molecules, namely CO, N_2O, NO, H_2O the spectroscopic entropy is found to exceed the calorimetric entropy by amounts $R \ln 2$, $R \ln 2$, $\frac{1}{2} R \ln 2$, $R \ln \frac{3}{2}$ so that o has values differing from unity. These non-zero values of $\ln o$ are understood and have been explained in §3.54. In each case they are due to metastability in the crystal. Finally for all other substances the value of o has not been determined experimentally. Its value is assumed to be unity. It is conceivable that there are other cases of $o > 1$ but they are likely to be few if any. The use of the assumption $o = 1$ can thus conceivably lead to equilibrium constants wrong by a factor such as 2. In the most extensive tables* $S^\ominus(298.15\text{ K})$ is denoted by S^0.

Tables[†] less extensive than those for 298.15 K exist relating to other temperatures. These tables give values of the purely calorimetric quantities $H^\ominus(T) - H^\ominus(298.15\text{ K})$ and $S^\ominus(T) - S^\ominus(298.15\text{ K})$.

§6.07 *Gaseous equilibria*

For every component B in a gas phase we have according to (4.08.1)

$$\lambda_B = \lambda_B^\ominus p_B / P^\ominus \qquad 6.07.1$$

where p_B is the fugacity of B, $P^\ominus$ is a standard pressure, normally one

---
* Rossini et al., Circular no. 500 of National Bureau of Standards, 1952.
† Kelley, U.S. Bureau of Mines, 1949, Bulletin 476.

atmosphere, and $\lambda_B^\ominus$ is the value of λ_B when p_B is equal to $P^\ominus$. Substituting (1) into (6.01.11) we obtain the equilibrium condition

$$\prod(p) = K \qquad 6.07.2$$

where K is a function of temperature only defined by

$$K = \prod(P^\ominus/\lambda^\ominus) \qquad 6.07.3$$

and is called the *equilibrium constant*. Lack of experimental data on virial coefficients in gaseous mixtures usually makes it impossible to correct for gas imperfection, even though the procedure is in principle straightforward. When we adopt the approximation

$$p_B = x_B P \quad \text{(perfect gas)} \qquad 6.07.4$$

(2) becomes

$$\prod(xP) = K. \qquad 6.07.5$$

We may rewrite (5) as

$$\prod(x) = K_x \quad \text{(perfect gas)} \qquad 6.07.6$$

where

$$K_x = K/\prod(P) \qquad 6.07.7$$

so that K_x as well as depending on the temperature is inversely proportional to $P^{\Sigma \nu_B}$. There is no advantage in using K_x instead of K.

§6.08 Equilibria between gases and solids

We turn now to a discussion of the equilibrium of reactions involving pure solids as well as gases. Examples are

$$CaCO_3(s) \to CaO(s) + CO_2(g) \qquad 6.08.1$$

$$NH_4Cl(s) \to NH_3(g) + HCl(g) \qquad 6.08.2$$

$$C(\text{graphite}) + CO_2(g) \to 2CO(g). \qquad 6.08.3$$

We have the general equilibrium condition (6.01.11)

$$\prod(\lambda) = 1 \qquad 6.08.4$$

where the λ of each gaseous species is related to its fugacity by (6.07.1). On the other hand we may regard the λ of each pure solid as a function of temperature only, since the effect of change of pressure on a solid is usually negligible.

We now extend our $\prod$ notation as follows. We write

$$\prod(l) = \prod_G(l) \prod_S(l) \qquad 6.08.5$$

where $\Pi_G(I)$ contains all the factors of $\Pi(I)$ relating to the gaseous species and $\Pi_S(I)$ all the factors relating to the solid species.

For example in the case of reaction (1)

$$\prod_G (\lambda) = \lambda_{CO_2} \qquad 6.08.6$$

$$\prod_S (\lambda) = \lambda_{CaO}/\lambda_{CaCO_3}. \qquad 6.08.7$$

Using this notation, the equilibrium condition (4) may be written

$$\prod_G (\lambda) \prod_S (\lambda) = 1. \qquad 6.08.8$$

Now substituting (6.07.1) into (8) we obtain

$$\prod_G (p) = K \qquad 6.08.9$$

where K is a function of temperature only given by

$$K = \prod_G (P^\ominus/\lambda^\ominus) \prod_S (1/\lambda^\ominus) \qquad 6.08.10$$

and is called the *equilibrium constant*. For example for reaction (3), we have

$$p_{CO}^2/p_{CO_2} = K \qquad 6.08.11$$

$$K = \lambda_{CO_2}^\ominus \lambda_C^\ominus P^\ominus / \lambda_{CO}^{\ominus 2}. \qquad 6.08.12$$

§6.09 *Temperature dependence*

For any reaction

$$0 = \sum_B v_B B \qquad 6.09.1$$

between gases and solids, or between gases only, the equilibrium constant K is given by (6.08.10)

$$K = \prod_G (P^\ominus/\lambda^\ominus) \prod_S (1/\lambda^\ominus). \qquad 6.09.2$$

For each species B whether gaseous or solid we have

$$d \ln \lambda_B^\ominus / dT = -H_B^\ominus / RT^2. \qquad 6.09.3$$

Taking logarithms and differentiating (2) with respect to T and using (3) we obtain

$$d \ln K/dT = \sum_B H_B^\ominus/RT^2 = \Delta H^\ominus/RT^2 \qquad 6.09.4$$

where $\Delta H^\ominus$ is the standard enthalpy of reaction. This may be written in the alternative form

$$d \ln K/d(1/T) = -\sum_B H_B^\ominus/R = -\Delta H^\ominus/R. \qquad 6.09.5$$

§6.10 Numerical example

We shall now illustrate the use of our formulae and of standard tables by a simple example. We choose the reaction

$$0 = 2\text{CO}(g) - \text{C(graphite)} - \text{CO}_2(\text{gas}) \quad \text{at } 1000 \text{ K}. \qquad 6.10.1$$

The equilibrium is determined by

$$p_{\text{CO}}^2 / p_{\text{CO}_2} = K \qquad 6.10.2$$

where

$$K = P^\ominus \lambda_{\text{CO}_2}^\ominus \lambda_{\text{C}}^\ominus / \lambda_{\text{CO}}^{\ominus 2} = (\lambda_{\text{CO}_2}^\ominus \lambda_{\text{C}}^\ominus / \lambda_{\text{CO}}^{\ominus 2}) \text{ atm}. \qquad 6.10.3$$

The tabulated experimental data* are as follows

B	C(graphite)	$CO_2(g)$	$CO(g)$	$\sum_B$
ν_B	-1	-1	$+2$	
$H_B^\ominus(25\,°\text{C})/\text{kcal mole}^{-1}$	0	-94.05	-26.42	
$S_B^\ominus(25\,°\text{C})/\text{cal K}^{-1}\text{ mole}^{-1}$	1.36	51.06	47.30	
$H_B^\ominus(1000\text{ K}) - H_B^\ominus(25\,°\text{C})/\text{kcal mole}^{-1}$	2.810	7.993	5.186	
$S_B^\ominus(1000\text{ K}) - S_B^\ominus(25\,°\text{C})/\text{cal K}^{-1}\text{ mole}^{-1}$	4.47	13.28	8.82	
$H_B^\ominus(1000\text{ K})/\text{kcal mole}^{-1}$	2.810	-86.059	-21.230	40.789
$S_B^\ominus(1000\text{ K})/\text{cal K}^{-1}\text{ mole}^{-1}$	5.83	64.34	56.12	42.07

From these we deduce

$$\ln(K/\text{atm}) = -\Delta H^\ominus / RT + \Delta S^\ominus / R = (42.07 - 40.79)/1.987 = 0.64$$
$$K = 1.9 \text{ atm}.$$

The accuracy of a calculation of this kind is at best about ± 0.05 in each term of $\ln K$. This usually leads to an uncertainty of at least 0.1 in $\ln K$ or 10% in K. In most cases the experimental uncertainty in a direct measurement of K is no less.

§6.11 Reactions between pure solids or liquids

We must now consider reactions between pure solid phases without any gases. Examples are

$$0 = \text{PbI}_2 - \text{Pb} - 2\text{I} \qquad 6.11.1$$
$$0 = \text{CuS} - \text{Cu} - \text{S}. \qquad 6.11.2$$

---

* Rossini et al., Circular no. 500 of National Bureau of Standards, 1952.

Incidentally, for the following considerations it is immaterial whether any of the phases is a pure liquid instead of a pure solid. As an example we may mention

$$0 = Ag + HgCl - Hg - AgCl. \qquad 6.11.3$$

The simplest type of reaction between solid phases is an allotropic change, such as

$$0 = \text{monoclinic sulphur} - \text{rhombic sulphur} \qquad 6.11.4$$

$$0 = \text{white tin} - \text{grey tin}. \qquad 6.11.5$$

The equilibrium condition for a reaction involving pure solid and liquid phases can still be expressed in the form (6.01.11)

$$\prod (\lambda) = 1 \quad \text{(equilibrium)} \qquad 6.11.6$$

but each λ is now a function of temperature only, if we disregard the small effect of changes of pressure. Hence the equilibrium condition (6) may be regarded as an equation determining the temperature of reversal of the change considered. This equation may or may not have a solution for T positive. Reactions (1), (2), (3) proceed naturally towards the right at all temperatures and there is no solution of (6). In point of fact very few reactions between pure solids and pure liquids have a reversal temperature. The most important exceptions are allotropic changes such as (4) and (5), among which we may, if we like, include simple fusion.

For reactions such as (1), (2), (3) at all temperatures we have

$$\prod (\lambda) < 1 \qquad 6.11.7$$

or taking logarithms and writing in full,

$$\sum_B v_B \ln \lambda_B < 0. \qquad 6.11.8$$

Another way of expressing the same thing is to state that the affinity, defined in §1.44, is positive at all temperatures. We shall see in chapter 8 how the affinity of some reactions can be accurately determined by measurements of electromotive force.

We shall now consider (6) in more detail and for this purpose we write it in the expanded form

$$\sum_B v_B \ln \lambda_B = 0. \qquad 6.11.9$$

But by definition

$$\ln \lambda_B = \mu_B/RT = H_B/RT - S_B/R. \qquad 6.11.10$$

Substituting (10) into (9), we obtain

$$T = \sum_B v_B H_B / \sum_B v_B S_B. \qquad 6.11.11$$

The numerator of (11) is the enthalpy of reaction ΔH and the denominator is the entropy of reaction ΔS. We now consider these separately.

For ΔH we write formally

$$\Delta H(T) = \Delta H(T') + \sum_B v_B \{H_B(T) - H_B(T')\}, \qquad 6.11.12$$

If for each of the substances the dependence of H on temperature has been determined calorimetrically and if in addition ΔH has been measured at any one temperature T', then by means of (12) ΔH can be calculated at any other temperature.

For ΔS we write formally

$$\Delta S(T) = \Delta S^0 + \sum_B v_B \{S_B(T) - S_B^0\} \qquad 6.11.13$$

where the superscript ^{0} denotes the value obtained by smooth extrapolation to $T = 0$. If now the dependence of H on temperature has been measured throughout the temperature range from T down to a temperature from which one can extrapolate to $T = 0$, then (13) determines ΔS for all temperatures apart from the constant ΔS^0. But S^0 is the quantity discussed in detail in §§ 3.51–3.57. It has the value zero except for a few well understood exceptions for which its value is known to be $R \ln o$, with o a small number such as 2 or $\frac{3}{2}$. With this knowledge of ΔS^0 or in the absence of evidence to the contrary assuming $\Delta S^0 = 0$, formula (13) determines ΔS for all temperatures.

Using (12) and (13) together, we can solve (11) for the *transition temperature T*. Alternatively using the experimental value of T, we can use (11), (12), (13) to determine an experimental value for ΔS^0.

§6.12 *Transition of sulphur*

We shall now illustrate the formulae of the preceding section by a numerical example. As already mentioned it is difficult to find an example of an equilibrium temperature for a reaction between solid phases except in the simplest case of an allotropic change. We accordingly choose as our example

$$0 = \text{monoclinic sulphur} - \text{rhombic sulphur} \qquad 6.12.1$$

and we shall use the subscripts $_R$ and $_M$ for the rhombic and monoclinic forms respectively. The transition temperature is

$$T = 368.6 \text{ K} \qquad \text{(transition).} \qquad 6.12.2$$

The enthalpy of transition at this temperature is given by

$$\Delta H/R = (H_M - H_R)/R = (47.5 \pm 5) \text{ K} \qquad (T = 368.6 \text{ K}). \qquad 6.12.3$$

Consequently the entropy of transition at this temperature is

$$\Delta S/R = (S_M - S_R)/R = 47.5/368.6 = 0.12 \pm 0.01 \qquad (T = 368.6 \text{ K}). \qquad 6.12.4$$

According to calorimetric measurements* on the two forms from 15 K to the transition temperature

$$\{S_R(368.6 \text{ K}) - S_R(15 \text{ K})\}/R = 4.38 \pm 0.03 \qquad 6.12.5$$

$$\{S_M(368.6 \text{ K}) - S_M(15 \text{ K})\}/R = 4.49 \pm 0.04. \qquad 6.12.6$$

Combining (4), (5), and (6) we obtain

$$\{S_M(15 \text{ K}) - S_R(15 \text{ K})\}/R = 0.12 - 4.49 + 4.38$$
$$= 0.01 \pm 0.05. \qquad 6.12.7$$

We conclude that well within the experimental accuracy

$$S_M^0 - S_R^0 = 0. \qquad 6.12.8$$

§6.13 Homogeneous equilibrium in solution

We turn now to homogeneous chemical equilibrium in a liquid solution. We again start from the general equilibrium condition (6.01.11)

$$\prod(\lambda) = 1 \qquad 6.13.1$$

and use

$$\lambda = \lambda^\ominus m\gamma. \qquad 6.13.2$$

Substituting (2) into (1) we obtain

$$\prod(m) \prod(\gamma) = K_m \qquad 6.13.3$$

where K_m is defined by

$$K_m = \prod(1/\lambda^\ominus) \qquad 6.13.4$$

and so depends only on the solvent and the temperature. K_m is called the *molality equilibrium constant*. Formula (3) tells us that the equilibrium molality product is inversely proportional to the equilibrium activity coefficient product.

In the special case of an ideal dilute solution (3) reduces to

$$\prod(m) = K_m. \qquad 6.13.5$$

* Eastman and McGavock, J. Amer. Chem. Soc. 1937 **59** 145.

§6.14 Temperature dependence

If we take logarithms of (6.13.4) we have

$$\ln K_m = -\sum_B \nu_B \ln \lambda_B^{\ominus}. \qquad 6.14.1$$

Differentiating with respect to T and using (5.15.2) we obtain

$$d \ln K_m/dT = \sum_B \nu_B H_B^{\infty}/RT^2 = \Delta H^{\infty}/RT^2. \qquad 6.14.2$$

where ΔH^{∞} is the *enthalpy of reaction at infinite dilution* in the given solvent.

§6.15 Use of volume concentrations

As mentioned in §5.28 volume concentrations are sometimes used instead of molalities but the practice is not recommended. In place of (6.13.3) one then obtains

$$\prod (c) \prod (y) = K_c \qquad 6.15.1$$

where c denotes volume concentration and y denotes a new kind of activity coefficient. We shall not go into details, but will only point out that the temperature dependence of K_c is given by*

$$\partial \ln K_c/\partial T = \Delta H^{\infty}/RT^2 - \alpha \sum_B \nu_B \qquad 6.15.2$$

where α denotes the thermal expansivity of the solvent. Spurious formulae for $\partial \ln K_c/\partial T$ obtained by false analogy with gaseous equilibria have sometimes been quoted, both in the past and recently[†].

§6.16 Heterogeneous equilibria involving solutions

We might also discuss equilibria involving solutions and vapour phases, or solutions and solids, or even solutions, solids, and vapour phases, but this is unnecessary because any equilibrium however complicated can be regarded as a superposition of a homogeneous equilibrium in a single phase, liquid or gaseous, and distribution equilibria of individual species between pairs of phases. Both these elementary types of equilibrium have been discussed in sufficient detail.

§6.17 Transitions of second order

This is perhaps the most convenient place to describe a phenomenon called

---

* Guggenheim, Trans. Faraday Soc. 1937 **33** 607.
† E.g. Clarke and Glew, Trans. Faraday Soc. 1966 **62** 547.

a *transition of the second order*. It is quite different from anything we have yet met, having some of the characteristics of phase changes and some of the characteristics of critical phenomena. We shall first show by a particular example how a transition of the second order arises from certain assumed properties of the thermodynamic functions. We shall then discuss briefly how and when such transitions occur.

As a preliminary step to our discussion, we shall consider the thermodynamic properties of the equilibrium between two isomers under the simplest conceivable conditions. Thus we consider the isomeric change

$$A \to B \qquad 6.17.1$$

occurring in a mixture of A and B in the absence of any other species. We further assume that the mixture is ideal. Finally we assume that the proper enthalpy of reaction has a value w independent of the temperature; in other words we assume that A and B have equal heat capacities. If then x denotes the mole fraction of B the proper Gibbs function G_m has the form

$$G_m = G_m^{\ominus}(T) + xw + RT\{(1-x)\ln(1-x) + x \ln x\} \qquad 6.17.2$$

where w is a constant and $G_m^{\ominus}(T)$ depends only on the temperature. From (2) we deduce

$$H_m = G_m^{\ominus} - T dG_m^{\ominus}/dT + xw \qquad 6.17.3$$

from which we verify that the proper enthalpy of reaction is w. We also deduce

$$S_m = -dG_m^{\ominus}/dT - R\{(1-x)\ln(1-x) + x \ln x\} \qquad 6.17.4$$

showing that the proper entropy of mixing has its ideal value

$$-R\{(1-x)\ln(1-x) + x \ln x\}. \qquad 6.17.5$$

The equilibrium value of x is obtained by minimizing G_m. We find

$$\partial G_m/\partial x = w + RT \ln\{x/(1-x)\} = 0 \qquad 6.17.6$$

so that

$$x/(1-x) = \exp(-w/RT). \qquad 6.17.7$$

Formula (7) is, of course, the simplest possible example of the equilibrium law. Before we dismiss this extremely simple system, there remains one important point to be investigated, namely the verification that (6) and (7) do correspond to a minimum of G_m, not to a maximum. We have

$$\partial^2 G_m/\partial x^2 = RT\{1/x + 1/(1-x)\} > 0 \qquad 6.17.8$$

thus verifying that we have found a minimum, not a maximum.

Let us now arbitrarily introduce a modification into the form of G_m assumed in (2), without at this stage enquiring into the physical significance of the change. We replace the term wx by $wx(1-x)$. We then have

$$G_m = G_m^\ominus(T) + x(1-x)w + RT\{(1-x)\ln(1-x) + x \ln x\} \quad \text{6.17.9}$$

$$H_m = G_m^\ominus - T\,dG_m^\ominus/dT + x(1-x)w \quad \text{6.17.10}$$

$$S_m = -dG_m^\ominus/dT - R\{(1-x)\ln(1-x) + x \ln x\} \quad \text{6.17.11}$$

from which we observe that the enthalpy is affected by the modification, but the entropy is not.

We now seek the equilibrium value of x by minimizing G_m. We find

$$\partial G_m/\partial x = -(2x-1)w + \ln\{x/(1-x)\} = 0 \quad \text{6.17.12}$$

so that

$$x/(1-x) = \exp\{(2x-1)w/RT\}. \quad \text{6.17.13}$$

One solution of (13) is obviously $x = \tfrac{1}{2}$, but this is not always the only solution. Nor is this solution necessarily a minimum rather than a maximum of G_m. We must investigate these points and shall do so in the first place graphically. Figure 6.1 shows $(G_m - G_m^\ominus)/RT$ plotted against $2x - 1$ for various values of $\tfrac{1}{2}w/RT$. Owing to the complete symmetry between x and $1-x$, we can without loss of generality assume that $x > 1-x$.

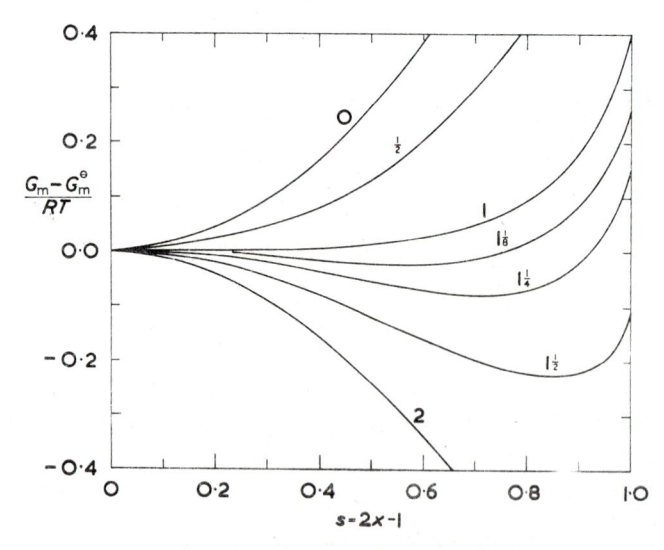

Fig. 6.1. Dependence of G_m on x for various values of $\tfrac{1}{2}w/RT$. The numbers attached to the curves are values of $\tfrac{1}{2}w/RT$ or T_λ/T

We see that if $w>0$ then at high temperatures, that is at small values of w/RT, the root $x=\tfrac{1}{2}$ is the only root and it corresponds to a minimum of G_m. At low enough temperatures, that is at large values of w/RT, there is another root $\tfrac{1}{2} \leq x \leq 1$ and this root corresponds to a minimum of G_m while the root $x=\tfrac{1}{2}$ now corresponds to a maximum. Thus there exists a temperature T_λ such that at temperatures below T_λ the equilibrium value of x is greater than $\tfrac{1}{2}$ and decreases as the temperature increases; the equilibrium value of x becomes $\tfrac{1}{2}$ at the temperature T_λ and remains $\tfrac{1}{2}$ at all high temperatures. The change occurring at the temperature T_λ is called a *transition of the second order* and the temperature T_λ is called a *lambda point* for a reason which will be explained later.

It is clear from figure 6.1 that T_λ is the temperature at which the two roots of (13) become equal, the root at $x=\tfrac{1}{2}$ changing from a minimum to a maximum. Thus there is a point of horizontal inflexion at $x=\tfrac{1}{2}$. We have then

$$\partial^2 G_m/\partial x^2 = -2w/RT_\lambda + 1/x(1-x) = 0 \qquad (x=\tfrac{1}{2}) \qquad 6.17.14$$

whence

$$w/RT_\lambda = 2. \qquad 6.17.15$$

It is clear from figure 6.1 that for negative values of w the minimum is always at $x=\tfrac{1}{2}$ and there can be no lambda point.

§6.18 Cooperative systems

Before proceeding to a more detailed examination of transitions of the second order, we shall explain in very general terms how they may arise. As a preliminary step, let us determine the enthalpy of change in the process (6.17.1). For the enthalpy H of the whole sytem, we have according to (6.17.10) changing to the variables n_A, n_B

$$H = (n_A + n_B)H^\ominus + n_A n_B w/(n_A + n_B) \qquad 6.18.1$$

where $H^\ominus$ is independent of n_A, n_B. Differentiating with respect to n_A, n_B in turn we obtain for the partial enthalpies

$$H_A = H^\ominus + n_B^2 w/(n_A + n_B)^2 = H^\ominus + x^2 w \qquad 6.18.2$$

$$H_B = H^\ominus + n_A^2 w/(n_A + n_B)^2 = H^\ominus + (1-x)^2 w \qquad 6.18.3$$

so that the proper *enthalpy of change* from A to B is

$$H_B - H_A = (1 - 2x)w. \qquad 6.18.4$$

Since we are considering a condensed phase, (4) is effectively equivalent to

$$U_B - U_A = (1-2x)w. \qquad 6.18.5$$

The outstanding characteristic of (5) is that the energy required to convert a molecule A into a molecule B depends in a marked degree on what fraction of all the molecules is present in each form. Such a characteristic would not be expected when the process

$$A \to B \qquad 6.18.6$$

represents a chemical change of one isomer to another, nor in such a case do we find a lambda point. It is however not difficult to mention other interpretations of (6) which might reasonably be expected to have the characteristic just mentioned. Suppose for example we consider a regular array of polar molecules or atoms in a lattice. Suppose further that each moleule or atom can point in either of two opposite directions. Suppose finally that we denote the molecules by A or B according to the direction in which they point. Then it is easily understandable that the energy required to turn round a molecule or atom may depend markedly on how many other molecules or atoms are pointing in either direction. This behaviour is typical of systems called *cooperative*. The significance of the name should be clear from this and the following examples.

Another more complicated, but possibly more important, interpretation of (6) is for A to represent a state of molecular libration and B a state of molecular rotation.

Another example occurring in certain alloys is the following. Suppose we have an alloy of the composition ZnCu containing N atoms of Zn and N atoms of Cu arranged on a regular lattice of $2N$ lattice points. We can picture two extreme arrangements of the two kinds of atoms on the lattice, one completely ordered, the other completely random. In the completely ordered arrangement every alternate lattice point A is occupied by a Zn atom and the remaining lattice points B are occupied by Cu atoms. In the opposite extreme arrangement every lattice point A or B is occupied by either Zn or Cu atoms arranged at complete random. We can moreover consider intermediate arrangements such that a fraction x of the Zn atoms occupy A lattice points and the fraction $1-x$ of Zn atoms occupy B lattice points. The remaining lattice points are of course occupied by the Cu atoms. We can then without loss of generality take $x \geq \frac{1}{2}$. In such a system the average energy required to move a Zn atom from an A point to a B point will depend markedly on how many A points are already occupied by Zn atoms. It is therefore at least conceivable that such a system might have a lambda point.

As a matter of fact the alloy having the composition ZnCu does have a lambda point and the thermodynamic properties of this system can be at least semi-quantitatively represented by a Gibbs function of the form (6.17.9). This form was first suggested by Gorsky* and later independently derived by approximate statistical considerations by Bragg and Williams[†]. It is outside the scope of this book to consider this aspect of the phenomenon and we shall accordingly confine ourselves to a purely phenomenological thermodynamic investigation of some of the general properties of lambda points, among others the property leading to the name.

§6.19 *Alternative notation*

The notation which we have used to introduce the subject of *transitions of the second order* seems natural. It is not however the notation most used. For the sake of completeness we describe briefly the alternative notation.

A quantity s called the *degree of order* is defined by [†]

$$s = 2x - 1 \qquad 6.19.1$$

or

$$x = \tfrac{1}{2}(1+s). \qquad 6.19.2$$

In this notation formula (6.17.9) becomes

$$G_m = G_m^\ominus + \tfrac{1}{4}(1-s^2)w + RT\{\tfrac{1}{2}(1+s)\ln(1+s) + \tfrac{1}{2}(1-s)\ln(1-s) - \ln 2\}. \qquad 6.19.3$$

The equilibrium value of s is determined according to (6.17.12) by

$$\ln\{(1+s)/(1-s)\} = sw/RT \qquad 6.19.4$$

which is equivalent to

$$\tanh(ws/2RT) = s. \qquad 6.19.5$$

Using (6.17.15) we can transform (5) to

$$T_\lambda/T = (\tanh^{-1} s)/s. \qquad 6.19.6$$

These formulae, of course, contain nothing which is not already contained in the formulae of §6.17. It is merely a historical accident that pioneer workers in this field used the variable s instead of x.

§6.20 *Lambda point*

We have seen how a Gibbs function of the form (6.17.9) leads without any

---

* Gorsky, Z. Phys. 1928 **50** 64.
† Bragg and Williams, Proc. Roy. Soc. A 1934 **145** 699.

further assumption to the occurrence of a *transition of the second order* and we have explained how this type of behaviour can occur in a *cooperative system*. We do not assert that a Gibbs function of approximately this form is the origin of all transitions of the second order. Still less do we assert that a Gibbs function of this form accounts accurately for any transition of the second order. We merely assert that the form (6.17.9) of the Gibbs function is one possible form which leads to the occurrence of a lambda point having certain general characteristics which we shall describe. We shall continue to make use of the particular forms of thermodynamic functions described in §6.17 merely for illustrative purposes.

From figure 6.1, or more accurately by calculation from (6.17.13), we can determine the equilibrium value of x as a function of T. The result is given in figure 6.2, where $s = 2x - 1$ is plotted against T/T_λ. We notice that at

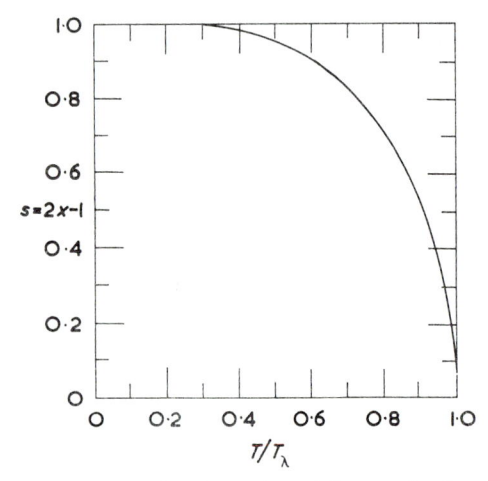

Fig. 6.2. Dependence of equilibrium value of degree of order on temperature

temperatures immediately below T_λ the equilibrium value of s changes rapidly with temperature and at temperatures below $\frac{1}{2}T_\lambda$ this equilibrium value differs hardly appreciably from unity. There is then a rapid change of the equilibrium value of s in the temperature range between T_λ and $\frac{1}{2}T_\lambda$. Associated with this change in s there is a rapid change in the part of the proper energy or enthalpy which depends on s namely the term

$$x(1-x)w = \tfrac{1}{4}(1-s^2)w. \qquad 6.20.1$$

This is shown in figure 6.3. The term (1) occurs in the energy additional to other terms due to the translational and internal degrees of freedom of the molecules. Thus as the temperature is decreased through the lambda point

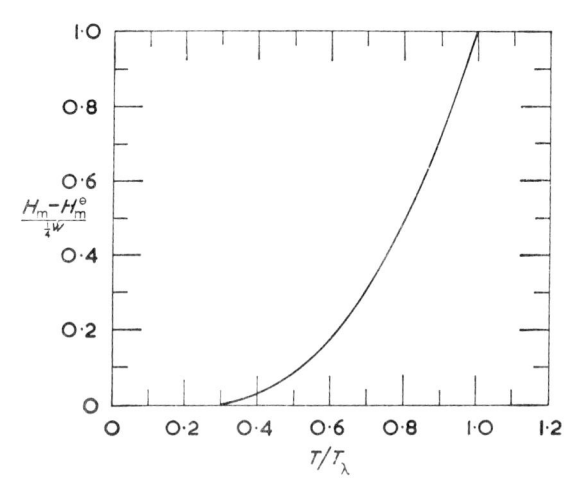

Fig. 6.3. Temperature dependence of enthalpy due to variation in degree of order

there is a sudden change in the temperature coefficient of the enthalpy, or in other words a discontinuity in the heat capacity C. This is shown in figure 6.4. The shape of the curve recalls a Greek capital Λ whence the name *lambda point* suggested by Ehrenfest.*

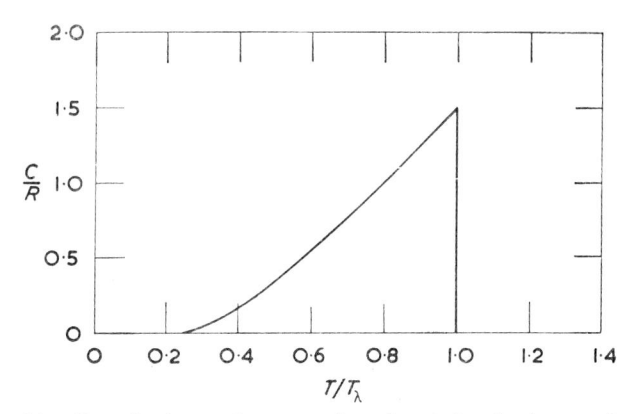

Fig. 6.4. Contribution to heat capacity of variation in degree of order

For the particular model considered in detail, we observe that in the immediate neighbourhood below the lambda temperature

$$\partial H/\partial s = 0 \qquad 6.20.2$$
$$\partial S/\partial s = 0 \qquad 6.20.3$$
$$\mathrm{d}s/\mathrm{d}T = \infty \qquad 6.20.4$$

* Keesom, Helium, Elsevier, 1942 p. 216.

in such a manner that

$$(\partial H/\partial s)(ds/dT) \text{ is finite} \qquad 6.20.5$$

$$(\partial S/\partial s)(ds/dT) \text{ is finite.} \qquad 6.20.6$$

The properties (5) and (6) are independent of the choice of s. On the other hand the relations (2), (3), (4) depend on the definition of s. For example if we replace s by $\sigma = s^2$, then

$$\partial H/\partial \sigma \text{ is finite} \qquad 6.20.7$$

$$\partial S/\partial \sigma \text{ is finite.} \qquad 6.20.8$$

We may then describe a transition of the second order as a discontinuity in C, with continuity of H, S, G, at a certain temperature T_λ called the *lambda point*.

The lambda point known longest is the one discovered by Curie and therefore called the *Curie point*, below which a substance such as iron has permanent magnetization and above which it has not. The Curie point will be referred to again in chapter 11.

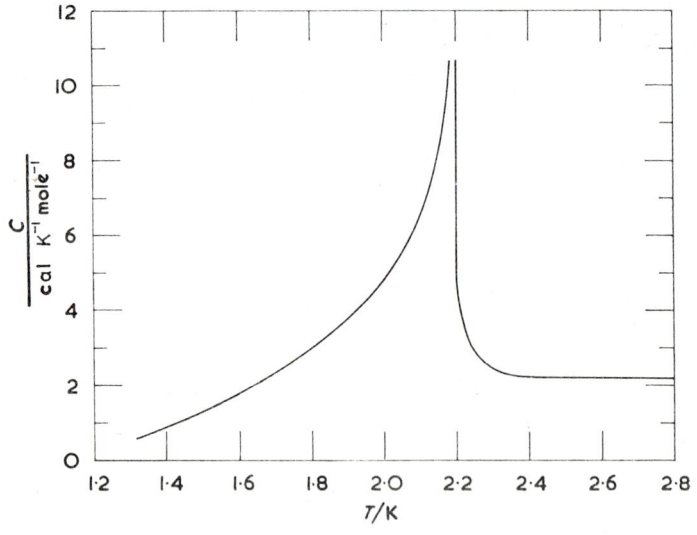

Fig. 6.5. Heat capacity of liquid helium near lambda point

Probably the most interesting, most studied, but perhaps least understood lambda point is that of helium at 2.2 K. The experimental data* for C plotted against T are shown in figure 6.5.

* Keesom, Helium, Elsevier, 1942 p. 215.

Many other lambda points are known to occur in crystals and are usually associated with a sudden change in the extent to which the molecules in the crystal can rotate freely. Few however, if any, have been studied in such detail as to be completely understood.

§6.21 *Comparison with phase change and critical point*

Since a substance has measurably different properties above and below the lambda point, there is a temptation to regard a *transition of the second order* as a kind of *phase change*. The expression *phase change of the second order* has been used, but as it has in the past led to considerable confusion it is better avoided.*

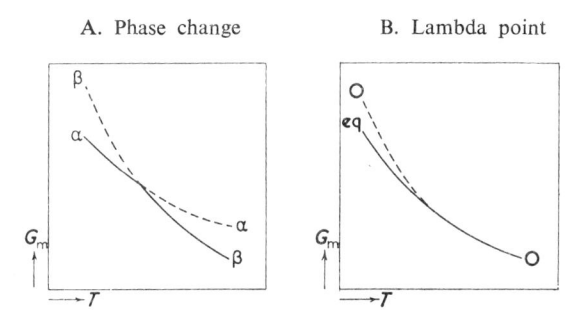

Fig. 6.6. Contrast between phase change and lambda point

The contrast between a lambda point and a phase change may be made clear by a plot of the proper Gibbs function against the temperature. This is shown in figure 6.6. Diagram A depicts a phase change. The curves of the two distinct phases α and β cut at the transition point, the dotted portions of the curves representing metastable states. Diagram B depicts a transition of the second order. The curve marked O represents the Gibbs function of a hypothetical phase with $s=0$, which is usually associated with complete randomness. The curve marked eq represents the Gibbs function of a phase in which at each temperature s has its equilibrium value. Below the lambda point the dotted curve marked O lies above the curve marked eq and consequently the former represents metastable states.

At the lambda point the two curves touch. We might ask what happens to the eq curve above the lambda point. If we extend the eq curve by the simplest analytical formula, ignoring physics, the curve would continue below the O curve, thus suggesting that it represents states more stable than the O curve. On further study we should however find that this hypothetical

* Guggenheim, Proc. Acad. Sci. Amst. 1934 37 3.

curve corresponds to negative values of s^2 and has therefore no physical meaning. It is therefore safer and more profitable to forget about such a curve.

On the other hand a comparison between a lambda point and a critical point, if not carried too far, is less dangerous. At temperatures below the lambda point there is a stable phase with a value of s determined by the temperature and there can also be a metastable phase with $s=0$; the latter can in fact sometimes be realized in practice by sudden chilling from a temperature above the lambda point. The difference between these two phases, measured by the values of s^2 gradually decreases as the temperature is raised and vanishes at the lambda point when the two phases become identical. This recalls the behaviour of liquid and vapour phases at the critical point, but here the resemblance ends.

§6.22 *Dependence of lambda point on pressure*

Up to this point we have considered how a transition of the second order occurs at a certain temperature, disregarding the pressure. This is in practice justifiable for most such transitions, but in principle there can be a dependence on the pressure. In practice the only known example where pressure changes are likely to be important is that of liquid helium. Let us then consider how the lambda point is affected when the pressure is changed.

In the particular model represented by (6.17.9) the dependence on pressure would result from the energy parameter w being a function of the pressure. We shall however not assume this model nor any other detailed model, but shall rather derive formulae of complete generality.

Regarding G as a function of s, as well as of T, P we have

$$\mathrm{d}G = -S\mathrm{d}T + V\mathrm{d}P + (\partial G/\partial s)\mathrm{d}s \qquad 6.22.1$$

and differentiating throughout with respect to s

$$\mathrm{d}(\partial G/\partial s) = -(\partial S/\partial s)\mathrm{d}T + (\partial V/\partial s)\mathrm{d}P + (\partial^2 G/\partial s^2)\mathrm{d}s. \qquad 6.22.2$$

Now the equilibrium value of s at each temperature is determined by

$$\partial G/\partial s = 0 \quad \text{(equilibrium)} \qquad 6.22.3$$

and in particular at the lambda point

$$s = 0 \quad \text{(lambda point)}. \qquad 6.22.4$$

If then we follow the lambda point at varying pressure we have (3) and owing to (4) we have

$$ds = 0 \quad \text{(lambda point)}. \qquad 6.22.5$$

Substituting (3) and (5) into (2) we obtain

$$-(\partial S/\partial s)dT + (\partial V/\partial s)dP = 0 \quad \text{(lambda point)} \qquad 6.22.6$$

or

$$\frac{dT_\lambda}{dP} = \frac{(\partial V/\partial s)_{s=0}}{(\partial S/\partial s)_{s=0}}. \qquad 6.22.7$$

Formula (7) describes in the most general way how the temperature of the lambda point depends on the pressure. The right side of (7) can however usefully be transformed into alternative forms more directly related to experimental quantities.

We accordingly multiply numerator and denominator of (7) by ds/dT, where s here denotes the equilibrium value. We obtain

$$\frac{dT_\lambda}{dP} = \frac{\partial V}{\partial s}\frac{ds}{dT} \bigg/ \frac{\partial S}{\partial s}\frac{ds}{dT} \qquad 6.22.8$$

where every quantity on the right side is given its equilibrium value at or immediately below T_λ. We shall now examine the physical significance of the numerator and denominator on the right of (8).

Let us use the superscripts $^-$ and $^+$ to denote the value of quantities immediately below and immediately above the temperature T_λ. Then we have

$$G^- = G^+ \qquad 6.22.9$$

$$H^- = H^+ \qquad 6.22.10$$

$$S^- = S^+ \qquad 6.22.11$$

$$C^- = C^+ + T_\lambda(\partial S/\partial s)(ds/dT) \qquad 6.22.12$$

so that the denominator on the right of (8) is $(C^- - C^+)/T_\lambda$.

Similarly if α denotes coefficient of thermal expansion

$$V^- = V^+ = V_\lambda \qquad 6.22.13$$

$$\alpha^- V_\lambda = \alpha^+ V_\lambda + (\partial V/\partial s)(ds/dT) \qquad 6.22.14$$

so that the numerator in (8) is $(\alpha^- - \alpha^+)V_\lambda$. Hence substituting (12) and (14) into (8) we obtain

$$dT_\lambda/dP = (\alpha^- - \alpha^+)V_\lambda T_\lambda/(C^- - C^+). \qquad 6.22.15$$

This formula shows how the effect of pressure on the lambda point is related to the discontinuities in C and in α.

Returning to (7), instead of multiplying numerator and denominator by ds/dT, we multiply by ds/dP, obtaining

$$\frac{dT_\lambda}{dP} = \frac{\partial V}{\partial s}\frac{ds}{dP} \bigg/ \frac{\partial S}{\partial s}\frac{ds}{dP}. \qquad 6.22.16$$

But if κ_T denotes isothermal compressibility, we have

$$\kappa_T^- V_\lambda = \kappa_T^+ V_\lambda - (\partial V/\partial s)(ds/dP). \qquad 6.22.17$$

Similarly

$$\left(\frac{\partial S}{\partial P}\right)^- = \left(\frac{\partial S}{\partial P}\right)^+ + \frac{\partial S}{\partial s}\frac{ds}{dP} \qquad 6.22.18$$

and so using Maxwell's relation (1.47.4) we have

$$\alpha^- V_\lambda = \alpha^+ V_\lambda - (\partial S/\partial s)(ds/dP). \qquad 6.22.19$$

Substituting (17) and (19) into (16) we obtain

$$dT_\lambda/dP = (\kappa_T^- - \kappa_T^+)/(\alpha^- - \alpha^+). \qquad 6.22.20$$

This formula relates the dependence of the lambda point on the pressure to the discontinuities in α and κ_T.

Formulae (15) and (20) are due to Ehrenfest.*

§6.23 *Transitions of higher order*

In an ordinary *phase change*, which we may call a *transition of the first order*, we have

$$\left.\begin{array}{ll} G & \text{continuous} \\ S = -\partial G/\partial T & \text{discontinuous} \end{array}\right\} \text{1st order transitions.}$$

In the *transitions of the second order*, which we have been discussing, we have

$$\left.\begin{array}{ll} G, \partial G/\partial T & \text{continuous} \\ C = -T\partial^2 G/\partial T^2 & \text{discontinuous} \end{array}\right\} \text{2nd order transitions.}$$

In a like manner we can define a *transition of the third order* by

$$\left.\begin{array}{ll} G, \partial G/\partial T, \partial^2 G/\partial T^2 & \text{continuous} \\ \partial^3 G/\partial T^3 & \text{discontinuous} \end{array}\right\} \text{3rd order transitions.}$$

* Ehrenfest, Proc. Acad. Sci. Amst. 1933 **36** 153.

SYSTEMS OF CHEMICALLY REACTING SPECIES 265

It is possible that transitions of the third order exist. It is further possible to extend the above definitions to transitions of still higher order. We shall however not pursue this matter any further.

§6.24 Components and degrees of freedom

Since the equilibrium condition for the chemical change

$$0 = \sum_B v_B B \qquad 6.24.1$$

is given by (6.01.7)

$$\sum_B v_B \mu_B = 0 \qquad 6.24.2$$

all variations of temperature, pressure, and composition consistent with chemical equilibrium must satisfy

$$\sum_B v_B d\mu_B = 0. \qquad 6.24.3$$

This is a relation between the chemical potentials additional to and independent of the Gibbs–Duhem relations. The existence of this relation reduces by one the number of degrees of freedom of the system.

Let us consider a particular example, namely a gaseous mixture of N_2, H_2, and NH_3, regarded as perfect. This single-phase system can be described by T, P, x_{N_2}, x_{H_2}, x_{NH_3} subject to the identity

$$x_{N_2} + x_{H_2} + x_{NH_3} = 1 \qquad 6.24.4$$

or alternatively by T, P, μ_{N_2}, μ_{H_2}, μ_{NH_3} subject to the Gibbs–Duhem relation

$$x_{N_2} d\mu_{N_2} + x_{H_2} d\mu_{H_2} + x_{NH_3} d\mu_{NH_3} = 0. \qquad 6.24.5$$

Hence in the absence of chemical reaction between the three components the system has four degrees of freedom. If however, for example by introducing a catalyst, we enable the process

$$N_2 + 3H_2 \rightleftharpoons 2NH_3 \qquad 6.24.6$$

to attain equilibrium, then there is the further restriction

$$\mu_{N_2} + 3\mu_{H_2} = 2\mu_{NH_3} \qquad 6.24.7$$

which reduces the number of degrees of freedom from four to three. This situation is sometimes described by saying that of the three species N_2, H_2, and NH_3 there are only two *independent components*. Whether or not this terminology is adopted the number of degrees of freedom is certainly three.

As a second example consider the system consisting of PCl_3, Cl_2, and PCl_5. There are three chemical species but owing to the equilibrium condition for the reaction

$$PCl_5 \rightleftharpoons PCl_3 + Cl_2 \qquad 6.24.8$$

there are only two independent components. For the gaseous phase alone the situation is similar to that in the system N_2, H_2, and NH_3. There are two independent components in one phase and so three degrees of freedom. If we consider the system consisting of the solid phase PCl_5 together with the gaseous phase we have two independent components in two phases and so two degrees of freedom. This means that there are two independent variables which we shall take to be the temperature T and the stoichiometric ratio r of Cl to P in the gas phase. The temperature determines the equilibrium constant K for the process

$$PCl_5(s) \rightleftharpoons PCl_3(g) + Cl_2(g) \qquad 6.24.9$$

and the fugacities are then determined by the simultaneous equations

$$p_{PCl_3} p_{Cl_2} = K \qquad 6.24.10$$

$$(3p_{PCl_3} + 2p_{Cl_2})/p_{PCl_3} = r. \qquad 6.24.11$$

In the particular case $r = 5$ the stoichiometric composition of the gas phase is the same as that of the solid phase PCl_5. In this case some authors go so far as to describe the system as of one component PCl_5. This attitude has nothing to recommend it. We have seen that the system as initially described has two degrees of freedom. These two degrees of freedom are of course reduced to one by specifying the value of r but the value $r = 5$ has no unique thermodynamic feature. The statement that the ratio of Cl to P in the gas phase is equal to its ratio in the solid phase is no different in kind from the statement that the ratio in the gas phase is one half, or double, the ratio in the solid phase. The distinction between $r = 5$ and other values of r is artificial and pointless. Furthermore it can lead to confusion. Suppose we are interested in the surface phase between solid and gas. Then although the ratio of Cl to P may be 5 in both gas phase and solid there is no reason to expect the ratio to be 5 in the surface phase. In other words there may well be preferential adsorption of either PCl_3 or Cl_2 and this can not be described in terms of the single component PCl_5. Again suppose there is a gravitational field. Then, as we shall see in chapter 9, the proportion of PCl_3 to Cl_2 will vary from layer to layer and can have the value unity at one height only. It is then essential to treat the system as of two components even though the overall stoichiometric composition may be that of PCl_5.

Now consider a system in which several chemical changes can take place. Some such changes may be expressible as linear combinations of others, but there will always be a definite number of chemical changes which are linearly independent. Consider for example a system consisting of solid graphite and a perfect gaseous mixture O_2, CO, CO_2. Then of the chemical changes

$$C + \tfrac{1}{2}O_2 \rightarrow CO \qquad 6.24.12$$
$$C + O_2 \rightarrow CO_2 \qquad 6.24.13$$
$$CO + \tfrac{1}{2}O_2 \rightarrow CO_2 \qquad 6.24.14$$
$$C + CO_2 \rightarrow 2CO \qquad 6.24.15$$

the third is obtained by subtracting the first from the second, while the fourth is obtained by subtracting the third from the first. Thus only two of these changes are independent. By a comparison of (1) and (2) it is clear that independent chemical processes have independent equilibrium conditions, whereas linearly related chemical processes have linearly related equilibrium conditions. Hence each linearly independent chemical equilibrium corresponds to a restrictive relation between the chemical potentials leading to a decrease by unity in the number of degrees of freedom. For example in the two-phase system consisting of solid graphite and a gaseous mixture of O_2, CO, CO_2 the effect of the two independent chemical equilibria is to reduce the number of degrees of freedom from four to two; thus the state of the system is completely determined by the temperature and the pressure. Incidentally in this particular system at equilibrium the amount of free O_2 is so small as to be undetectable. The system may therefore be more simply described as a two-phase system containing the three species C, CO, and CO_2 between which there is a single chemical reaction

$$C + CO_2 \rightarrow 2CO. \qquad 6.24.16$$

The equilibrium condition for this process reduces the number of degrees of freedom from three to two. Whichever way we consider the system we find that the number of degrees of freedom is two. Whether we regard the system as consisting of four components with two independent chemical processes, of three components with one independent chemical process, or of two independent components is a mere difference of terminology without practical importance.

CHAPTER 7

SOLUTIONS OF ELECTROLYTES

§7.01 Characteristics of strong electrolytes

When certain substances such as common salt are dissolved in water, the solution has a comparatively high conductivity showing that charged ions must be present. We owe to Arrhenius the suggestion that for these substances, called *strong electrolytes*, the solute is composed largely of the free ions, such as Na^+ and Cl^- in the case of common salt. Study of the optical properties by Bjerrum* led him in 1909 to the conclusion that at least in dilute solutions there are at most very few undissociated molecules and in many such cases the properties of the solution can be accurately accounted for on the assumption that no undissociated molecules are present.

It would be outside the province of this book to discuss whether a dilute solution of a strong electrolyte contains a small fraction of undissociated molecules or none at all. All that matters is that the description of a salt solution as completely dissociated into independent ions, though admittedly an oversimplification, is at least an incomparably better model than any other of equal simplicity. We shall therefore compare the properties of every real solution of strong electrolytes with an idealized solution containing independent ions.

§7.02 Ionic mole ratios and ionic molalities

In accordance with the programme outlined in the previous section, we describe the composition of solutions of electrolytes in terms of the ions, not in terms of the undissociated molecules. We accordingly describe the composition of a solution containing one or more electrolytes by the *mole ratio* r_i of each ionic species i defined by

$$r_i = n_i/n_1 . \qquad 7.02.1$$

* Bjerrum, Proc. 7th Int. Cong. Pure and Appl. Chem. London 1909 Sect. 10 p. 58; Z. Elektrochem. 1918 **24** 321.

In practice it is customary instead of *mole ratios* to use *molalities* m_i defined by

$$m_i = r_i/r^{\ominus} \qquad 7.02.2$$

where $r^{\ominus}$ is a standard value of r corresponding to one mole of the ionic species per kilogramme of solvent and equal to $M_1/\text{kg mole}^{-1}$.

§7.03 *Electrical neutrality*

When we carry out our intention of describing the properties of electrolyte solutions in terms of the ionic species, we shall find that most of the formulae have a close resemblance to those for non-electrolytes. There is however one important difference, namely that the molalities m_i of all the ionic species are not independent because the solution as a whole is electrically neutral. We now proceed to express this condition mathematically.

We use the symbol z to denote the charge on an ion measured in units of the charge of a proton, so that for example

For
$\quad$ Na^+ $\quad$ $z=1$
$\quad$ Ba^{2+} $\quad$ $z=2$
$\quad$ La^{3+} $\quad$ $z=3$
$\quad$ Cl^- $\quad$ $z=-1$
$\quad$ SO_4^{2-} $\quad$ $z=-2$
$\quad$ PO_4^{3-} $\quad$ $z=-3$
$\quad$ $FeC_6N_6^{4-}$ $\quad$ $z=-4$.

We call z the *charge number* of the ion.

If then m_i denotes the molality of the ionic species i having a charge number z_i, the condition for *electrical neutrality* of the solution may be written

$$\sum_i z_i m_i = 0. \qquad 7.03.1$$

Alternatively if we use the subscript $_+$ to denote positively charged ions or *cations* and $_-$ to denote negatively charged ions or *anions*, then we may write (1) in the form

$$\sum_+ z_+ m_+ = \sum_- |z_-| m_- \qquad 7.03.2$$

wherein $|z_-| = -z_-$ is a positive integer.

Owing to the condition of *electrical neutrality* (1) or (2), a solution containing c ionic species, as well as the solvent, has c not c+1, independent components.

§7.04 Ionic absolute activities

Since most equilibrium conditions are expressible in a general, yet convenient, form in terms of absolute activities we shall make continual use of the absolute activity λ_i of each ionic species i. By following this procedure we shall in fact obtain formulae closely resembling those already obtained for non-electrolytes. There is however one important difference. We saw in the previous section that if there are c ionic species i and so c ionic molalities m_i, then only $c-1$ are independent. There must clearly be some analogous or related property of the set of c quantities λ_i. We shall now discover this property by considering the physical significance of the λ_i's first in particular cases and then in general.

Let us consider the distribution of NaCl between two phases, of which at least one α is a solution; the other β may be a solution in a different solvent or the solid phase. We shall now determine the equilibrium condition for NaCl ab initio on the same lines as in §1.39 but in terms of Na$^+$ and Cl$^-$. We assume the temperature, but not necessarily the pressure, to be the same in the two phases. Suppose now a small quantity dn_{Na^+} of Na$^+$ and a small quantity dn_{Cl^-} of Cl$^-$ to pass from the phase α to the phase β, the temperature of the whole system being kept constant. Then the increase in the Helmholtz function is given by

$$dF = -P^\alpha dV^\alpha - \mu^\alpha_{Na^+} dn_{Na^+} - \mu^\alpha_{Cl^-} dn_{Cl^-}$$
$$- P^\beta dV^\beta + \mu^\beta_{Na^+} dn_{Na^+} + \mu^\beta_{Cl^-} dn_{Cl^-}. \qquad 7.04.1$$

By an argument analogous to that of §1.39, if the two phases are in mutual equilibrium with respect to the NaCl, the process being considered must be reversible and so the increase in the free energy must be equal to the work done on the system. Thus

$$dF = -P^\alpha dV^\alpha - P^\beta dV^\beta. \qquad 7.04.2$$

Subtracting (2) from (1) we obtain

$$(\mu^\beta_{Na^+} - \mu^\alpha_{Na^+})dn_{Na^+} + (\mu^\beta_{Cl^-} - \mu^\alpha_{Cl^-})dn_{Cl^-} = 0. \qquad 7.04.3$$

The condition for electrical neutrality (7.03.2) in this case takes the simple form

$$dn_{Na^+} = dn_{Cl^-} = dn. \qquad 7.04.4$$

Substituting (4) into (3) we have

$$(\mu^\beta_{Na^+} - \mu^\alpha_{Na^+} + \mu^\beta_{Cl^-} - \mu^\alpha_{Cl^-})dn = 0 \qquad 7.04.5$$

or dividing by dn

$$\mu^\alpha_{Na^+} + \mu^\alpha_{Cl^-} = \mu^\beta_{Na^+} + \mu^\beta_{Cl^-}. \qquad 7.04.6$$

Since according to the definition of λ_i

$$\mu_i = RT \ln \lambda_i \qquad 7.04.7$$

we may rewrite (6) as

$$\ln \lambda^\alpha_{Na^+} + \ln \lambda^\alpha_{Cl^-} = \ln \lambda^\beta_{Na^+} + \ln \lambda^\beta_{Cl^-} \qquad 7.04.8$$

or

$$\lambda^\alpha_{Na^+} \lambda^\alpha_{Cl^-} = \lambda^\beta_{Na^+} \lambda^\beta_{Cl^-}. \qquad 7.04.9$$

We thus see that any phase equilibrium relating to NaCl involves only the sum

$$\mu_{Na^+} + \mu_{Cl^-} \qquad 7.04.10$$

or the product

$$\lambda_{Na^+} \lambda_{Cl^-}. \qquad 7.04.11$$

In the same way an equilibrium relating to $BaCl_2$ would involve only the sum

$$\mu_{Ba^{2+}} + 2\mu_{Cl^-} \qquad 7.04.12$$

or the product

$$\lambda_{Ba^{2+}} \lambda^2_{Cl^-} \qquad 7.04.13$$

and an equilibrium relating to $LaCl_3$, only the sum

$$\mu_{La^{3+}} + 3\mu_{Cl^-} \qquad 7.04.14$$

or the product

$$\lambda_{La^{3+}} \lambda^3_{Cl^-} \qquad 7.04.15$$

and so on.

But it might be asked what about an equilibrium relating to the chloride ion by itself? The answer is that the transfer of a chloride ion, or any other ion alone from one phase to another involves a transfer of electrical charge, that is to say an electric current. We shall consider such processes in detail in the following chapter on *electrochemical systems*. Meanwhile as long as we exclude processes involving an electric current, and in this chapter we do so, we shall meet the μ_i's and λ_i's only in combinations corresponding to zero net electric charge. We can express this mathematically by stating that the only linear combinations

$$\sum_i v_i \mu_i \qquad 7.04.16$$

and the only products

$$\prod_i (\lambda_i)^{v_i} \qquad 7.04.17$$

which will occur will be those in which the v_i's satisfy the relation

$$\sum_i v_i z_i = 0. \qquad 7.04.18$$

This means that, apart from electrochemical flow of charge, with which we are not concerned in this chapter, we could in each phase assign an arbitrary value to the absolute activity λ_i of one ionic species, for instance the chloride ion. The λ_i's of the remaining ions would then be unambiguously determined. Nothing is however gained by thus arbitrarily fixing the values of the λ_i's. We can just as well leave the arbitrary factor in the λ_i's undetermined, knowing that only those combinations (17) of the λ_i's satisfying (18) will ever occur and that in these combinations the arbitrary factors cancel.

§7.05 *Ideal dilute and real solutions*

It would be rational, as in the case of non-electrolytes, first to define an ideal dilute solution of electrolytes and thereafter to compare the properties of real solutions with ideal dilute solutions. Since however no solution of a strong electrolyte is even approximately ideal dilute even at the highest dilution at which accurate measurements can be made, there seems no point in devoting space to such solutions. We therefore pass straight to real solutions, of which ideal dilute solutions constitute an idealized limiting case.

§7.06 *Osmotic coefficient of the solvent*

We can define the *osmotic coefficient* ϕ of the solvent in complete analogy with the case where the solute species are non-electrolytes merely replacing r_s by r_i the mole ratio of an ionic species, or m_s by m_i the molality of an ionic species. For electrolyte solutions (5.10.1) becomes

$$-\ln(\lambda_1/\lambda_1^0) = -\ln(p_1/p_1^0) = \phi \sum_i r_i = \phi r^\ominus \sum_i m_i. \qquad 7.06.1$$

We shall use (1) to describe the several equilibrium properties of the solvent. Before doing so we however point out that if the solution contains non-electrolytes as well as electrolytes, the former may be included formally inside the summation Σ_i. We merely treat an electrically uncharged species as if it were an ionic species with $z = 0$.

§7.07 *Freezing point and boiling point*

Formula (5.17.11) relating the freezing point T of a solution to the

SOLUTIONS OF ELECTROLYTES 273

freezing point T^0 of the pure solvent, becomes

$$\phi \sum_i r_i = \phi r^\ominus \sum_i m_i = \phi(M_1/\text{kg mole}^{-1}) \sum_i m_i = \langle \Delta_f H_1^0 \rangle (1/RT - 1/RT^0) \quad 7.07.1$$

where $\langle \Delta_f H_1^0 \rangle$ denotes the value of the proper enthalpy of fusion $\Delta_f H_1^0$ of the pure solvent averaged over the reciprocal temperature range $1/T^0$ to $1/T$. In (1) the value of ϕ is that at the freezing point of the solution.

The relation (5.18.1) between the boiling point T of a solution of involatile solutes and the boiling point T^0 of the pure solvent becomes

$$\phi \sum_i r_i = \phi r^\ominus \sum_i m_i = \phi(M_1/\text{kg mole}^{-1}) \sum_i m_i = \langle \Delta_e H_1^0 \rangle (1/RT^0 - 1/RT) \quad 7.07.2$$

where $\langle \Delta_e H_1^0 \rangle$ denotes the value of the proper enthalpy of evaporation $\Delta_e H_1^0$ of the pure solvent averaged over the reciprocal temperature range $1/T$ to $1/T^0$. In (2) the value of ϕ is that at the boiling point of the solution.

§7.08 *Osmotic pressure*

Formula (5.16.2) for the osmotic pressure Π becomes for a solution of electrolytes

$$\Pi \langle V_1 \rangle / RT = \phi \sum_i r_i. \quad 7.08.1$$

§7.09 *Ionic activity coefficient*

In analogy with (5.09.1) and (5.09.2) the activity coefficient γ_i of the ionic species i is related to the absolute activity λ_i by

$$\lambda_i = \lambda_i^\ominus m_i \gamma_i \quad 7.09.1$$

$$\gamma_i \to 1 \quad \text{as} \quad \sum_i m_i \to 0. \quad 7.09.2$$

The proportionality constant $\lambda_i^\ominus$ depends on the solvent and the temperature. Furthermore, as explained in §7.04 *in each solution* an arbitrary value may be assigned to λ_i for any one ionic species; the values for the remaining ionic species are then determined in that solution.

§7.10 *Mean activity coefficient of electrolyte*

Let us consider an electrolyte which consists of ν_+ cations R of charge number z_+ and ν_- anions X of charge number z_- so that according to the condition of electrical neutrality $\nu_+ z_+ + \nu_- z_- = 0$. The absolute activity $\lambda_{R,X}$

of the electrolyte $R_{\nu_+} X_{\nu_-}$ is then related to the absolute activities of the two ionic species by

$$\lambda_{R,X} = \lambda_R^{\nu_+} \lambda_X^{\nu_-}. \qquad 7.10.1$$

Substituting (7.09.1) into (1) we have

$$\lambda_{R,X} = (\lambda_R^\ominus m_R \gamma_R)^{\nu_+} (\lambda_X^\ominus m_X \gamma_X)^{\nu_-} \qquad 7.10.2$$

and in the limit of infinite dilution

$$\lambda_{R,X} \to (\lambda_R^\ominus m_R)^{\nu_+} (\lambda_X^\ominus m_X)^{\nu_-} \quad \text{as} \quad \sum_i m_i \to 0. \qquad 7.10.3$$

Since $\lambda_{R,X}$ and m_R and m_X are all well defined quantities it is clear from (3) that in spite of the indefiniteness in $\lambda_R^\ominus$ and $\lambda_X^\ominus$ separately, the product $(\lambda_R^\ominus)^{\nu_+} (\lambda_X^\ominus)^{\nu_-}$ is completely defined. Returning now to (2) since $\lambda_{R,X}$, m_R, m_X, and, as we have just seen, the product $(\lambda_R^\ominus)^{\nu_+} (\lambda_X^\ominus)^{\nu_-}$ are all well defined, it follows that the product $\gamma_R^{\nu_+} \gamma_X^{\nu_-}$ is also well defined.

We now introduce a quantity $\gamma_{R,X}$ called the *mean activity coefficient* of the electrolyte, related to γ_R and γ_X by

$$\gamma_{R,X}^{\nu_+ + \nu_-} = \gamma_R^{\nu_+} \gamma_X^{\nu_-}. \qquad 7.10.4$$

Substituting (4) into (2) we have

$$\lambda_{R,X} = (\lambda_R^\ominus)^{\nu_+} (\lambda_X^\ominus)^{\nu_-} m_R^{\nu_+} m_X^{\nu_-} \gamma_{R,X}^{\nu_+ + \nu_-}. \qquad 7.10.5$$

Since $\gamma_{R,X}$ is well defined, while γ_R and γ_X individually are not, it would be wrong to regard (4) as a definition of $\gamma_{R,X}$ in terms of γ_R and γ_X. Nevertheless formula (4) does contain something of physical significance. For let us consider a solution containing two cations R, R' and two anions X, X' from which we can form four different electrolytes, for each of which we can write a relation of the form (4). What these relations together tell us is that the four mean activity coefficients are not independent. We can best illustrate the point by a simple example. Let us consider the two cations Na^+, K^+ and the two anions Cl^-, NO_3^-. Then we have formally

$$\gamma_{Na,Cl}^2 = \gamma_{Na^+} \gamma_{Cl^-} \qquad 7.10.6$$

$$\gamma_{K,Cl}^2 = \gamma_{K^+} \gamma_{Cl^-} \qquad 7.10.7$$

$$\gamma_{Na,NO_3}^2 = \gamma_{Na^+} \gamma_{NO_3^-} \qquad 7.10.8$$

$$\gamma_{K,NO_3}^2 = \gamma_{K^+} \gamma_{NO_3^-}. \qquad 7.10.9$$

In a given solution each of the quantities on the left of formulae (6) to (9) is well defined, while the individual factors on the right are not. But these four formulae together lead to the physically significant result

$$\gamma_{Na, Cl}/\gamma_{K, Cl} = \gamma_{Na, NO_3}/\gamma_{K, NO_3}. \qquad 7.10.10$$

§7.11 Temperature dependence

Just as for non-ionic species, we have according to (5.03.10)

$$\partial \ln \lambda_i/\partial T = -H_i/RT^2 \qquad 7.11.1$$

so that according to (7.09.1)

$$\partial \ln(\lambda_i^\ominus \gamma_i)/\partial T = -H_i/RT^2. \qquad 7.11.2$$

Proceeding to the limit of infinite dilution (2) becomes

$$\partial \ln \lambda_i^\ominus/\partial T = -H_i^\infty/RT^2 \qquad 7.11.3$$

where H_i^∞ denotes the limiting value of H_i when $\Sigma_i m_i \to 0$. Now subtracting (3) from (2) we find

$$\partial \ln \gamma_i/\partial T = -(H_i - H_i^\infty)/RT^2. \qquad 7.11.4$$

For reasons previously given, only linear combinations of these formulae will occur of the type defined by (7.04.18). In particular for an electrolyte composed of v_+ cations R and v_- anions X, we have according to (1)

$$\partial \ln \lambda_{R, X}/\partial T = -H_{R, X}/RT^2 \qquad 7.11.5$$

where

$$H_{R, X} = v_+ H_R + v_- H_X \qquad 7.11.6$$

is the *partial enthalpy* of the electrolyte. Similarly from (4) we deduce

$$(v_+ + v_-)\partial \ln \gamma_{R, X}/\partial T = -(H_{R, X} - H_{R, X}^\infty)/RT^2 \qquad 7.11.7$$

where $H_{R, X}^\infty$ denotes the limiting value of $H_{R, X}$ as $\Sigma_i m_i \to 0$.

§7.12 Distribution of electrolyte between two solvents

The equilibrium condition for the distribution of an electrolyte consisting of v_+ cations R and v_- anions X between two solvents α and β can be written either in terms of the electrolytes as

$$\lambda_{R, X}^\alpha = \lambda_{R, X}^\beta \qquad 7.12.1$$

or in terms of the ions as

$$(\lambda_R^\alpha)^{v_+}(\lambda_X^\alpha)^{v_-} = (\lambda_R^\beta)^{v_+}(\lambda_X^\beta)^{v_-}. \qquad 7.12.2$$

According to (7.10.1) the two conditions are equivalent. Substituting (7.10.5) into (1) we obtain

$$\frac{[m_R^{\nu_+} m_X^{\nu_-} \gamma_{R,X}^{\nu_++\nu_-}]^\beta}{[m_R^{\nu_+} m_X^{\nu_-} \gamma_{R,X}^{\nu_++\nu_-}]^\alpha} = l_{R,X}^{\alpha\beta} \qquad 7.12.3$$

where

$$l_{R,X}^{\alpha\beta} = (\lambda_R^{\ominus\alpha}/\lambda_R^{\ominus\beta})^{\nu_+} (\lambda_X^{\ominus\alpha}/\lambda_X^{\ominus\beta})^{\nu_-} \qquad 7.12.4$$

and according to (7.11.3) we have

$$\partial \ln l_{R,X}^{\alpha\beta}/\partial T = (H_{R,X}^{\infty\beta} - H_{R,X}^{\infty\alpha})/RT^2 \qquad 7.12.5$$

We notice that the numerator of the right side is the limiting value as $\Sigma_i m_i \to 0$ of the *partial enthalpy of transfer of the electrolyte* from the solvent α to the solvent β.

§7.13 Solubility

For the equilibrium between the solid electrolyte composed of the ions R, X and a solution containing R, X and possibly other electrolytes, we have

$$\lambda_{R,X} = \lambda_{R,X}^S \qquad 7.13.1$$

where we denote the solid phase by the superscript ^{S} and the solution by no superscript.

Substituting from (7.10.5) into (1) we obtain

$$m_R^{\nu_+} m_X^{\nu_-} \gamma_{R,X}^{\nu_++\nu_-} = s_{R,X}^{\nu_++\nu_-} \qquad 7.13.2$$

where

$$s_{R,X}^{\nu_++\nu_-} = \lambda_{R,X}^S / (\lambda_R^{\ominus})^{\nu_+} (\lambda_X^{\ominus})^{\nu_-} \qquad 7.13.3$$

is called the *solubility product* of the electrolyte and $s_{R,X}$ is called the *mean solubility* of the electrolyte. Since

$$\partial \ln \lambda_{R,X}^S / \partial T = -H_{R,X}^S/RT^2 \qquad 7.13.4$$

we have, using this and (7.11.3) in (3)

$$\partial \ln s_{R,X}/\partial T = (H_{R,X}^\infty - H_{R,X}^S)/(\nu_+ + \nu_-)RT^2. \qquad 7.13.5$$

We notice that the numerator on the right of (5) is the limiting value as $\Sigma_i m_i \to 0$ of the *enthalpy of dissolution of the solid electrolyte* in the given solvent.

§7.14 Chemical reactions

If we consider the chemical reaction

$$0 = \sum_i v_i B_i \qquad 7.14.1$$

where some or all of the species B_i may be ionic, the condition of equilibrium, in the notation defined in §6.01 is according to (6.13.3)

$$\prod(m_i) \prod(\gamma_i) = K_m(T). \qquad 7.14.2$$

The fact that some or all of the reacting species may be ions has no effect on the form of (2). It is however of interest to notice that, owing to the conservation of net electric charge, it follows from (1) that

$$\sum_i v_i z_i = 0 \qquad 7.14.3$$

and so

$$\ln \prod(\gamma_i) \qquad 7.14.4$$

conforms to the type of product which is physically well defined according to the condition (7.04.18).

We shall illustrate the point by an example. Consider the reaction

$$0 = 2\mathrm{Fe}^{2+} + \mathrm{Sn}^{4+} - 2\mathrm{Fe}^{3+} - \mathrm{Sn}^{2+}. \qquad 7.14.5$$

According to (2) the equilibrium condition is

$$\frac{m_{\mathrm{Fe}^{2+}}^2 \, m_{\mathrm{Sn}^{4+}} \, \gamma_{\mathrm{Fe}^{2+}}^2 \, \gamma_{\mathrm{Sn}^{4+}}}{m_{\mathrm{Fe}^{3+}}^2 \, m_{\mathrm{Sn}^{2+}} \, \gamma_{\mathrm{Fe}^{3+}}^2 \, \gamma_{\mathrm{Sn}^{2+}}} = K_m \qquad 7.14.6$$

wherein the *activity coefficients product* is well defined. It can in fact be expressed in terms of mean activity coefficients as follows

$$\frac{\gamma_{\mathrm{Fe}^{2+}}^2 \, \gamma_{\mathrm{Sn}^{4+}}}{\gamma_{\mathrm{Fe}^{3+}}^2 \, \gamma_{\mathrm{Sn}^{2+}}} = \frac{\gamma_{\mathrm{Fe}^{2+}}^2 \, \gamma_{\mathrm{Sn}^{4+}} \, \gamma_{\mathrm{Cl}^-}^8}{\gamma_{\mathrm{Fe}^{3+}}^2 \, \gamma_{\mathrm{Sn}^{2+}} \, \gamma_{\mathrm{Cl}^-}^8} = \frac{\gamma_{\mathrm{Fe}^{2+},\mathrm{Cl}^-}^6 \, \gamma_{\mathrm{Sn}^{4+},\mathrm{Cl}^-}^5}{\gamma_{\mathrm{Fe}^{3+},\mathrm{Cl}^-}^8 \, \gamma_{\mathrm{Sn}^{2+},\mathrm{Cl}^-}^3}. \qquad 7.14.7$$

For the temperature dependence of K_m we have (6.14.2)

$$\mathrm{d}\ln K_m / \mathrm{d}T = \Delta H^\infty / RT^2 \qquad 7.14.8$$

where ΔH^∞ denotes the *enthalpy of reaction* at infinite dilution.

§7.15 Gibbs–Duhem relation for electrolyte solutions

For any phase whatever we have the Gibbs–Duhem relation (1.30.2). For a solution of electrolytes in a solvent 1 this becomes

$$SdT - VdP + n_1 d\mu_1 + \sum_i n_i d\mu_i = 0 \qquad 7.15.1$$

or considering variations of composition at constant temperature and pressure

$$n_1 d\mu_1 + \sum_i n_i d\mu_i = 0 \qquad (T, P \text{ const.}). \qquad 7.15.2$$

We may rewrite (2) in terms of absolute activities λ_i as

$$n_1 d \ln \lambda_1 + \sum_i n_i d \ln \lambda_i = 0 \qquad (T, P \text{ const.}). \qquad 7.15.3$$

According to the definition of mole ratios and molalities given in §5.02 and extended to ions in §7.02 we have

$$n_i/n_1 = r_i = r^\ominus m_i = (M_1/\text{kg mole}^{-1}) m_i. \qquad 7.15.4$$

If then we divide (2) and (3) throughout by n_1 and use (4) we obtain

$$d\mu_1 = -\sum_i r_i d\mu_i = -(M_1/\text{kg mole}^{-1}) \sum_i m_i d\mu_i \qquad (T, P \text{ const.}) \qquad 7.15.5$$

$$d \ln \lambda_1 = -\sum_i r_i d \ln \lambda_i = -(M_1/\text{kg mole}^{-1}) \sum_i m_i d \ln \lambda_i \qquad (T, P \text{ const.}).$$
$$7.15.6$$

As explained in §7.03 all variations of composition of an electrolyte solution are subject to the condition for electrical neutrality

$$\sum_i z_i m_i = 0 \qquad 7.15.7$$

so that

$$\sum_i z_i dm_i = 0. \qquad 7.15.8$$

The variations in formulae (1), (2), (3), (5), (6) are all subject to the condition (8); but for variations satisfying (8) these formulae hold just as well for electrolyte solutions as for other solutions.

We now recall the definition of the osmotic coefficient (7.06.1)

$$-\ln(\lambda_1/\lambda_1^0) = \phi \sum_i r_i = \phi (M_1/\text{kg mole}^{-1}) \sum_i m_i \qquad 7.15.9$$

and the definition of ionic activity coefficients γ_i by (7.09.1)

$$\lambda_i = \lambda_i^\ominus m_i \gamma_i. \qquad 7.15.10$$

Differentiating (9) with respect to changes of composition at constant temperature and pressure we obtain

$$d \ln \lambda_1 = -(M_1/\text{kg mole}^{-1}) d(\phi \sum_i m_i) \qquad (T, P \text{ const.}). \qquad 7.15.11$$

Taking logarithms of (10) and differentiating we obtain

$$d \ln \lambda_i = d \ln m_i + d \ln \gamma_i \quad (T, P \text{ const.}). \quad 7.15.12$$

Now substituting (11) and (12) into (6) we obtain

$$d\{(\phi-1)\sum_i m_i\} = \sum_i m_i \, d \ln \gamma_i \quad (T, P \text{ const.}) \quad 7.15.13$$

of the same form as formula (5.11.2) and due to Bjerrum*.

In particular for a solution of a single electrolyte having v_+ cations R and v_- anions X, formula (13) becomes

$$(v_+ + v_-)\partial\{(1-\phi)m\}/\partial m = -v_+ m \partial \ln \gamma_R/\partial m - v_- m \partial \ln \gamma_X/\partial m \quad 7.15.14$$

where m denotes the *molality of the electrolyte*. The mean activity coefficient $\gamma_{R,X}$ of the electrolyte is related to the ionic activity coefficients γ_R and γ_X by (7.10.4)

$$\gamma_{R,X}^{v_+ + v_-} = \gamma_R^{v_+} \gamma_X^{v_-}. \quad 7.15.15$$

We now divide (14) throughout by $(v_+ + v_-)m$ and use (15) obtaining

$$-\partial \ln \gamma_{R,X}/\partial m = m^{-1} \partial\{(1-\phi)m\}/\partial m \quad 7.15.16$$

or integrating from 0 to m

$$-\ln \gamma_{R,X} = \int_0^m \frac{\partial\{(1-\phi)m\}}{\partial m} \frac{dm}{m}. \quad 7.15.17$$

Just as in a solution of a single solute non-electrolyte, formula (16) or (17) may be used to determine either of the quantities γ or ϕ if the other is known as a function of composition at all molalities less than m. On the other hand the more general relation (13) should not be used in this manner, but rather as a check on the self-consistency of assumed formulae for ϕ and the γ_i's because it is also necessary for the γ_i's to satisfy the relations of the type

$$\partial \ln \gamma_i/\partial m_k = \partial \ln \gamma_k/\partial m_i. \quad 7.15.18$$

As an example of (17) suppose

$$1 - \phi = am^t \quad (a, t \text{ const.}). \quad 7.15.19$$

Then substituting (19) into (17) we obtain

$$-\ln \gamma_{R,X} = (1+t^{-1})am^t = (1+t^{-1})(1-\phi). \quad 7.15.20$$

---

* Bjerrum, Z. Physik. Chem. 1923 **104** 406.

§7.16 Limiting behaviour at high dilutions

It was already proved* over fifty years ago that deviations from ideality due to the long-range electrostatic interactions between ions in highly dilute electrolyte solutions are quite different from the deviations in non-electrolyte solutions.

The distinction can for a single solute be expressed in the form

$$1-\phi \propto m \quad \text{as} \quad m \to 0 \quad \text{(non-electrolyte)} \qquad 7.16.1$$

$$1-\phi \propto m^t \quad \text{as} \quad m \to 0 \quad (t<1) \quad \text{(electrolyte)}. \qquad 7.16.2$$

This distinction is most strikingly expressed in the form

$$\mathrm{d}(1-\phi)/\mathrm{d}m \to \text{finite limit} \quad \text{as} \quad m \to 0 \quad \text{(non-electrolyte)} \qquad 7.16.3$$

$$\mathrm{d}(1-\phi)/\mathrm{d}m \to \infty \quad \text{as} \quad m \to 0 \quad \text{(electrolyte)}. \qquad 7.16.4$$

The latter behaviour is shown graphically in figure 7.1 which is of historical interest being taken from a paper by Bjerrum[†] written as early as 1916.

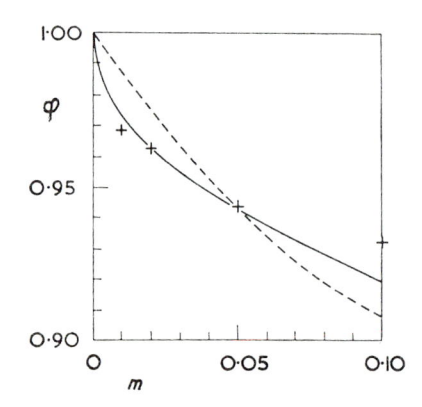

Fig. 7.1. Osmotic coefficients in aqueous solutions of potassium chloride. ++ Freezing-point measurements. ——— Electrostatic interaction according to Milner. ---- Incomplete dissocation ignoring electrostatic interaction

Milner* in 1912 had shown by statistical methods that the theoretical value of t is near $\frac{1}{2}$. Various values of t in (2) were used empirically in the period around 1922, some authors using different values of t for different electrolytes. Brønsted[‡] pointed out in 1922 that in the limit of high dilutions

---

* Milner, Phil. Mag. 1912 **23** 551.
† Bjerrum, 16te Skand. Naturforskermøte 1916 p. 229.
‡ Brønsted, J. Amer. Chem. Soc. 1922 **44** 938.

the value of t and likewise the proportionality factor in (2) must be the same for all electrolytes of the same charge type. For 1–1 electrolytes Brønsted proposed $t=\frac{1}{2}$. Finally in 1923 Debye and Hückel* determined by a statistical treatment the theoretical law valid in the limit $m\to 0$. According to this law $t=\frac{1}{2}$ and the proportionality constant in (2) is also determined by the theory.

§7.17 *Limiting law of Debye and Hückel*

As already mentioned the behaviour of a strong electrolyte in the limit of high dilution is given quantitatively by the formulae due to Debye and Hückel.* We shall now specify these formulae. All the deviations from ideality are most concisely expressed in terms of two characteristic lengths denoted by s and κ^{-1}. The definition of s is

$$s = e^2/4\pi\varepsilon_0\varepsilon_r kT \qquad 7.17.1$$

where e is the elementary charge, ε_0 is the rationalized permittivity of empty space (so that $4\pi\varepsilon^0$ is the unrationalized permittivity of empty space), ε_r is the relative permittivity ('dielectric constant') of the solvent and k is the Boltzmann constant. The other characteristic length κ^{-1} is defined by

$$\kappa^2 = 8\pi L\varrho s \sum_i \tfrac{1}{2} z_i^2 n_i/n_1 M_1 \qquad 7.17.2$$

where L is the Avogadro constant, ϱ is the density of the solvent, and M_1 is the proper mass of the solvent.

If G denotes the Gibbs function of the solution and G^{id} denotes the Gibbs function of an ideal dilute solution of the same composition, then at high dilutions

$$(G-G^{\mathrm{id}})/RT = -\tfrac{1}{3}\sum_i n_i z_i^2 \kappa s. \qquad 7.17.3$$

Differentiating (3) with respect to n_1 and noting that $\kappa \propto n_1^{-\frac{1}{2}}$ we obtain

$$(1-\phi)\sum_i n_i/n_1 = (\mu_1-\mu_1^{\mathrm{id}})/RT = \tfrac{1}{6}\sum_i n_i z_i^2 \kappa s/n_1 \qquad 7.17.4$$

so that

$$1-\phi = \tfrac{1}{6}(\sum_i n_i z_i^2/\sum_i n_i)\kappa s. \qquad 7.17.5$$

Differentiating (3) with respect to n_i and noting that $\kappa \propto (\Sigma_i n_i z_i^2)^{\frac{1}{2}}$ we obtain

$$-\ln \gamma_i = \tfrac{1}{2} z_i^2 \kappa s. \qquad 7.17.6$$

* Debye and Hückel, Phys. Z. 1923 **24** 185.

We readily verify that Bjerrum's condition (7.15.13)

$$\mathrm{d}\{(1-\phi)\sum_i m_i\} = -\sum_i m_i \mathrm{d} \ln \gamma_i \qquad (T, P \text{ const.}) \qquad 7.17.7$$

is satisfied when used with (7.15.20) and $t = \frac{1}{2}$.

We now rewrite the above formulae for numerical calculations. We recall that the molality m_i of an ionic species i is defined by

$$m_i = r_i/r_i^{\ominus} = r_i/(M_1/\text{kg mole}^{-1}) = n_i/n_1 M_1 \text{ mole kg}^{-1}. \qquad 7.17.8$$

We also define the *ionic strength*, following Lewis and Randall*, by

$$I = \tfrac{1}{2} \sum_i z_i^2 m_i. \qquad 7.17.9$$

We further define a dimensionless parameter α depending on the nature of the solvent and on the temperature, by

$$\alpha = (2\pi L \varrho s^3 \text{ mole kg}^{-1})^{\frac{1}{2}} \qquad 7.17.10$$

and we observe that

$$\tfrac{1}{2}\kappa s = \alpha I^{\frac{1}{2}}. \qquad 7.17.11$$

We can now rewrite (3) as

$$(G - G^{\mathrm{id}})/RT = -\tfrac{2}{3}\alpha \sum_i z_i^2 n_i I^{\frac{1}{2}} \qquad 7.17.12$$

and (5) as

$$1 - \phi = \tfrac{1}{3}\alpha (\sum_i z_i^2 m_i / \sum_i m_i) I^{\frac{1}{2}} \qquad 7.17.13$$

and (6) as

$$-\ln \gamma_i = \alpha z_i^2 I^{\frac{1}{2}}. \qquad 7.17.14$$

In the simple case of a single electrolyte composed of ν_+ cations of charge number z_+ and ν_- anions of charge number z_- formula (13) becomes

$$1 - \phi = \tfrac{1}{3}\alpha\{(\nu_+ z_+^2 + \nu_- z_-^2)/(\nu_+ + \nu_-)\} I^{\frac{1}{2}}. \qquad 7.17.15$$

Using the condition for electrical neutrality

$$\nu_+ z_+ + \nu_- z_- = 0 \qquad 7.17.16$$

we can rewrite (15) as

$$1 - \phi = \tfrac{1}{3}\alpha z_+ |z_-| I^{\frac{1}{2}}. \qquad 7.17.17$$

From (14) it follows that the mean activity coefficient $\gamma_\pm$ of an electrolyte composed of ν_+ cations of charge number z_+ and ν_- anions of charge

---

* Lewis and Randall, J. Amer. Chem. Soc. 1921 **43** 1141.

number z_- is given by

$$-\ln \gamma_\pm = \alpha\{(v_+ z_+^2 + v_- z_-^2)/(v_+ + v_-)\}I^{\frac{1}{2}}. \qquad 7.17.18$$

Using (16) we can rewrite (18) as

$$-\ln \gamma_\pm = \alpha z_+ |z_-| I^{\frac{1}{2}}. \qquad 7.17.19$$

§7.18 *Aqueous solutions*

We now illustrate the formulae of the previous section by giving numerical values for water. At 0 °C we have

$$L = 0.60225 \times 10^{24} \text{ mole}^{-1}$$
$$\varrho = 0.9999 \times 10^3 \text{ kg m}^{-3}$$
$$e = 1.6021 \times 10^{-19} \text{ C}$$
$$4\pi\varepsilon_0 = 8.85416 \times 10^{-12} \text{ C V}^{-1} \text{ m}^{-1}$$
$$k = 1.3805 \times 10^{-23} \text{ J K}^{-1}$$
$$T = 273.15 \text{ K}$$
$$\varepsilon_r = 88.23.$$

As usual we take $r^\ominus = M_1/\text{kg mole}^{-1}$. These values lead to

$$s = 6.935 \text{ Å} \qquad \alpha = 1.123. \qquad 7.18.1$$

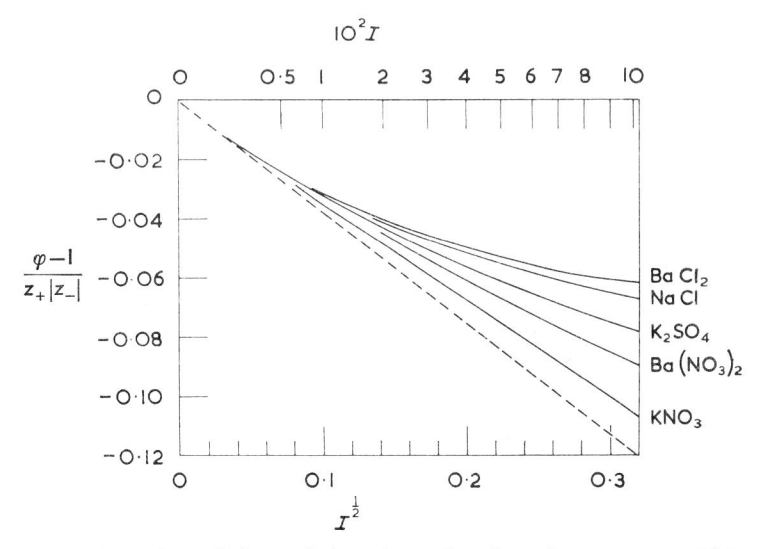

Fig. 7.2. Osmotic coefficients of electrolytes of various charge types at 0 °C

Similarly for water at 25 °C we have

$$\varrho = 0.9971 \times 10^3 \text{ kg m}^{-3}$$
$$T = 298.15 \text{ K}$$
$$\varepsilon_r = 78.54.$$

These lead to

$$s = 7.134 \text{ Å} \quad \alpha = 1.171. \qquad 7.18.2$$

Figure 7.2 gives a plot of $(\phi - 1)/|z_+ z_-|$, determined by freezing-point measurements, against $I^{\frac{1}{2}}$ for several electrolytes of various charge types. The tangent at the origin shown as a broken line has the theoretical slope $\frac{1}{3}\alpha$ according to formula (7.17.17).

§7.19 Less dilute solutions

The limiting law of Debye and Hückel described in the previous two sections is most valuable in providing a reliable means of extrapolating experimental data to infinite dilution, since experimental measurements determine only ratios of the values of γ in the several solutions. To determine values of γ itself in the several solutions some assumption has to be made concerning the value of γ in at least one such solution, for example the most dilute. The limiting law of Debye and Hückel provides the necessary assumption.

On the other hand this limiting law is accurate only at very high dilutions. For example when the solvent is water it is accurate enough at $I = 10^{-3}$, but already at $I = 10^{-2}$ deviations are experimentally detectable and at $I = 10^{-1}$ deviations are serious. In other solvents having smaller permittivities deviations from the limiting law appear at correspondingly lower ionic strengths.

For less dilute solutions various formulae can be used, all reducing to the limiting formula of Debye and Hückel at high dilutions and all more or less empirical at less high dilutions. Two of these will be described in the succeeding sections.

§7.20 Formulae of Debye and Hückel

We saw in §7.17 how the limiting law of Debye and Hückel can be conveniently and succinctly expressed in terms of two lengths s and κ^{-1}, the former being completely determined by the solvent and the temperature, and the latter being inversely proportional to the square root of the ionic strength. The formulae for less dilute solutions contain another characteristic length a

representing an average distance of closest approach between pairs of ions and therefore called the *ionic diameter*.

Formula (7.17.3) of the limiting law is replaced by*

$$(G-G^{id})/RT = -\tfrac{1}{3}\sum_i n_i z_i^2 \kappa s\{\tfrac{3}{2}(1+\kappa a)^{-1} - \tfrac{1}{2}\sigma(\kappa a)\} \qquad 7.20.1$$

where $\sigma(y)$ is a tabulated function[†] defined by

$$\sigma(y) = 3y^{-3}\{1+y-(1+y)^{-1} - 2\ln(1+y)\} \qquad 7.20.2$$

or when $y<1$

$$\sigma(y) = 1 + 3\sum_{t=1}^{\infty}(t+1)(t+3)^{-1}(-y)^t \qquad (y<1). \qquad 7.20.3$$

By differentiating (1) with respect to n_1 and noting that $\kappa \propto n_1^{-\frac{1}{2}}$ we deduce for the osmotic coefficient

$$(1-\phi)\sum_i n_i/n_1 = (\mu_1 - \mu_1^{id})/RT = -\tfrac{1}{2}\kappa d\{(G-G^{id})/RT\}/d\kappa$$
$$= \tfrac{1}{6}\sum_i n_i z_i^2 \kappa s n_1^{-1}\sigma(\kappa a) \qquad 7.20.4$$

so that

$$1-\phi = \tfrac{1}{6}(\sum_i n_i z_i^2/\sum_i n_i)\kappa s\sigma(\kappa a). \qquad 7.20.5$$

By differentiating (1) with respect to n_i and noting that $\kappa^2 \propto \Sigma_i n_i z_i^2$ we obtain

$$-\ln \gamma_i = \tfrac{1}{2}z_i^2 \kappa s(1+\kappa a)^{-1}. \qquad 7.20.6$$

As in §7.17 we now rewrite the above formulae for numerical calculations in terms of the ionic strength I and the dimensionless parameter α. Formula (1) becomes

$$(G-G^{id})/RT = -\tfrac{2}{3}\alpha \sum_i z_i^2 n_i I^{\frac{1}{2}}\{\tfrac{3}{2}(1+2\alpha I^{\frac{1}{2}}a/s)^{-1} - \tfrac{1}{2}\sigma(2\alpha I^{\frac{1}{2}}a/s)\}. \qquad 7.20.7$$

Formulae (5) and (6) become respectively

$$1-\phi = \tfrac{1}{3}\alpha(\sum_i z_i^2 m_i/\sum_i m_i)I^{\frac{1}{2}}\sigma(2\alpha I^{\frac{1}{2}}a/s) \qquad 7.20.8$$

$$-\ln \gamma_i = \alpha z_i^2 I^{\frac{1}{2}}(1+2\alpha I^{\frac{1}{2}}a/s)^{-1}. \qquad 7.20.9$$

For solutions containing only a single electrolyte composed of ν_+ cations of charge number z_+ and ν_- anions of charge number z_- these formulae, when we use the condition of electrical neutrality,

$$\nu_+ z_+ + \nu_- z_- = 0 \qquad 7.20.10$$

---

* Debye and Hückel, Phys. Z. 1923 **24** 185.
† Harned and Owen, Physical Chemistry of Electrolyte Solutions, Reinhold 2nd ed. 1950 p. 597.

reduce to

$$1-\phi = \tfrac{1}{3}\alpha z_+|z_-|I^{\frac{1}{2}}\sigma(2\alpha I^{\frac{1}{2}}a/s) \qquad 7.20.11$$

$$-\ln \gamma_\pm = \alpha z_+|z_-|I^{\frac{1}{2}}(1+2\alpha I^{\frac{1}{2}}a/s)^{-1}. \qquad 7.20.12$$

§7.21 *Specific interactions*

The formulae of Debye and Hückel enunciated in the previous section contain a single adjustable parameter a. By ad hoc adjustment of the value assigned to a it is usually possible to account for the behaviour of a solution containing a single electrolyte at ionic strengths not exceeding 0.1. In a mixture of several electrolytes these formulae predict the same value of $\gamma_\pm$ for all electrolytes of the same charge type and this contradicts the experimental facts. To conform with these facts further adjustable parameters are essential. We shall describe the use of a simple and convenient set of such parameters. For the sake of brevity and simplicity we shall here consider solutions containing only ions of charge number ± 1.

We begin by choosing a real or hypothetical single electrolyte as a standard with which to compare other electrolytes either alone or in a mixed solution. We shall find that it does not matter how this standard electrolyte is prescribed provided it resembles a typical real electrolyte. We shall mention alternative convenient choices of the standard electrolyte:

1. Some real electrolyte, say NaCl.
2. A hypothetical electrolyte accurately described by the formulae of Debye and Hückel with $2\alpha a/s = 1$ which corresponds to $a \approx 3\text{Å}$ in water.
3. A hypothetical electrolyte accurately described by the formulae of Debye and Hückel with $2a/s = 1$ which corresponds to $a \approx 3.5\text{Å}$ in water.

We use the superscript $^\ominus$ to denote the standard electrolyte and we repeat that what follows is independent of the choice of standard.

We assume that the Gibbs function G of a solution containing n_R cations R and n_X anions X is related to the Gibbs function $G^\ominus$ of a solution of the standard electrolyte at the same total molality m by

$$(G-G^\ominus)/RT = n_1 \sum_R \sum_X 2\beta_{R,X} m_R m_X. \qquad 7.21.1$$

The outstanding feature of this formula is that there is a single *interaction parameter* $\beta_{R,X}$ for every combination of a cation R and an anion X, that is to say one parameter for each electrolyte. The formula contains no parameter for interaction between two cations, nor between two anions. This is the essence of a principle enunciated by Brønsted* in 1921 and called by him the *principle of specific interaction of ions*. According to this principle two ions of the same sign will so rarely come close to each other in dilute solution that their mutual interactions may be assumed to be determined by their charges, but otherwise to be non-specific. Ions of the opposite sign on the other hand often come close to each other and their mutual interactions are therefore specific depending on their sizes, shapes, polarizabilities, and so on. When this principle is introduced into a statistical treatment[†] it leads to parameters of the type $\beta_{R,X}$ but none of the type $\beta_{R,R'}$ or $\beta_{X,X'}$.

By differentiating (1) with respect to n_1 noting that $m_R \propto n_1^{-1}$ and $m_X \propto n_1^{-1}$ we obtain

$$\left(\sum_R n_R + \sum_X n_X\right)(\phi^\ominus - \phi) = -\sum_R \sum_X 2\beta_{R,X} m_R m_X \qquad 7.21.2$$

so that

$$\phi - \phi^\ominus = \sum_R \sum_X \beta_{R,X} m_R m_X / m \qquad 7.21.3$$

where

$$m = \sum_R m_R = \sum_X m_X. \qquad 7.21.4$$

By differentiating (1) with respect to n_R noting that $m_R \propto n_R$ we obtain

$$\ln(\gamma_R/\gamma^\ominus) = 2\sum_{X'} \beta_{R,X'} m_{X'}. \qquad 7.21.5$$

Similarly

$$\ln(\gamma_X/\gamma^\ominus) = 2\sum_{R'} \beta_{R',X} m_{R'}. \qquad 7.21.6$$

Consequently the mean activity coefficient $\gamma_{R,X}$ of RX is given by

$$\ln(\gamma_{R,X}/\gamma^\ominus) = \sum_{X'} \beta_{R,X'} m_{X'} + \sum_{R'} \beta_{R',X} m_{R'}. \qquad 7.21.7$$

In a solution of a single electrolyte RX these formulae reduce to

$$\phi - \phi^\ominus = \beta_{R,X} m$$
$$\ln(\gamma_{R,X}/\gamma^\ominus) = 2\beta_{R,X} m. \qquad 7.21.8$$

We see that every parameter $\beta_{R,X}$ can be determined by measurements on solutions of the single electrolyte RX. Thus the properties of all solutions of mixed electrolytes can be predicted from the properties of solutions of single electrolytes.

* Brønsted, J. Amer. Chem. Soc. 1922 **44** 938.
† Guggenheim, Phil. Mag. 1935 **19** 588; Guggenheim, Applications of Statistical Mechanics, Clarendon Press 1966 chapter 9.

The *principle of specific interaction* leads to a number of conclusions concerning mixtures of electrolytes which have been confirmed experimentally by Brønsted*. We shall not give details, but shall merely mention one illustrative example of the usefulness of the principle.

From formula (7) it follows that the mean activity of NaCl present as a trace in a solution of HCl at $m = 10^{-1}$ is equal to that of HCl present as a trace in a solution of NaCl at $m = 10^{-1}$. The latter can be measured electrometrically, as we shall see in the next chapter, while there is no convenient experimental method for determining the former. Hence the former is best determined by measuring the latter.

TABLE 7.1

Interaction coefficients

Values of $\beta_{R,X} - \beta_{Na,Cl}$ at 25 °C

		NaF	−0.08	KF	−0.02		
HCl	0.12	NaCl	0.00	KCl	−0.05	RbCl	−0.09
HBr	0.18	NaBr	0.02	KBr	−0.04	RbBr	−0.10
HI	0.21	NaI	0.06	KI	0.00	RbI	−0.11
HClO$_4$	0.15	NaClO$_3$	−0.05	KClO$_3$	−0.19	RbNO$_3$	−0.29
		NaClO$_4$	−0.02			RbAc	0.11
LiCl	0.07	NaBrO$_3$	−0.14	KBrO$_3$	−0.22	CsCl	−0.15
LiBr	0.11			KIO$_3$	−0.22	CsBr	−0.15
LiI	0.20	NaNO$_3$	−0.11	KNO$_3$	−0.26	CsI	−0.16
						CsNO$_3$	−0.30
LiClO$_4$	0.19	NaAc	0.08	KAc	0.11	CsAc	0.13
LiNO$_3$	0.06	NaCNS	0.05	KCNS	−0.06	AgNO$_3$	−0.29
		NaH$_2$PO$_4$	−0.21	KH$_2$PO$_4$	−0.31	TlClO$_4$	−0.32
LiAc	0.03					TlNO$_3$	−0.51
						TlAc	−0.19

Values of the parameter $\beta_{R,X}$ are known for a large number of 1 : 1 electrolytes in water. The values of $\beta_{R,X}$ of course depend on the choice of standard electrolyte, but the difference between the values for any two electrolytes is almost independent of this choice. For this reason we give[†] in table 7.1 values of $\beta_{R,X} - \beta_{Na,Cl}$ at 25 °C. The values for HCl, NaCl, and KCl are obtained from electromotive-force measurements by use of the theory given in the following chapter. These values are probably accurate to ±0.02. The remaining values are obtained by the isopiestic measurements

---

* Brønsted, Kgl. Danske Videnskab. Selskab Mat.-Fys. Medd. 1921 **4**(4); J. Amer. Chem. Soc. 1922 **44** 877; 1923 **45** 2898.

† Guggenheim and Turgeon, Trans. Faraday Soc. 1955 **51** 747.

relative to NaCl or KCl of Robinson and Stokes*. Some of the β values may be uncertain by as much as ± 0.1 but most of them are probably more accurate than this.

§7.22 Chemical reactions involving solvent

In §7.14 we obtained the condition for equilibrium in a chemical reaction between solute ionic species, including non-ionic species as if they were ionic with $z=0$. We shall now consider how in dilute solution this condition can be extended to include chemical reactions involving the solvent.

A reaction involving the solvent is called *solvolysis* with the exception of simple addition called *solvation*. In particular if the solvent is water it is called *hydrolysis*.

For the sake of brevity we shall consider not the general case, but a specific example. We choose the hydrolysis of chlorine

$$Cl_2 + 2H_2O \rightarrow H_3O^+ + Cl^- + HOCl. \qquad 7.22.1$$

The equilibrium condition in its most general form is

$$\lambda_{H_3O^+} \lambda_{Cl^-} \lambda_{HOCl} / \lambda_{Cl_2} \lambda_{H_2O}^2 = 1 \qquad 7.22.2$$

which we rewrite as

$$\lambda_{H_3O^+} \lambda_{Cl^-} \lambda_{HOCl} / \lambda_{Cl_2} = \lambda_{H_2O}^2. \qquad 7.22.3$$

According to (7.06.1) we have

$$\lambda_{H_2O}/\lambda_{H_2O}^0 = \exp(-\phi \sum_i r_i). \qquad 7.22.4$$

In dilute solution it is sufficiently accurate to replace (4) by the approximation

$$\lambda_{H_2O}/\lambda_{H_2O}^0 = 1 \qquad 7.22.5$$

with an accuracy depending on the composition of the solution. As a typical example in an aqueous solution of 1:1 electrolytes at a total molality one tenth, we have approximately

$$\sum_i r_i \approx 3.6 \times 10^{-3}$$

$$\phi \approx 0.92$$

so that

$$\lambda_{H_2O}/\lambda_{H_2O}^0 \approx \exp(-3.3 \times 10^{-3}) \approx 0.997. \qquad 7.22.6$$

Hence we may usually replace (3) by

$$\lambda_{H_3O^+} \lambda_{Cl^-} \lambda_{HOCl} / \lambda_{Cl_2} = \lambda_{H_2O}^{0\,2}. \qquad 7.22.7$$

* Robinson and Stokes, Trans. Faraday Soc. 1949 **45** 612.

Using the relation (7.09.1) for each reacting species other than the solvent H_2O we obtain

$$(m_{H_3O^+} m_{Cl^-} m_{HOCl}/m_{Cl_2})(\gamma_{H_3O^+}\gamma_{Cl^-}\gamma_{HOCl}/\gamma_{Cl_2}) = K_m(T) \qquad 7.22.8$$

the constant $\lambda_{H_2O}^{02}$ being absorbed as a factor of K_m.

From this typical example we see that for a chemical reaction involving the solvent, the equilibrium condition takes the approximate form in dilute solution

$$\prod'(m_i \gamma_i) = K_m(T) \qquad 7.22.9$$

where $\prod'$ differs from $\prod$ by the omission of factors relating to the solvent.

§7.23 Acid–base equilibrium

One of the most important classes of chemical processes between ions in solution, is that of the transfer of a proton from one ion or molecule to another. Any ion or molecule capable of losing a proton is called an *acid*; any ion or molecule capable of gaining a proton is called a *base*. These definitions due to Brønsted* are simpler and more rational than earlier definitions which they supersede. The acid and base which differ from each other by one proton are called a *conjugate pair*†. Obviously the electric charge number of any acid exceeds by unity that of its conjugate base. Table 7.2 gives examples of well-known conjugate pairs of acids and bases. It is clear from several examples in table 7.2 that an ion or a molecule may be both an acid and a base.

TABLE 7.2

Typical conjugate acids and bases

Acid	Base
CH_3CO_2H	$CH_3CO_2^-$
NH_4^+	NH_3
H_2O	OH^-
H_3O^+	H_2O
H_3PO_4	$H_2PO_4^-$
$H_2PO_4^-$	HPO_4^{2-}
HPO_4^{2-}	PO_4^{3-}
$H_3N^+ \cdot CH_2 \cdot CO_2H$	$H_3N^+ \cdot CH_2 \cdot CO_2^-$
$H_3N^+ \cdot CH_2 \cdot CO_2^-$	$H_2N \cdot CH_2 \cdot CO_2^-$

* Brønsted, Rec. Trav. Chim. Pays-Bas 1923 **42** 718.
† Brønsted and Guggenheim, J. Amer. Chem. Soc. 1927 **49** 2554.

If A and B denote an acid and its conjugate base, while A' and B' denote another conjugate pair then the chemical reaction

$$A + B' \to B + A' \qquad 7.23.1$$

is typical of acid–base reactions. The equilibrium condition is

$$(m_B m_{A'}/m_A m_{B'})(\gamma_B \gamma_{A'}/\gamma_A \gamma_{B'}) = K \qquad 7.23.2$$

where K depends on the solvent and the temperature, but not on the composition of the solution. As a typical example we have, using Ac as an abbreviation for CH_3CO_2

$$NH_4^+ + Ac^- \to NH_3 + HAc \qquad 7.23.3$$

$$(m_{NH_3} m_{HAc}/m_{NH_4^+} m_{Ac^-})(\gamma_{NH_3}\gamma_{HAc}/\gamma_{NH_4^+}\gamma_{Ac^-}) = K. \qquad 7.23.4$$

Since water is both a base and an acid it can react with either an acid or a base dissolved in it. As examples of acids reacting with water, we mention

$$HAc + H_2O \to Ac^- + H_3O^+ \qquad 7.23.5$$

$$NH_4^+ + H_2O \to NH_3 + H_3O^+ \qquad 7.23.6$$

$$H_2PO_4^- + H_2O \to HPO_4^{2-} + H_3O^+ \qquad 7.23.7$$

and as examples of bases reacting with water

$$H_2O + Ac^- \to OH^- + HAc \qquad 7.23.8$$

$$H_2O + NH_3 \to OH^- + NH_4^+ \qquad 7.23.9$$

$$H_2O + H_2PO_4^- \to OH^- + H_3PO_4. \qquad 7.23.10$$

We note that according to the definition of *hydrolysis* given in the preceding section, reactions (5) to (10) are all examples of *hydrolysis*. On the other hand reaction (3) does not involve the solvent H_2O and is therefore not a *hydrolysis*.

The reactions (5), (6), and (7) are all examples of the general type

$$A + H_2O \to B + H_3O^+ \qquad 7.23.11$$

of which the equilibrium condition in dilute solution becomes

$$(m_B m_{H_3O^+}/m_A)(\gamma_B \gamma_{H_3O^+}/\gamma_A) = K_A \qquad 7.23.12$$

where K_A is called the *acidity constant* of A in water at the given temperature. K_A is a measure of the strength of the acid A relative to water. The reciprocal of K_A may likewise be regarded as a measure of the strength of the conjugate base B. For example the acidity constants K_{HAc} of HAc and $K_{NH_4^+}$ of NH_4^+ have the values at 25 °C

$$K_{HAc} = 1.75 \times 10^{-5} \qquad 7.23.13$$

$$K_{NH_4^+} = 6.1 \times 10^{-10}. \qquad 7.23.14$$

Two molecules of H_2O can react together, the one acting as an acid, the other as a base, thus:

$$H_2O + H_2O \rightarrow OH^- + H_3O^+.$$

The equilibrium is determined by

$$m_{H_3O^+} m_{OH^-} \gamma_{H_3O^+} \gamma_{OH^-} = K_w \qquad 7.23.15$$

where K_w is called the *ionization product* of water. Its values at various temperatures are as follows:

$$0\,°C \quad K_w = 0.115 \times 10^{-14} \qquad 7.23.16$$

$$20\,°C \quad K_w = 0.68 \times 10^{-14} \qquad 7.23.17$$

$$25\,°C \quad K_w = 1.01 \times 10^{-14}. \qquad 7.23.18$$

The equilibrium constants for reactions of the type

$$H_2O + B \rightarrow OH^- + A \qquad 7.23.19$$

can always be expressed in terms of an acidity constant and the ionization constant of water. For example for reaction (8) we have

$$(m_{OH^-} m_{HAc}/m_{Ac^-})(\gamma_{OH^-} \gamma_{HAc}/\gamma_{Ac^-}) = K_w/K_{HAc} \qquad 7.23.20$$

where K_{HAc} is the acidity constant K_A of HAc. Similarly for reaction (9) we have

$$(m_{OH^-} m_{NH_4^+}/m_{NH_3})(\gamma_{OH^-} \gamma_{NH_4^+}/\gamma_{NH_3}) = K_w/K_{NH_4^+} \qquad 7.23.21$$

where $K_{NH_4^+}$ denotes the acidity constant of NH_4^+.

If we apply the definition (12) of an acidity constant to H_3O^+ we obtain

$$K_{H_3O^+} = (m_{H_2O} m_{H_3O^+}/m_{H_3O^+})(\gamma_{H_2O} \gamma_{H_3O^+}/\gamma_{H_3O^+})$$
$$= m_{H_2O} \gamma_{H_2O}$$
$$\approx m_{H_2O} \approx 55.5. \qquad 7.23.22$$

From (12) and (22) we see that no molecule or ion which is a much stronger acid than H_3O^+ can exist in appreciable quantity in water. For example HCl is a much stronger acid than H_3O^+. Consequently when dissolved in water it is almost completely changed to H_3O^+ and Cl^-. Similarly H_2SO_4

is a much stronger acid than H_3O^+ and is therefore almost completely changed to H_3O^+ and HSO_4^-. On the other hand $K_{HSO_4^-} = 1.0 \times 10^{-2}$ so that HSO_4^- being a much weaker acid than H_3O^+ can exist in appreciable amount in water.

Similarly no base much stronger than OH^- can exist in appreciable quantity in water, since it would be hydrolysed to its conjugate acid and OH^-. Examples of bases too strong to exist in water are O^{2-} and NH_2^- which are hydrolysed as follows

$$H_2O + O^{2-} \rightarrow OH^- + OH^- \qquad 7.23.23$$

$$H_2O + NH_2^- \rightarrow OH^- + NH_3. \qquad 7.23.24$$

Examples of very strong bases, but not so strong that they cannot exist at all in water are S^{2-} and CN^-.

When a strongly alkaline substance such as NaOH is dissolved in water, the base present in the solution is OH^-. Often NaOH is itself referred to loosely as a base.

Similar relations hold in other solvents which can react as both base and acid. Reactions of an ion or molecule with the solvent are called *solvolysis*.

§7.24 Weak electrolytes

An electrically neutral molecule, not itself an electrolyte, which by hydrolysis or other reaction is partly changed into ions is often called a *weak electrolyte*. In particular an electrically neutral acid such as HAc which is partly hydrolysed according to

$$HAc + H_2O \rightarrow Ac^- + H_3O^+ \qquad 7.24.1$$

and an electrically neutral base such as NH_3 which is partly hydrolysed according to

$$H_2O + NH_3 \rightarrow OH^- + NH_4^+ \qquad 7.24.2$$

are by this definition *weak electrolytes*. For these substances the names *electrically neutral acids* and *electrically neutral bases* are sufficient and more informative.

§7.25 Surface phases

The formulae previously derived for surface phases apply just as well to solutions of electrolytes as to solutions of non-electrolytes. In particular for variations of composition at constant temperature formula (5.26.1) becomes

$$-\mathrm{d}\gamma = \sum_i (\Gamma_i - r_i\Gamma_1)\mathrm{d}\mu_i \quad (T \text{ const.}) \quad 7.25.1$$

where the summation Σ_i extends over all ions and other solute species. Expressed in terms of absolute activities (1) becomes

$$-\mathrm{d}\gamma = RT\sum_i (\Gamma_i - r_i\Gamma_1)\,\mathrm{d}\ln\lambda_i \quad (T \text{ const.}). \quad 7.25.2$$

Even if the solution is extremely dilute the term $r_i\Gamma_1$ must not be omitted for although $r_i \ll 1$ at the same time $|\Gamma_1| \gg |\Gamma_i|$.

The above relations, and in fact all the relations, for the surface of an electrolyte solution are formally analogous to those for the surface of a non-electrolyte solution. There is however a significant difference requiring careful treatment, namely counting the number of independent components. Let us consider some typical examples beginning with the simplest.

A solution of hydrochloric acid in water contains the species H_2O, H_3O^+, and Cl^-. We omit OH^-, not so much because it is present in negligible amount as because it is in any case not an independent component, since

$$OH^- = 2H_2O - H_3O^+. \quad 7.25.3$$

Of the three species H_2O, H_3O^+, and Cl^- the condition for electrical neutrality imposes the restrictions

$$m_{H_3O^+} = m_{Cl^-} \quad 7.25.4$$

$$\Gamma_{H_3O^+} = \Gamma_{Cl^-} \quad 7.25.5$$

so that there are only two independent components. We may take these to be H_2O on the one hand and $(H_3O^+ + Cl^-)$ on the other. More simply we may choose as independent components H_2O and HCl.

Similarly a solution of sodium hydroxide in water contains the species H_2O, OH^-, and Na^+. We omit H_3O^+ not so much because it is present in negligible amount as because it is in any case not an independent component owing to (3). The condition for electrical neutrality imposes the restrictions

$$m_{OH^-} = m_{Na^+} \quad 7.25.6$$

$$\Gamma_{OH^-} = \Gamma_{Na^+} \quad 7.25.7$$

so that there are only two independent components which we may take to be H_2O and $NaOH$. Thus for the surface tension of the solution of $NaOH$ we have at constant temperature

$$-\mathrm{d}\gamma = RT(\Gamma_{Na^+} - r_{Na^+}\Gamma_{H_2O})\,\mathrm{d}\ln\lambda_{Na^+} + RT(\Gamma_{OH^-} - r_{OH^-}\Gamma_{H_2O})\,\mathrm{d}\ln\lambda_{OH^-}$$

$$= 2RT(\Gamma_{NaOH} - r_{NaOH}\Gamma_{H_2O})\,\mathrm{d}\ln(m_{NaOH}\gamma_{Na,OH}) \quad 7.25.8$$

where m_{NaOH} and Γ_{NaOH} are defined by, respectively,

$$m_{\text{Na}^+} = m_{\text{OH}^-} = m_{\text{NaOH}} \qquad 7.25.9$$

$$\Gamma_{\text{Na}^+} = \Gamma_{\text{OH}^-} = \Gamma_{\text{NaOH}}. \qquad 7.25.10$$

The reader should have no difficulty in distinguishing between γ without any subscript denoting surface tension and $\gamma_{\text{Na,OH}}$ denoting the mean activity coefficient of NaOH.

Let us now consider a solution made by dissolving both hydrogen chloride and sodium hydroxide in water. Of the five species H_2O, H_3O^+, OH^-, Na^+, and Cl^- in the system only three are independent. For the equilibrium

$$H_2O + H_2O \rightarrow H_3O^+ + OH^- \qquad 7.25.11$$

imposes the restriction

$$m_{H_3O^+} m_{OH^-} \gamma_{H_3O^+} \gamma_{OH^-} = K_w \qquad 7.25.12$$

and the condition for electrical neutrality imposes the restrictions

$$m_{\text{Na}^+} + m_{H_3O^+} = m_{Cl^-} + m_{OH^-} \qquad 7.25.13$$

$$\Gamma_{\text{Na}^+} + \Gamma_{H_3O^+} = \Gamma_{Cl^-} + \Gamma_{OH^-}. \qquad 7.25.14$$

If the hydrogen chloride is in excess, it is natural to choose as the three independent components H_2O, NaCl, and HCl. In this case m_{OH^-} is negligible compared with all the other terms in (13). If on the contrary the sodium hydroxide is in excess, it is natural to choose as the three independent components H_2O, NaCl, and NaOH. In this case $m_{H_3O^+}$ is negligible compared with the other terms in (13). These remarks apply equally to the bulk of the solution and to the surface layer.

Suppose now we stipulate that precisely equivalent amounts of hydrogen chloride and sodium hydroxide are contained in the solution. Then the relation (13) is replaced by the two relations

$$m_{\text{Na}^+} = m_{Cl^-} \qquad 7.25.15$$

$$m_{H_3O^+} = m_{OH^-} \qquad 7.25.16$$

so that the solution contains only two independent components, which we naturally take to be H_2O and NaCl. But the restriction (15) which reduces by one the number of independent components in the bulk of the solution, does not imply any analogous restriction on the Γ's. In other words the surface layer can contain as well as H_2O and NaCl either an excess of HCl or an excess of NaOH. Thus the number of components necessary to describe the composition of the surface phase is still three, not two.

We shall now analyse this problem, beginning with unspecified quantities of NaCl and NaOH dissolved in water, introducing the restriction that the quantity of NaOH is zero only at a later stage. There are four ionic species Na^+, Cl^-, H_3O^+, and OH^- in the solvent H_2O. These are not independent, but are subject to the conditions for electrical neutrality

$$m_{Na^+} + m_{H_3O^+} = m_{Cl^-} + m_{OH^-} \qquad 7.25.17$$

$$\Gamma_{Na^+} + \Gamma_{H_3O^+} = \Gamma_{Cl^-} + \Gamma_{OH^-} \qquad 7.25.18$$

and to the condition for ionization equilibrium of the solvent water

$$\lambda_{H_3O^+} \lambda_{OH^-} = \lambda_{H_2O}^2 = \text{const.} \qquad 7.25.19$$

For variations of the surface tension with composition at constant temperature we have the general relation of the form (2)

$$-d\gamma/RT = (\Gamma_{Na^+} - r_{Na^+} \Gamma_{H_2O}) \, d \ln \lambda_{Na^+}$$
$$+ (\Gamma_{Cl^-} - r_{Cl^-} \Gamma_{H_2O}) \, d \ln \lambda_{Cl^-}$$
$$+ (\Gamma_{H_3O^+} - r_{H_3O^+} \Gamma_{H_2O}) \, d \ln \lambda_{H_3O^+}$$
$$+ (\Gamma_{OH^-} - r_{OH^-} \Gamma_{H_2O}) \, d \ln \lambda_{OH^-}. \qquad 7.25.20$$

Using (17), (18), and (19) we can replace (20) by

$$-d\gamma/RT = (\Gamma_{Cl^-} - r_{Cl^-} \Gamma_{H_2O}) \, d \ln(\lambda_{Na^+} \lambda_{Cl^-})$$
$$+ ([\Gamma_{Na^+} - \Gamma_{Cl^-}] - [r_{Na^+} - r_{Cl^-}]\Gamma_{H_2O}) \, d \ln(\lambda_{Na^+} \lambda_{OH^-})$$
$$= (\Gamma_{Cl^-} - r_{Cl^-} \Gamma_{H_2O}) \, d \ln(m_{Na^+} m_{Cl^-} \gamma_{Na,Cl}^2)$$
$$+ ([\Gamma_{Na^+} - \Gamma_{Cl^-}] - [r_{Na^+} - r_{Cl^-}]\Gamma_{H_2O}) \, d \ln(m_{Na^+} m_{OH^-} \gamma_{Na,OH}^2). \qquad 7.25.21$$

Thus by studying the dependence of the surface tension on the composition by variations of the molalities of NaCl and of NaOH, provided the activity coefficients are known, we can determine the separate values of

$$\Gamma_{Cl^-} - r_{Cl^-} \Gamma_{H_2O} \qquad 7.25.22$$

and of

$$[\Gamma_{Na^+} - \Gamma_{Cl^-}] - [r_{Na^+} - r_{Cl^-}]\Gamma_{H_2O}. \qquad 7.25.23$$

The expression (22) is a measure of the adsorption of NaCl relative to H_2O, while the expression (23) is a measure of the adsorption of NaOH relative to H_2O. In particular as the molality of NaOH is made to tend to zero, so the quantity (23) tends to

$$\Gamma_{Na^+} - \Gamma_{Cl^-}. \qquad 7.25.24$$

The value of (24) then becomes the surface concentration of NaOH in a solution which in the bulk contains only NaCl and H_2O.

To recapitulate, by varying the molalities of both NaCl and NaOH and measuring surface tension we can determine separately the coefficients of the two terms on the right of (21), namely the quantities (22) and (23) of which the latter reduces to (24) in a solution containing no excess NaOH. By measuring the surface tension of solutions containing varying amounts of NaCl only without any NaOH it is not possible to separate the two terms on the right of (21) and consequently the quantity (22) can not be determined in this way.

CHAPTER 8

ELECTROCHEMICAL SYSTEMS

§8.01 *Electrically charged phases*

In the previous chapter we saw how a solution containing ions can be treated by means of the same formulae as one containing only electrically neutral molecules. In particular the formula

$$dG = -S\,dT + V\,dP + \sum_i \mu_i\,dn_i \qquad 8.01.1$$

from which follows

$$\mu_i = (\partial G/\partial n_i)_{T,\,P,\,n_j} \qquad 8.01.2$$

are applicable. The only significant difference in our treatment of ions was the imposition of the condition for electrical neutrality

$$\sum_i n_i z_i = 0 \qquad 8.01.3$$

where z_i is the charge number of the ionic species i. We shall now consider what happens if we try to relax the condition (3).

To obtain a clear picture of what happens it is useful to begin with some simple numerical calculations. The charge e on a proton is given by

$$e = 1.6021 \times 10^{-19} \text{ C}. \qquad 8.01.4$$

Consequently the proper electric charge, associated with an ionic species having a charge number 1, called the *Faraday constant* and denoted by F is given by

$$F = Le = 0.96487 \times 10^5 \text{ C mole}^{-1}. \qquad 8.01.5$$

Let us now consider a single phase surrounded by a vacuum and thus electrically insulated. Let us further imagine that this phase, instead of satisfying the condition of electrical neutrality (3), contains an excess of 10^{-10} moles of an ionic species with charge number $+1$. Then most, if not all, the excess electrical charge will accumulate at the surface of the phase. For simplicity let us suppose that the phase is spherical with a radius one centimetre. The electrical potential ψ of a charged sphere of radius r in vacuo is determined by

$$\psi = Q/4\pi\varepsilon_0 r \qquad 8.01.6$$

where Q is the charge on the sphere and ε_0 is the rational permittivity of a vacuum. Substituting the values

$$Q = 10^{-10} \; F \; \text{mole} = 0.96 \times 10^{-5} \; C$$
$$4\pi\varepsilon_0 = 1.11 \times 10^{-10} \; C \; V^{-1} \; m^{-1}$$
$$r = 10^{-2} \; m \qquad 8.01.7$$

into (6), we obtain

$$\psi = (0.96 \times 10^{-5}/1.11 \times 10^{-10} \times 10^{-2})V = 0.86 \times 10^7 \; V. \qquad 8.01.8$$

From this example we have reached the striking conclusion that a departure from the condition of electrical neutrality corresponding to a quantity of ions far too small to be detected chemically corresponds to an electrostatic potential which could be encountered only in specialized high tension laboratories. Any other numerical example would lead to the same conclusion.

§8.02 *Phases of identical composition*

The above general result leads to the use of the following terminology. We speak of two phases having the *same chemical content*, but *different electrical potentials*. Actually two such phases differ in chemical content but the difference is too small to be detectable by chemical means, or any other means, except electrical. For example suppose we mention two spheres of copper each containing precisely one gramme differing in electrical potential by 200 V. If this electrical potential difference is ascribed to an excess of copper ions Cu^{2+} with a charge number $+2$, then the amount of this excess is about 3.5×10^{-16} moles or 2×10^{-14} grammes. This excess is so small as to be entirely negligible except in its electrical effect. Consequently it is of no importance or interest whether the electrical charge is in fact due to an excess of Cu^{2+} ions or to an equivalent deficiency of electrons or even to some extraneous kind of ion such as H_3O^+, present as an impurity.

Similar considerations apply to a pair of phases of different size but of the same chemical composition.

§8.03 *Electrochemical potentials*

Having agreed as to what we mean when we speak of two phases having the same chemical composition but different electrical potentials, we see that the μ_i's occurring in the formulae mentioned in §8.01 have values depending on the electrical state of the phase as well as on its chemical composition. To stress this fact we call the μ_i of an ionic species its *electrochemical potential*.*

The difference of the electrochemical potential μ_i between two phases of *identical chemical composition* will clearly be proportional to the proper electrical charge $z_i F$ associated with the species in question but independent of all its other individual characteristics. Hence for any two phases α and β of identical chemical composition we may write for any ionic species i

$$\mu_i^\beta - \mu_i^\alpha = z_i F(\psi^\beta - \psi^\alpha) \qquad 8.03.1$$

where $\psi^\beta - \psi^\alpha$ is the *electrical potential difference* between the two phases. Formula (1) may be regarded as the thermodynamic definition of the electrical potential difference between two phases of *identical chemical composition*. The equilibrium condition for a given ionic species between two phases of identical composition is that the two phases should be at the same electrical potential. In fact the laws of mathematical electrostatics are applicable to any ionic species, in particular to electrons, only in so far as differences in chemical composition between several phases are excluded or ignored.

For the distribution of the ionic species i between two phases α, β of different chemical composition the equilibrium condition is equality of the electrochemical potential μ_i, that is to say

$$\mu_i^\alpha = \mu_i^\beta. \qquad 8.03.2$$

Any splitting of $\mu_i^\beta - \mu_i^\alpha$ into a chemical part and an electrical part is in general arbitrary and without physical significance.

As long ago as 1899 Gibbs wrote:[†] 'Again, the consideration of the electrical potential in the electrolyte, and especially the consideration of the difference of potential in electrolyte and electrode, involve the consideration of quantities of which we have no apparent means of physical measurement, while the difference of potential in pieces of metal of the same kind attached to the electrodes is exactly one of the things which we can and do measure.' This principle was however ignored or forgotten until rediscovered and reformulated thirty years later as follows:* 'The electric potential difference between two points in different media can never be measured and has not yet been defined in terms of physical realities. It is therefore a conception which has no physical significance.' The electrostatic potential difference

---

* Guggenheim, J. Phys. Chem. 1929 **33** 842.
† Gibbs, Collected Works, Longmans, vol. **1**, p. 429.

between two points is admittedly defined in *electrostatics*, the mathematical theory of an imaginary fluid *electricity*, whose equilibrium or motion is determined entirely by the electric field. *Electricity* of this kind does not exist. Only electrons and ions have physical existence and these differ fundamentally from the hypothetical fluid *electricity* in that their equilibrium is *thermodynamic* not *electrostatic*.

Although the above considerations seem almost obvious to anyone who has thought about the matter, there has in the past been considerable confusion due to misleading terminology. It therefore seems worth while considering in more detail some simple examples. Consider a potentiometer wire made of copper and in particular two sections of the wire α' and α'' between which the electrical potential difference $\psi'' - \psi'$ is say 2 V. Since α' and α'' are both in copper, there is no ambiguity in the meaning of $\psi'' - \psi'$. If two pieces of copper wire are attached to α' and α'', then the electrical potential difference between these two is also $\psi'' - \psi' = 2$ V. If instead of copper wire we attach two pieces of silver wire β' and β'' to α' and α'' respectively, then the difference of electrical potential between β' and β'' is likewise 2 V. The electrical potential difference between a piece of copper and a piece of silver is however not defined. The silver wire β' and the copper wire α' are in equilibrium with respect to electrons, so that

$$\mu_{\text{el}-}^{\alpha'} = \mu_{\text{el}-}^{\beta'} \qquad 8.03.3$$

where the subscript $_{\text{el}-}$ denotes electrons. Likewise

$$\mu_{\text{el}-}^{\alpha''} = \mu_{\text{el}-}^{\beta''}. \qquad 8.03.4$$

Thus the situation is completely described by (3) or (4) together with

$$\mu_{\text{el}-}^{\alpha''} - \mu_{\text{el}-}^{\alpha'} = \mu_{\text{el}-}^{\beta''} - \mu_{\text{el}-}^{\beta'} = -F(\psi'' - \psi'). \qquad 8.03.5$$

Suppose further that the two pieces of silver wire β', β'' be dipped respectively into two solutions γ', γ'' both having the same composition and both containing a silver salt. Then between each piece of silver wire and the solution with which it is in contact there will be equilibrium with respect to silver ions Ag^+. Hence

$$\mu_{Ag^+}^{\gamma'} = \mu_{Ag^+}^{\beta'} \qquad 8.03.6$$

$$\mu_{Ag^+}^{\gamma''} = \mu_{Ag^+}^{\beta''}. \qquad 8.03.7$$

At the same time

$$\mu_{Ag^+}^{\gamma''} - \mu_{Ag^+}^{\gamma'} = \mu_{Ag^+}^{\beta''} - \mu_{Ag^+}^{\beta'} = F(\psi'' - \psi') = F \times 2 \text{ V}. \qquad 8.03.8$$

If the two solutions γ', γ'' are contained in insulating vessels and the silver wires are removed without otherwise touching or disturbing the two solutions then the relations (8) remain valid until one solution is touched by some other electrically charged or electrically conducting body. From this it is clear that the value of μ_{Ag^+} in a solution of a silver salt depends not only on the composition of the solution but also on its, usually accidentally determined, electrical state. If the solutions also contain nitrate ions NO_3^- then, since both solutions have the same composition,

$$\mu_{NO_3^-}^{\gamma''} - \mu_{NO_3^-}^{\gamma'} = -F(\psi'' - \psi'). \qquad 8.03.9$$

Adding (8) and (9), we obtain

$$\mu_{Ag^+}^{\gamma'} + \mu_{NO_3^-}^{\gamma'} = \mu_{Ag^+}^{\gamma''} + \mu_{NO_3^-}^{\gamma''} \qquad 8.03.10$$

the electrical terms cancelling. We accordingly speak of the *chemical potential* of a salt, for example $\mu_{AgNO_3} = \mu_{Ag^+} + \mu_{NO_3^-}$, but of the *electrochemical potentials* of ions, for example μ_{Ag^+} and $\mu_{NO_3^-}$.

§8.04 Absolute activities of ions

Since the absolute activity λ_i is related to μ_i by

$$\mu_i = RT \ln \lambda_i \qquad 8.04.1$$

it is clear that the absolute activity of an ionic species contains a factor depending on the, usually accidentally determined, electrical state of the system. The same applies to the activity coefficient of an ionic species. As already emphasized in the previous chapter all such indeterminacy disappears in formulae relating to electrically neutral combinations of ions, in particular to salts.

§8.05 Dilute solutions in common solvent

According to (7.09.1) the absolute activity λ_i of an ionic species i is related to its molality m_i and its activity coefficient γ_i by

$$\lambda_i = \lambda_i^\ominus m_i \gamma_i \qquad 8.05.1$$

where $\lambda_i^\ominus$ depends on the solvent and temperature and moreover contains a partly undetermined factor, which however cancels in all applications to processes not involving a net transfer of electric charge. Correspondingly the electrochemical potential μ_i has the form

$$\mu_i = RT \ln \lambda_i^\ominus + RT \ln m_i + RT \ln \gamma_i \qquad 8.05.2$$

and includes an undetermined additive term which cancels in all applications to processes not involving a net transfer of electrical charge. We shall investigate this term in greater detail.

Let us formally write

$$\mu_i = z_i F\psi + RT \ln l_i^\ominus + RT \ln m_i + RT \ln \gamma_i \qquad 8.05.3$$

where $l_i^\ominus$ is independent of the electrical state of the phase and ψ is the electrical potential of the phase. Let us now apply (3) to two phases denoted by a single and a double dash respectively and then subtract. We obtain

$$\mu_i'' - \mu_i' = z_i F(\psi'' - \psi') + RT \ln(l_i^{\ominus''}/l_i^{\ominus'}) + RT \ln(m_i''/m_i') + RT \ln(\gamma_i''/\gamma_i'). \qquad 8.05.4$$

We now re-examine the condition for the term containing $\psi'' - \psi'$ to be physically defined.

The easiest case is when the two phases have the same chemical composition so that

$$l_i^{\ominus''} = l_i^{\ominus'} \qquad 8.05.5$$

$$m_i'' = m_i' \qquad 8.05.6$$

$$\gamma_i'' = \gamma_i'. \qquad 8.05.7$$

Formula (4) then reduces to

$$\mu_i'' - \mu_i' = z_i F(\psi'' - \psi'). \qquad 8.05.8$$

Since $\mu_i'' - \mu_i'$ is always well defined, formula (8) in this special case defines $\psi'' - \psi'$.

We now consider the extreme opposite case of two solutions in different solvents or two different pure phases. In this case there is no means of distinguishing in (4) between the term containing $\psi'' - \psi'$ and the term containing $\ln(l_i^{\ominus''}/l_i^{\ominus'})$. The splitting into these two terms has in this case no physical significance. These remarks merely repeat and confirm what has already been stated in the preceding sections.

We have still to consider the intermediate case of two solutions of different composition in the same solvent, of course at the same temperature. We then have

$$l_i^{\ominus''} = l_i^{\ominus'} \qquad 8.05.9$$

so that (4) reduces to

$$\mu_i'' - \mu_i' = z_i F(\psi'' - \psi') + RT \ln(m_i''/m_i') + RT \ln(\gamma_i''/\gamma_i'). \qquad 8.05.10$$

Since $\mu_i'' - \mu_i'$ is well defined and m_i'', m_i' are measurable, the question whether

$\psi'' - \psi'$ is determinate depends on our knowledge of γ_i''/γ_i'. If both the solutions are so dilute that we can evaluate γ_i by an explicit formula such as (7.20.9), then we may consider that (10) defines $\psi'' - \psi'$. If on the other hand either solution is so concentrated that our knowledge of the value of γ_i is incomplete, then the value of $\psi'' - \psi'$ becomes correspondingly indefinite.

§8.06 Volta potentials

It is outside the province of this book to consider thermionic phenomena. In case however any reader may be puzzled by the fact that the so-called *Volta potential difference* or *contact potential difference* between two metals can be determined, it seems worth while stressing that the only measurable potential difference of this kind is that between two regions in free space immediately *outside* the two metals respectively.

§8.07 Membrane equilibrium (non-osmotic)

Suppose two solutions α and β at the same temperature and pressure in the same solvent be separated by a membrane permeable to some ions, but not to others, nor to the solvent. We call this a *non-osmotic membrane equilibrium*. Then for every permeant ion we have the equilibrium condition

$$\mu_i^\alpha = \mu_i^\beta. \qquad 8.07.1$$

If for example one of the permeant ions is the Ag^+ ion, we have

$$\mu_{Ag^+}^\alpha = \mu_{Ag^+}^\beta. \qquad 8.07.2$$

If then we place in each of the two solutions a piece of silver wire since each piece of wire is, with respect to Ag^+, in equilibrium with the solution in which it dips, the equality of μ_{Ag^+} also holds between the two pieces of silver wire. Hence the two pieces of silver wire have equal electrical potentials, as could be verified by connecting them to a potentiometer or electrometer.

We have yet to consider what, if anything, can be said concerning the electrical potential difference between the two solutions. Since the solvent is the same in both solutions, we may in accordance with (8.05.10) replace (2) by

$$F(\psi^\beta - \psi^\alpha) + RT \ln(m_{Ag^+}^\beta / m_{Ag^+}^\alpha) + RT \ln(\gamma_{Ag^+}^\beta / \gamma_{Ag^+}^\alpha) = 0. \qquad 8.07.3$$

Supposing that m_{Ag^+} has been measured in both solutions, the determination of $\psi^\beta - \psi^\alpha$ reduces to that of the values of γ_{Ag^+} in the two solutions. If the solutions are so dilute that accurate or at least approximate, formulae for the activity coefficients γ are available then the electrical potential difference $\psi^\beta - \psi^\alpha$ can be evaluated with greater or less accuracy as the case may be. If either solution is so concentrated that γ_{Ag^+} cannot be evaluated, then no more can $\psi^\beta - \psi^\alpha$.

If there are several permeant ions, then the relations of the form (1) can be combined into relations corresponding to processes involving no net flow of electric charge. For example for a salt composed of v_+ cations R of charge number z_+ and v_- anions X of charge number z_-, both permeant, the equilibrium condition is

$$v_+ \mu_R^\alpha + v_- \mu_X^\alpha = v_+ \mu_R^\beta + v_- \mu_X^\beta \qquad 8.07.4$$

which can be written in the equivalent form

$$(m_R^\alpha)^{v_+}(m_X^\alpha)^{v_-}(\gamma_{R,X}^\alpha)^{v_++v_-} = (m_R^\beta)^{v_+}(m_X^\beta)^{v_-}(\gamma_{R,X}^\beta)^{v_++v_-}. \qquad 8.07.5$$

§8.08 Osmotic membrane equilibrium

In the preceding section we assumed that the membrane was impermeable to the solvent. The more usual case when the membrane is permeable to the solvent, called *osmotic membrane equilibrium*, is less simple. In this case equilibrium as regards the solvent between two phases separated by the membrane, will generally require a pressure difference between the two phases, the *osmotic pressure difference*, and this pressure difference complicates the exact conditions of equilibrium for the solute ions. We shall consider only the case of one and the same solvent on both sides of the membrane.

The conditions for membrane equilibrium can be written in the general form

$$\mu_1^\alpha = \mu_1^\beta \qquad 8.08.1$$

for the solvent and

$$\mu_i^\alpha = \mu_i^\beta \qquad 8.08.2$$

for each permeant ionic species.

We have now to take account of how each μ depends on the pressure, but for the sake of brevity we shall neglect the compressibility of the solutions. We have then in accordance with (7.06.1)

$$\mu_1(P) = \mu_1^0(P) - RT\phi \sum_i r_i$$

$$= \mu_1^{0\ominus} + PV_1^0 - RT\phi \sum_i r_i \qquad 8.08.3$$

where $\mu_1^0(P)$ is the value of μ_1 for the pure solvent at the pressure P, while $\mu_1^{0\ominus}$ is the value for the pure solvent in the limit of zero pressure.

Similarly for each ionic species i we replace (8.05.3) by

$$\mu_i = PV_i + z_i F\psi + RT\ln(l_i^\ominus m_i \gamma_i) \qquad 8.08.4$$

where $l_i^\ominus$ is independent of the pressure.

Using (3) in (1) we obtain for the equilibrium value of the pressure difference

$$P^\beta - P^\alpha = RT(\phi^\beta \sum_i r_i^\beta - \phi^\alpha \sum_i r_i^\alpha)/V_1^0. \qquad 8.08.5$$

Using (4) in (2) we obtain

$$RT\ln(m_i^\beta \gamma_i^\beta / m_i^\alpha \gamma_i^\alpha) + z_i F(\psi^\beta - \psi^\alpha) = (P^\alpha - P^\beta)V_i. \qquad 8.08.6$$

Finally substituting (5) into (6) we have

$$\ln(m_i^\beta \gamma_i^\beta / m_i^\alpha \gamma_i^\alpha) + z_i F(\psi^\beta - \psi^\alpha)/RT = V_i(\phi^\alpha \sum_i r_i^\alpha - \phi^\beta \sum_i r_i^\beta)/V_1^0. \qquad 8.08.7$$

Whether formula (7) by itself has any physical significance depends, as explained in §8.05, on whether values of γ_i can be computed. If they can, then from formula (7) the value of $\psi^\beta - \psi^\alpha$ can be computed, since all the other quantities occurring in (7) are measurable. In any case the term containing $\psi^\beta - \psi^\alpha$ can be eliminated by applying (7) to several ionic species together forming an electrically neutral combination. Thus for the equilibrium distribution of a permeant electrolyte consisting of ν_+ cations R and ν_- anions X we obtain

$$\frac{(m_R^\beta)^{\nu_+}(m_X^\beta)^{\nu_-}(\gamma_{R,X}^\beta)^{\nu_+ + \nu_-}}{(m_R^\alpha)^{\nu_+}(m_X^\alpha)^{\nu_-}(\gamma_{R,X}^\alpha)^{\nu_+ + \nu_-}} = \exp\{(\nu_+ V_R + \nu_- V_X)(\phi^\alpha \sum_i r_i^\alpha - \phi^\beta \sum_i r_i^\beta)/V_1^0\}.$$

$$8.08.8$$

At high dilutions when all $r_i \ll 1$ the quantity within the $\{\ \}$ on the right side of (8) may be so small that it can be neglected. Under such conditions (8) reduces to

$$(m_R^\beta)^{\nu_+}(m_X^\beta)^{\nu_-}(\gamma_{R,X}^\beta)^{\nu_+ + \nu_-} = (m_R^\alpha)^{\nu_+}(m_X^\alpha)^{\nu_-}(\gamma_{R,X}^\alpha)^{\nu_+ + \nu_-} \qquad 8.08.9$$

of the same form as (8.07.5) for a non-osmotic membrane equilibrium.

The thermodynamic methods of Gibbs were first applied to osmotic membrane equilibria by Donnan. Such an equilibrium is accordingly called a *Donnan membrane equilibrium*.

§8.09 Contact equilibrium

The most important and simplest example of non-osmotic equilibrium is that of two phases with one common ion, the surface of separation being in effect a membrane permeable to the common ion but impermeable to all others. This may be called *contact equilibrium*.

We have already met several examples of contact equilibrium. For example, for two metals say Cu and Ag in contact there is equilibrium between the metals as regards electrons, but not as regards the positive ions Cu^{2+} or Ag^+. This equilibrium is expressed by

$$\mu_{el^-}^{Cu} = \mu_{el^-}^{Ag} \qquad 8.09.1$$

the subscript $_{el^-}$ denoting electrons and the superscripts denoting the two phases.

Likewise for a piece of metal M of say Cu dipping into a solution S containing ions of the metal, in this case Cu^{2+}, the contact equilibrium is completely described by

$$\mu_{Cu^{2+}}^{M} = \mu_{Cu^{2+}}^{S} \qquad 8.09.2$$

the metal and solution being in mutual equilibrium as regards the metallic ions only.

In neither of these cases is any contact electrical potential difference thermodynamically definable.

§8.10 Examples of galvanic cell

We shall now introduce the subject of galvanic cells by the detailed study of a simple example in terms of the electrochemical potentials. At a later stage we shall proceed to derive more general formulae applicable to all galvanic cells.

We describe a cell symbolically by writing down in order a number of phases separated by vertical lines, each phase being in contact with the phases written down immediately to its left and right. For example

$$Cu \mid Zn \mid \begin{array}{c} \text{Solution I} \\ \text{containing } Zn^{2+} \end{array} \mid \begin{array}{c} \text{Solution II} \\ \text{containing } Ag^+ \end{array} \mid Ag \mid Cu \qquad 8.10.1$$

may be regarded as denoting a copper *terminal* attached to a zinc *electrode* dipping into a solution I containing zinc ions; this solution is in contact with another solution II containing silver ions in which there is dipping a silver electrode attached to another copper terminal. We shall use the following superscripts to denote the several phases:

'   the copper terminal on the left
Zn   the zinc electrode
I   the solution on the left
II   the solution on the right
Ag   the silver electrode
''   the copper terminal on the right.

In the metal phases, since there is equilibrium between electrons, metallic ions, and the metal atoms, we have

$$\tfrac{1}{2}\mu'_{Cu^{2+}} + \mu'_{el^-} = \tfrac{1}{2}\mu''_{Cu^{2+}} + \mu''_{el^-} = \tfrac{1}{2}\mu^{Cu}_{Cu} \qquad 8.10.2$$

$$\tfrac{1}{2}\mu^{Zn}_{Zn^{2+}} + \mu^{Zn}_{el^-} = \tfrac{1}{2}\mu^{Zn}_{Zn} \qquad 8.10.3$$

$$\mu^{Ag}_{Ag^+} + \mu^{Ag}_{el^-} = \mu^{Ag}_{Ag}. \qquad 8.10.4$$

The contact equilibrium conditions are

$$\mu'_{el^-} = \mu^{Zn}_{el^-} \qquad 8.10.5$$

$$\mu^{Zn}_{Zn^{2+}} = \mu^{I}_{Zn^{2+}} \qquad 8.10.6$$

$$\mu^{II}_{Ag^+} = \mu^{Ag}_{Ag^+} \qquad 8.10.7$$

$$\mu^{Ag}_{el^-} = \mu''_{el^-}. \qquad 8.10.8$$

From (5) and (8) we deduce

$$\mu''_{el^-} - \mu'_{el^-} = \mu^{Ag}_{el^-} - \mu^{Zn}_{el^-} \qquad 8.10.9$$

and so using (3) and (4)

$$\mu''_{el^-} - \mu'_{el^-} = \mu^{Ag}_{Ag} - \tfrac{1}{2}\mu^{Zn}_{Zn} - \mu^{Ag}_{Ag^+} + \tfrac{1}{2}\mu^{Zn}_{Zn^{2+}} \qquad 8.10.10$$

and then using (6) and (7)

$$\mu''_{el^-} - \mu'_{el^-} = \mu^{Ag}_{Ag} - \tfrac{1}{2}\mu^{Zn}_{Zn} - \mu^{II}_{Ag^+} + \tfrac{1}{2}\mu^{I}_{Zn^{2+}}. \qquad 8.10.11$$

We may further write

$$\mu''_{el^-} - \mu'_{el^-} = -F(\psi'' - \psi') \qquad 8.10.12$$

where $\psi'' - \psi'$ denotes the electrical potential difference between the two copper terminals. It is evident from relations (5) and (8) that the value of (12) would be the same if both copper terminals were replaced by any other metal provided both were of the same metal. Thus $\psi'' - \psi'$ is determined by the nature of the two electrodes and of the two solutions. The electric potential difference $\psi'' - \psi'$ is called the *electromotive force* of the cell and is denoted by E. We accordingly replace (12) by

$$\mu''_{el^-} - \mu'_{el^-} = -FE. \qquad 8.10.13$$

Substituting (13) into (11) we then obtain

$$-FE = \mu_{Ag}^{Ag} - \tfrac{1}{2}\mu_{Zn}^{Zn} - \mu_{Ag^+}^{II} + \tfrac{1}{2}\mu_{Zn^{2+}}^{I}. \qquad 8.10.14$$

We shall now assume that there is at least one anion, say NO_3^- present in both solutions I and II so that

$$\tfrac{1}{2}\mu_{Zn^{2+}}^{I} = \tfrac{1}{2}\mu_{Zn(NO_3)_2}^{I} - \mu_{NO_3^-}^{I} \qquad 8.10.15$$

$$\mu_{Ag^+}^{II} = \mu_{AgNO_3}^{II} - \mu_{NO_3^-}^{II}. \qquad 8.10.16$$

Using (15) and (16) we can rewrite (14) as

$$FE = \{\tfrac{1}{2}\mu_{Zn}^{Zn} - \tfrac{1}{2}\mu_{Zn(NO_3)_2}^{I} + RT \ln m_{NO_3^-}^{I}\}$$
$$+ \{\mu_{NO_3^-}^{I} - \mu_{NO_3^-}^{II} + RT \ln(m_{NO_3^-}^{II}/m_{NO_3^-}^{I})\}$$
$$- \{\mu_{Ag}^{Ag} - \mu_{AgNO_3}^{II} + RT \ln m_{NO_3^-}^{II}\}. \qquad 8.10.17$$

We have now a formula for E containing three terms in $\{\ \}$ of which the first relates only to the Zn electrode and the solution around this electrode and the last relates only to the Ag electrode and the solution around this electrode. The middle term on the other hand is independent of the nature of the electrodes and relates to an anion present in both solutions. One might be inclined to call the first of these three terms the *electrode potential* of the Zn electrode, the second the *liquid–liquid junction potential*, and the last the *electrode potential* of the silver. Such a procedure is harmless provided it is realized that

(a) this decomposition of E into three terms is affected by our arbitrary choice of the anion NO_3^- for use in our formulae;
(b) other alternative decompositions of E into three terms can be obtained by the arbitrary choice of some other ion instead of NO_3^- in our formulae;
(c) any such decomposition of E is no more nor less fundamental than another;
(d) there is in general no means of decomposing E into three terms which is less arbitrary than the one described.

In view of some inevitable arbitrariness in the decomposition of the electromotive force of a cell into two *electrode potentials* and a *liquid–liquid junction potential*, we shall for the most part abandon any attempt at such a decomposition. We shall accordingly in the next section derive a general formula for the electromotive force of any cell by a more powerful method which makes no reference at all to the localization of separate terms in the electromotive force. Before proceeding to this general treatment, we shall however draw attention to a special case where the arbitrariness referred to above effectively disappears.

Reverting to formula (17), let us now consider the particular case where the molalities of Zn^{2+} and Ag^+ in the two electrode solutions are extremely small compared with the molalities of other ions in these solutions and the compositions of the two electrode solutions are apart from their content of Zn^{2+} and Ag^+ nearly identical. Under these particular conditions we may regard the two electrode solutions as effectively identical except with regard to the equilibrium between solution and electrode. We may accordingly drop the superscripts ^{I} and ^{II} so that (17) reduces to

$$FE = \{\tfrac{1}{2}\mu_{Zn}^{Zn} - \tfrac{1}{2}\mu_{Zn(NO_3)_2}\} - \{\mu_{Ag}^{Ag} - \mu_{AgNO_3}\} \qquad 8.10.18$$

where the μ's without superscripts refer to the solution. We may then regard the cell (1) under consideration as

$$Cu \mid Zn \mid \begin{array}{c} \text{Solution containing} \\ Zn^{2+} \text{ and } Ag^+ \end{array} \mid Ag \mid Cu \qquad 8.10.19$$

bearing in mind that in reality the Ag^+ must be kept away from the Zn electrode to avoid irreversible dissolution of Zn with plating out of Ag.

It is usual to describe certain cells in this manner as if containing only one solution, but in reality there must always be some real, though possibly small difference between the composition of the two electrode solutions. Consider for example the cell commonly described as

$$Pt, H_2 \mid \text{Aqueous solution of } HCl \mid AgCl \mid Ag \mid Pt \qquad 8.10.20$$

This description implies that an electrode consisting of platinum in contact with hydrogen and another electrode consisting of a mixture of AgCl and Ag are dipping into the same solution. In fact the platinum dips into a solution saturated with H_2, but containing no AgCl, while the silver is immersed in a solution saturated with AgCl but containing no hydrogen. If in fact any part of the solution contained both hydrogen and silver chloride, these might* react irreversibly to give silver and hydrogen chloride. Thus the cell is more accurately described by

$$Pt, H_2 \left| \begin{array}{c} \text{Solution I} \\ \text{Aqueous solution} \\ \text{of HCl saturated} \\ \text{with } H_2 \end{array} \right| \begin{array}{c} \text{Solution II} \\ \text{Aqueous solution} \\ \text{of HCl saturated} \\ \text{with AgCl} \end{array} \left| AgCl \right| Ag \left| Pt \right. \qquad 8.10.21$$

* Actually in the case of this cell the irreversible process will usually be too slow to affect the accuracy of the electromotive force measurements.

By an analysis of (21) similar to that applied to (1) it can be shown that the electromotive force E is given accurately by

$$FE = \tfrac{1}{2}\mu_{H_2}^G - \mu_{H^+}^I + \mu_{AgCl}^{AgCl} - \mu_{Ag}^{Ag} - \mu_{Cl^-}^{II} \qquad 8.10.22$$

where the superscript ^{G} denotes the gas phase. Since however as far as the HCl is concerned we may regard the solutions I and II as essentially identical we may drop these superscripts and (22) reduces to

$$FE = \tfrac{1}{2}\mu_{H_2}^G + \mu_{AgCl}^{AgCl} - \mu_{Ag}^{Ag} - \mu_{HCl} \qquad 8.10.23$$

where μ_{HCl} denotes the chemical potential of HCl in the solution.

§8.11 General treatment of electromotive force

We now proceed to a more general treatment applicable to any galvanic cell. We begin by describing the characteristics common to all such cells. In so doing it is convenient to assume that the system to which we refer as the *cell* is terminated at both ends by terminals of the same metal. The essential characteristic of the galvanic cell is that a process involving ions can take place in it in such a manner that the process is necessarily accompanied by a transfer of electric charge from one terminal to the other without building up any charge in any of the intermediate phases of the cell. Moreover the charge which flows from the one terminal to the other is directly proportional to the change in the extent of the process.

For example in the cell, already discussed in the previous section,

$$\text{Pt, } H_2 \,|\, \text{Aqueous solution of HCl} \,|\, \text{AgCl} \,|\, \text{Ag} \,|\, \text{Pt} \qquad 8.11.1$$

the process accompanying the flow of one mole of positive charge from the left to the right is

$$\tfrac{1}{2}H_2(g) + AgCl(s) \to Ag(s) + HCl(aq) \qquad 8.11.2$$

where (g) denotes gas, (s) denotes solid, and (aq) denotes aqueous solution.

We now suppose the two terminals of the cell to be put into contact respectively with two points of a potentiometer bridge so placed that the electric potential of the right contact exceeds that of the left contact by an amount E'. Then in general an electric current will flow through the cell and between the two points of contact with the potentiometer bridge. If either of the points of contact is moved along the bridge the current will increase or decrease and it will change sign when E' has a certain value E. When E' is slightly less than E there will be a flow of current from left to right in the cell and from right to left in the potentiometer bridge; this flow of current will be accompanied by a well-defined chemical change in the cell. When E' is slightly greater than E there will be a flow of current in the opposite direction and the accompanying chemical change in the cell will also be reversed. When E' is equal to E there will be no flow of current and no chemical change, but by a small shift in the point of contact between cell terminal and potentiometer bridge a small current can be made to flow in either direction. This is a typical and a particularly realistic example of a *reversible process*. The value E of E' at which the current changes sign is the *electromotive force* of the cell. We note that a positive value of E means that the electrode on the right is positive.

We now stipulate that $E' = E$ so that the electromotive force of the cell is balanced against the potential difference in the potentiometer bridge and we consider the flow of one mole of positive charge from left to right in the cell, the temperature being maintained constant throughout and the pressure on every phase being kept constant. The pressures on different phases will usually, but not necessarily always, be all equal. Then since, as we have seen, this process is reversible and isothermal, it follows from (1.33.5) that the work w done on the cell is equal to the increase in the Helmholtz function, that is to say

$$w = \Delta F. \qquad 8.11.3$$

In the present case w consists of two distinct parts, namely

(a) the work $-\sum_\alpha P^\alpha \Delta V^\alpha$ done by the pressures P^α acting on the several phases α,
(b) the electrical work $-FE$ done by the potentiometer on the cell in transferring one mole of positive charge through a potential difference E.

We may therefore replace (3) by

$$-FE = \Delta F + \sum_\alpha P^\alpha \Delta V^\alpha = \Delta G. \qquad 8.11.4$$

It must be emphasized that the symbol Δ in both (3) and (4) denotes the increase of a function when the process taking place is that associated with the flow in the cell of *one mole of positive charge from the left to the right*.

From (4) we see that the electrical work obtainable from a reversible isothermal process, at constant pressure on each phase, is equal to the decrease in the Gibbs function G.

§8.12 Temperature dependence

By combining (8.11.4) with the Gibbs–Helmholtz relation (1.49.5) we obtain*

$$F(E - T\partial E/\partial T) = -\Delta H. \qquad 8.12.1$$

By subtracting (1) from (8.11.4), or by a more direct method, we obtain

$$F(\partial E/\partial T) = \Delta S. \qquad 8.12.2$$

In both (1) and (2) the symbol Δ denotes increase when the chemical change takes place which accompanies the flow of one mole of positive charge from left to right in the cell.

It is perhaps worth while drawing attention to the physical meaning of ΔH and ΔS. If the cell is kept in a thermostat and balanced against a potentiometer so that any flow of current is reversible, then when one mole of positive charge flows from left to right in the cell

(a) the work done on the cell by the potentiometer is $-FE$
(b) the work done on the cell by external pressures is $-\Sigma_\alpha P^\alpha \Delta V^\alpha$
(c) the heat absorbed is $T\Delta S = FT(\partial E/\partial T)$
(d) the increase in the energy of the cell is the sum of the above three terms namely $\Delta U = -FE - \Sigma_\alpha P^\alpha \Delta V^\alpha + FT(\partial E/\partial T)$
(e) the increase in the enthalpy is $\Delta H = \Delta U + \Sigma_\alpha P^\alpha \Delta V^\alpha = -F\{E - T(\partial E/\partial T)\}$.

If, on the other hand, the cell is kept in a thermostat and short-circuited so that the process takes place irreversibly without the performance of electrical work, then when the process takes place to the same extent as before,

(a) the electrical work done on the cell is zero
(b) the work done on the cell by external pressures is $-\Sigma_\alpha P^\alpha \Delta V^\alpha$
(c) the heat absorbed is ΔH.

§8.13 Application of Nernst's heat theorem

The measurement of electromotive force provides a method of determining ΔG for the accompanying chemical reaction; this can be combined with a value of ΔH, determined calorimetrically, so as to obtain the value of ΔS. Since however the magnitude of $T\Delta S$ is often small compared with those of ΔG and ΔH, the relative error in ΔS determined in this way can be large.

---

* Although formula (1.49.5) is generally called the Gibbs–Helmholtz relation, it is in fact due to Gibbs, while its corollary (8.12.1) was derived by Helmholtz.

If on the other hand accurate measurements of electromotive force are made over a range of temperatures so as to give an accurate value of the temperature coefficient of the electromotive force, this provides directly the value of ΔS for the cell reaction. Values of ΔS thus obtained for any chemical reaction between only solid phases may be used to test Nernst's heat theorem, provided heat capacity data down to low temperatures are available for each substance. The procedure is illustrated by the following example.*

In the cell

$$\text{Pt} \mid \text{Pb}^{(\text{Hg})} \mid \begin{array}{c} \text{Solution of Pb salt} \\ \text{saturated with PbI}_2 \end{array} \mid \text{PbI}_2 \mid \text{I} \mid \text{Pt} \qquad 8.13.1$$

where the superscript $^{(\text{Hg})}$ denotes that the lead is in the form of an amalgam, the chemical process when one mole of positive charge flows from left to right is

$$\tfrac{1}{2}\text{Pb}^{(\text{Hg})} + \text{I} \rightarrow \tfrac{1}{2}\text{PbI}_2. \qquad 8.13.2$$

For the cell at 25 °C it is found that

$$E = 893.62 \text{ mV} \qquad 8.13.3$$

$$\partial E/\partial T = (-0.042 \pm 0.005) \text{ mV K}^{-1}. \qquad 8.13.4$$

In the cell

$$\text{Pt} \mid \text{Pb}^{(\text{Hg})} \mid \text{PbI}_2 \mid \text{Solution of KI} \mid \text{AgI} \mid \text{Ag} \mid \text{Pt} \qquad 8.13.5$$

where $\text{Pb}^{(\text{Hg})}$ denotes the same lead amalgam as in (1), the cell process accompanied by the flow of one mole of positive charge from left to right is

$$\tfrac{1}{2}\text{Pb}^{(\text{Hg})} + \text{AgI} \rightarrow \tfrac{1}{2}\text{PbI}_2 + \text{Ag}. \qquad 8.13.6$$

For this cell at 25 °C it is found that[†]

$$E = (207.8 \pm 0.2) \text{ mV} \qquad 8.13.7$$

$$\partial E/\partial T = (-0.188 \pm 0.002) \text{ mV K}^{-1}. \qquad 8.13.8$$

The data for neither of these cells can be used directly for testing Nernst's heat theorem owing to lack of calorimetric data for PbI_2 down to low temperatures. However by subtracting (7) from (3) and (8) from (4) we obtain for a cell at 25 °C in which the cell process is

* Webb, J. Phys. Chem. 1925 **29** 827.
† Gerke, J. Amer. Chem. Soc. 1922 **44** 1703.

$$Ag + I \to AgI \qquad 8.13.9$$
$$E = (685.8 \pm 0.2)\,\text{mV} \qquad 8.13.10$$
$$\partial E/\partial T = (0.146 \pm 0.004)\,\text{mV K}^{-1}. \qquad 8.13.11$$

Multiplying (10) and (11) by

$$F = 0.9649 \times 10^5 \text{ C mole}^{-1}$$
$$= 0.09649 \text{ kJ mV}^{-1} \text{ mole}^{-1} \qquad 8.13.12$$

and using (8.11.4) and (8.12.2), we obtain for the process (9) at 298 K

$$\Delta G = -66.17 \text{ kJ mole}^{-1} \qquad (T = 298 \text{ K}) \qquad 8.13.13$$
$$\Delta S = (14.06 \pm 0.4) \text{ J K}^{-1} \text{ mole}^{-1} \qquad (T = 298 \text{ K}). \qquad 8.13.14$$

From (13) and (14) we derive incidentally

$$\Delta H = \Delta G + T\Delta S$$
$$= (-66.17 + 4.22) \text{ kJ mole}^{-1}$$
$$= -61.95 \text{ kJ mole}^{-1}$$
$$= -14.81 \text{ kcal mole}^{-1} \qquad 8.13.15$$

with which may be compared the calorimetrically measured value* -14.97 kcal mole^{-1}.

We must now convert the value of ΔS at 298 K given by (14) to the corresponding value in the limit $T \to 0$. The following calorimetric data are available[†] for $S(298 \text{ K}) - S(0)$.

AgI	(115.5 ± 1.2) J K^{-1} mole^{-1}	8.13.16
Ag	(42.5 ± 0.4) J K^{-1} mole^{-1}	8.13.17
I	58.4 J K^{-1} mole^{-1}.	8.13.18

Although accurate calorimetric data for AgI are available down to $T = 15$ K, at this temperature C/R has the exceptionally high value 1.45 which leads to the rather high uncertainty, due to the extrapolation to $T = 0$, shown in (16).

Combining (16), (17), and (18) we obtain for the process (9)

$$\Delta S(298 \text{ K}) - \Delta S(0) = (14.6 \pm 1.2) \text{ J K}^{-1} \text{ mole}^{-1}. \qquad 8.13.19$$

* Webb, J. Phys. Chem. 1925 **29** 827.
† AgI, see Pitzer, J. Amer. Chem. Soc. 1941 **63** 516; Ag, see Griffiths and Griffiths, Proc. Roy. Soc. A 1914 **90** 557; I, see Lange, Z. Physik. Chem. 1924 **110** 343. Experimental data for Ag and I recomputed by Kelley, U.S. Bureau of Mines 1932 Bulletin 350.

Now comparing (19) with (14) we obtain

$$\Delta S(0) = (-0.5 \pm 1.3) \text{ J K}^{-1} \text{ mole}^{-1}. \qquad 8.13.20$$

so that within the experimental accuracy $\Delta S(0) = 0$ in agreement with Nernst's heat theorem.

§8.14 Cells without transference

When a galvanic cell contains only two solutions, one surrounding each electrode, and these two solutions are so nearly alike in composition that they may be regarded as identical except with respect to the reactions at the electrodes, the cell is called a *cell without transference*. When a current flows through the cell there is in fact necessarily transference of some electrolyte from the one electrode to the other, but if the two electrode solutions are of nearly identical composition the changes in the chemical potentials of the electrolytes transferred are negligible and so this transference is without importance.

As a typical example of a *cell without transference* we again consider the cell

| Pt, H_2 | Solution I
Aqueous HCl
saturated with H_2 | Solution II
Aqueous HCl
saturated with AgCl | AgCl | Ag | Pt | 8.14.1 |

When one mole of positive charge flows from the left to the right, the following changes take place:

(a) at the left electrode

$$\tfrac{1}{2}H_2(g) \rightarrow H^+(\text{aq I}) \qquad 8.14.2$$

(b) at the right electrode

$$AgCl(s) \rightarrow Ag(s) + Cl^-(\text{aq II}) \qquad 8.14.3$$

(c) there is a simultaneous transfer of some H^+ ions from left to right and of Cl^- ions from right to left such that the net transfer of charge from left to right is one mole and that electrical neutrality is preserved in both electrode solutions.

Since however we ignore the effect on the properties of the HCl of saturating the solution with either H_2 or AgCl, we need not distinguish between the two electrode solutions. We may therefore replace (1) by

$$\text{Pt, } H_2 \mid \text{Aqueous HCl} \mid \text{AgCl} \mid \text{Ag} \mid \text{Pt} \qquad 8.14.4$$

Correspondingly (a) and (b) reduce to

(a) $\quad \tfrac{1}{2}H_2(g) \to H^+(aq)$ $\qquad$ 8.14.5

(b) $\quad AgCl(s) \to Ag(s) + Cl^-(aq)$ $\qquad$ 8.14.6

and (c) may be ignored. Thus the process accompanying the flow of one mole of charge reduces to the chemical change

$$\tfrac{1}{2}H_2(g) + AgCl(s) \to Ag(s) + H^+(aq) + Cl^-(aq) \qquad 8.14.7$$

for which

$$\Delta G = \mu_{Ag}^{Ag} + \mu_{HCl} - \tfrac{1}{2}\mu_{H_2}^G - \mu_{AgCl}^{AgCl} \qquad 8.14.8$$

where the superscript ^{G} denotes the gas phase and μ_{HCl} denotes the chemical potential of HCl in the solution.

Substituting (8) into (8.11.4) we obtain

$$FE = -\mu_{Ag}^{Ag} - \mu_{HCl} + \tfrac{1}{2}\mu_{H_2}^G + \mu_{AgCl}^{AgCl} \qquad 8.14.9$$

in agreement with (8.10.23).

Explicit formulae for all cells without transference can be obtained similarly. We shall merely quote, without detailed derivation, one other example

$$Pt \left| \begin{array}{c} \text{Solution containing} \\ Sn^{2+} \text{ and } Sn^{4+} \end{array} \right| \left. \begin{array}{c} \text{Solution containing} \\ Fe^{2+} \text{ and } Fe^{3+} \end{array} \right| Pt \qquad 8.14.10$$

Provided that both electrode solutions contain a preponderating excess of other electrolytes and have nearly the same composition so that we may regard them as a single solution, the process accompanying the flow of one mole of positive charge from left to right is the chemical change

$$\tfrac{1}{2}Sn^{2+} + Fe^{3+} \to \tfrac{1}{2}Sn^{4+} + Fe^{2+} \qquad 8.14.11$$

for which

$$\Delta G = \tfrac{1}{2}\mu_{Sn^{4+}} + \mu_{Fe^{2+}} - \tfrac{1}{2}\mu_{Sn^{2+}} - \mu_{Fe^{3+}}. \qquad 8.14.12$$

Consequently

$$\begin{aligned} FE &= \tfrac{1}{2}\mu_{Sn^{2+}} + \mu_{Fe^{3+}} - \tfrac{1}{2}\mu_{Sn^{4+}} - \mu_{Fe^{2+}} \\ &= \tfrac{1}{2}\mu_{SnCl_2} + \mu_{FeCl_3} - \tfrac{1}{2}\mu_{SnCl_4} - \mu_{FeCl_2} \end{aligned} \qquad 8.14.13$$

provided there is some Cl^- ion in the solutions.

§8.15 Standard electromotive force

Let us return to the cell described by (8.14.4) and rewrite formula (8.14.9) for its electromotive force in terms of absolute activities. We have

$$FE/RT = -\ln \lambda_{Ag}^S - \ln \lambda_{H^+} - \ln \lambda_{Cl^-} + \tfrac{1}{2}\ln \lambda_{H_2}^G + \ln \lambda_{AgCl}^S \qquad 8.15.1$$

where we have denoted the solid phase by the superscript ^{S}, the gas phase by the superscript ^{G}, and the liquid phase by no superscript. Using (7.09.1) and (3.17.1) we can rewrite (1) as

$$FE/RT = FE^\ominus/RT - \ln(m_{H^+} m_{Cl^-} \gamma_{H,Cl}^2) + \tfrac{1}{2}\ln(p_{H_2}/P^\ominus) \qquad 8.15.2$$

where $P^\ominus$ denotes the standard pressure, taken to be one atmosphere, and $E^\ominus$ is defined by

$$FE^\ominus/RT = -\ln \lambda_{Ag}^S - \ln \lambda_{H^+}^\ominus - \ln \lambda_{Cl^-}^\ominus + \tfrac{1}{2}\ln \lambda_{H_2}^\ominus + \ln \lambda_{AgCl}^S. \qquad 8.15.3$$

This quantity $E^\ominus$ is independent of the composition of the solution and independent of the pressure of the gaseous hydrogen and is called the *standard electromotive force* of cells having the specified electrodes in the specified solvent at a specified temperature.

Similarly the electromotive force of the cell described by (8.14.10) can be expressed by the formula

$$\frac{FE}{RT} = \frac{FE^\ominus}{RT} - \tfrac{1}{2}\ln \frac{m_{Sn^{4+}} \gamma_{Sn^{4+}}}{m_{Sn^{2+}} \gamma_{Sn^{2+}}} + \ln \frac{m_{Fe^{3+}} \gamma_{Fe^{3+}}}{m_{Fe^{2+}} \gamma_{Fe^{2+}}} \qquad 8.15.4$$

in which the standard electromotive force $E^\ominus$ is defined by

$$\frac{FE^\ominus}{RT} = -\tfrac{1}{2}\ln \frac{\lambda_{Sn^{4+}}^\ominus}{\lambda_{Sn^{2+}}^\ominus} + \ln \frac{\lambda_{Fe^{3+}}^\ominus}{\lambda_{Fe^{2+}}^\ominus}. \qquad 8.15.5$$

It can readily be verified that in all these formulae only such combinations of $\lambda^\ominus$'s and of γ's occur as satisfy the condition (7.04.17) with (7.04.18).

Formulae such as the above have two applications. Firstly by making measurements of E over a range of molalities as low as possible and extrapolation with the help of formulae such as those of §7.20 we can obtain the value of the standard electromotive force $E^\ominus$. Secondly having determined the value of $E^\ominus$ by extrapolation we can insert this value into the formulae and so obtain information about certain combinations of activity coefficients in solutions of given composition.

§8.16 Numerical example

We shall illustrate the procedure described in the previous section by an example. We choose the cell

$$\text{Pt, H}_2 \left| \begin{array}{c} \text{Aqueous HCl} \\ \text{molality } m \end{array} \right| \text{HgCl} \left| \text{Hg} \right. \qquad 8.16.1$$

for which there exist measurements* at 25 °C of exceptionally high accuracy. The electromotive force of this cell is given by

$$FE = FE^\ominus - 2RT \ln m - 2RT \ln \gamma \qquad 8.16.2$$

where m is the molality and γ the mean activity coefficient of HCl. We assume that at sufficiently high dilutions γ can be represented by

$$\ln \gamma = -\alpha m^{\frac{1}{2}}(1+m^{\frac{1}{2}})^{-1} + 2\beta m \qquad 8.16.3$$

where α has the value determined by (7.17.10) and β is an adjustable constant. We now define the experimental quantity $E^{\ominus\prime}$ by

$$E^{\ominus\prime} = E + (2RT/F)\ln m - (2RT/F)\alpha m^{\frac{1}{2}}(1+m^{\frac{1}{2}})^{-1}. \qquad 8.16.4$$

Using (3) and (4) in (2) we obtain

$$E^{\ominus\prime} = E^\ominus - (4RT/F)\beta m. \qquad 8.16.5$$

If then we plot $E^{\ominus\prime}$ against m, in so far as γ can be represented by formula (3), we shall obtain a straight line of slope $-4RT\beta/F$ and of intercept at $m=0$ equal to $E^\ominus$. This plot is shown in figure 8.1 from which we find that $E^\ominus = 267.96$ mV and $\beta = 0.270$. We further see from the diagram that with

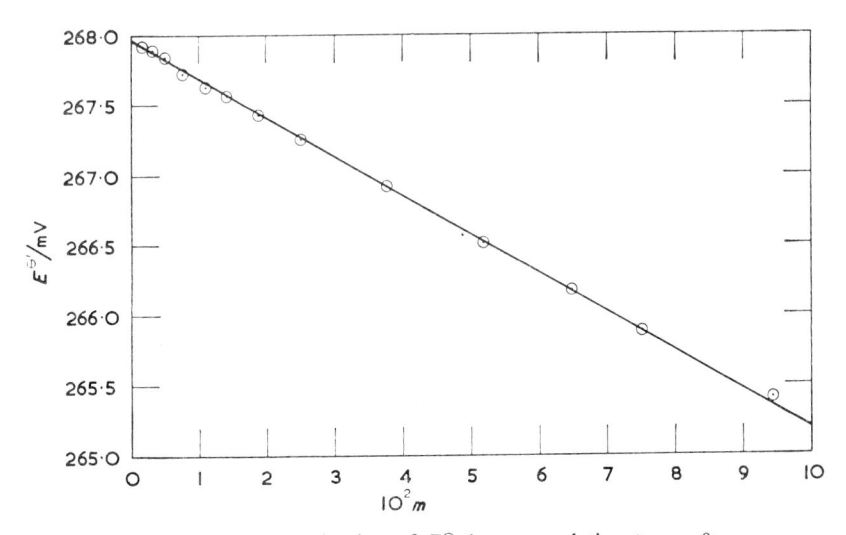

Fig. 8.1. Determination of $E^\ominus$ by extrapolation to $m=0$

* Hills and Ives, J. Chem. Soc. 1951 315.

this value of β formula (3) is as accurate as the experimental measurements for all values of m up to about 0.08.

§8.17 Standard electromotive force of half-cell

Let us consider the three cells

$$\text{Pt, H}_2 \mid \text{Solution containing H}^+ \mid \text{Solution containing Cl}^- \mid \text{AgCl} \mid \text{Ag} \mid \text{Pt} \qquad 8.17.1$$

$$\text{Pt, H}_2 \mid \text{Solution containing H}^+ \mid \text{Solution containing Sn}^{2+} \text{ and Sn}^{4+} \mid \text{Pt} \qquad 8.17.2$$

$$\text{Pt} \mid \text{Solution containing Sn}^{2+} \text{ and Sn}^{4+} \mid \text{Solution containing Cl}^- \mid \text{AgCl} \mid \text{Ag} \mid \text{Pt} \qquad 8.17.3$$

In each cell we assume that the two electrode solutions have nearly the same composition. The standard electromotive forces of the three cells are given respectively by

$$FE^\ominus/RT = \tfrac{1}{2}\ln\lambda_{H_2}^\ominus - \ln\lambda_{H^+}^\ominus + \ln\lambda_{AgCl}^S - \ln\lambda_{Ag}^S - \ln\lambda_{Cl^-}^\ominus \qquad 8.17.4$$

$$FE^\ominus/RT = \tfrac{1}{2}\ln\lambda_{H_2}^\ominus - \ln\lambda_{H^+}^\ominus + \tfrac{1}{2}\ln(\lambda_{Sn^{4+}}^\ominus/\lambda_{Sn^{2+}}^\ominus) \qquad 8.17.5$$

$$FE^\ominus/RT = -\tfrac{1}{2}\ln(\lambda_{Sn^{4+}}^\ominus/\lambda_{Sn^{2+}}^\ominus) + \ln\lambda_{AgCl}^S - \ln\lambda_{Ag}^S - \ln\lambda_{Cl^-}^\ominus \qquad 8.17.6$$

and we observe that the value of $E^\ominus$ for the third cell is equal to the difference between the values of $E^\ominus$ for the first and second cells. It is clear from this example that if there are available n different kinds of electrodes, although these can be paired to give $\tfrac{1}{2}n(n-1)$ different cells, only $n-1$ of these $E^\ominus$ values are independent. For example if we know the $E^\ominus$ values for all cells in which one of the electrodes is the Pt, H$_2$ electrode, then the $E^\ominus$ values of all other combinations can be obtained by adding and subtracting.

The $E^\ominus$ value of a cell consisting of an electrode α and a Pt,H$_2$ electrode is called the *standard electromotive force of the half-cell* α. We recall the convention that a positive value of E means that the electrode on the right is positive. We shall now illustrate by an example how this convention is extended to the electromotive force of half-cells.

We may state that the right hand half-cell

$$\text{Cl}^- \mid \text{AgCl} \mid \text{Ag} \qquad 8.17.7$$

has the standard electromotive force $E_r^\ominus = 222.5$ mV at 25 °C. This means that the cell

$$\text{Pt, H}_2 \left| \begin{array}{c} \text{Solution} \\ \text{containing H}^+ \end{array} \right| \begin{array}{c} \text{Solution} \\ \text{containing Cl}^- \end{array} \left| \text{AgCl} \right| \text{Ag} \qquad 8.17.8$$

has $E^\ominus = 222.5$ mV with the electrode on the right positive. Alternatively we may state that the left hand half-cell

$$\text{Ag} \mid \text{AgCl} \mid \text{Cl}^- \qquad 8.17.9$$

has the standard electromotive force $E_1^\ominus = -222.5$ mV at 25 °C. This means that the cell

$$\text{Ag} \left| \text{AgCl} \right| \begin{array}{c} \text{Solution} \\ \text{containing Cl}^- \end{array} \left| \begin{array}{c} \text{Solution} \\ \text{containing H}^+ \end{array} \right| \text{Pt, H}_2 \qquad 8.17.10$$

has $E^\ominus = -222.5$ mV with the electrode on the right negative. These conventions, which are unambiguous, have now been internationally agreed.*

§8.18 *Cells with transference with two similar electrodes*

Any cell which does not satisfy the conditions in the definition of a cell without transference, is called a *cell with transference*. The detailed discussion of a *cell with transference* is more involved than that of a *cell without transference*. We shall initially restrict ourselves to the case that the two electrodes are of the same chemical nature so that the chemical processes taking place at the electrodes are the converse of each other. For example we may consider the cell

$$\text{Ag} \left| \text{AgCl} \right| \begin{array}{c} \text{Solution I} \\ \text{containing} \\ \text{Cl}^- \end{array} \left| \begin{array}{c} \text{Bridge} \\ \text{solutions} \end{array} \right| \begin{array}{c} \text{Solution II} \\ \text{containing} \\ \text{Cl}^- \end{array} \left| \text{AgCl} \right| \text{Ag} \qquad 8.18.1$$

We assume that the two electrode solutions I and II are connected by *bridge solutions* in which the composition varies continuously. It is essential to exclude any discontinuity of composition, for in that case the passage of an infinitesimal current would not be reversible and it would not then be possible to apply thermodynamic equations. Suppose for example in two solutions in contact the cation Na$^+$ were present in that on the left but not that on the right, while the cation K$^+$ were present in the solution on the right but not that on the left. Then an infinitesimal current from left to right would transfer Na$^+$ from the left solution to the right solution. Reversal of the current would on the other hand transfer K$^+$ from the right solution to the left solution. If however any two solutions in contact differ only

---

* I.U.P.A.C., C.R. XVII Conference 1953 p. 83; I.U.P.A.C. Manual of Physicochemical Symbols and Terminology, Butterworths 1959.

infinitesimally in composition, the passage of current will be reversible. It is true that simultaneously there is taking place an irreversible diffusion between the two solutions tending to equalize their compositions.

This condition of continuity of composition is the only condition imposed on the nature of the bridge solutions. In view of this condition the compositions of the outermost bridge solutions are identical respectively with those of the electrode solutions. If the bridge solutions are formed by natural mixing or interdiffusion of the two electrode solutions, then their compositions throughout will be intermediate between those of the two electrode solutions. On the other hand the middle portion of the bridge solutions may consist of a solution of entirely different composition from either electrode solution, but such solution must be connected to each electrode solution through solutions of continuously varying composition. The formulae which we are about to derive are applicable to all cases, but we begin by considering the more elementary cell

$$\text{Ag} \mid \text{AgCl} \mid \begin{array}{c} \text{Solution} \\ \text{molalities } m \end{array} \mid \begin{array}{c} \text{Solution} \\ \text{molalities } m+\text{d}m \end{array} \mid \text{AgCl} \mid \text{Ag} \qquad 8.18.2$$

where the two electrode solutions differ only infinitesimally in composition.

Even in the elementary cell (2) there is not thermodynamic equilibrium and there is inevitably a state of interdiffusion between the two solutions. We are consequently compelled to introduce some assumption extraneous to classical thermodynamics which applies strictly only to equilibrium conditions. Initially we make the simplest, but not the least restrictive, assumption leading to correct conclusions. In chapter 13 it will be shown how the same conclusions can be reached by a less restrictive assumption. We here assume that the flow J_i per unit area per unit time of the ionic species i is directly proportional to the gradient of its electrochemical potential μ_i. We may, for simplicity and without any loss of generality in our conclusions, assume that the gradients are in the y-direction. Our assumption thus becomes

$$J_i = -L_i \text{d}\mu_i/\text{d}y \qquad 8.18.3$$

where L_i may depend on the composition of the solution but is independent of the gradient of the composition and independent of the flow. We consider the condition of zero electric current which exists when the cell circuit is open or alternatively when the cell is exactly balanced against a potentiometer bridge. At each electrode we have the equilibrium

$$\text{Cl}^- + \text{Ag} \rightleftharpoons \text{AgCl} + \text{el}^- \qquad 8.18.4$$

where Cl^- denotes chloride ion in solution and el^- denotes an electron in the silver. The condition for this equilibrium is

$$z_{Cl^-}^{-1} d\mu_{Cl^-} = -d\mu_{el^-} \qquad 8.18.5$$

since the silver has the same chemical potential at both ends and likewise the silver chloride. The charge number of the chloride ion is of course -1, but we have deliberately displayed it as z_{Cl^-} in order to facilitate generalization to other cells having electrodes reversible with respect to ions other than the chloride ion. The electromotive force dE of the cell (2) is then given by

$$F dE = -d\mu_{el^-} = z_{Cl^-}^{-1} d\mu_{Cl^-} \qquad 8.18.6$$

which we may also write as

$$F dE/dy = z_{Cl^-}^{-1} d\mu_{Cl^-}/dy. \qquad 8.18.7$$

The condition for zero electric current is

$$\sum_i z_i J_i = 0. \qquad 8.18.8$$

Substituting (3) into (8) we have

$$\sum_i z_i L_i d\mu_i/dy = 0. \qquad 8.18.9$$

We now multiply (7) by $\Sigma_i z_i^2 L_i$ and subtract (9) obtaining

$$\sum_i z_i^2 L_i F dE/dy = \sum_i z_i^2 L_i (-z_i^{-1} d\mu_i/dy + z_{Cl^-}^{-1} d\mu_{Cl^-}/dy) \qquad 8.18.10$$

and consequently

$$F dE/dy = \sum_i z_i^2 L_i (-z_i^{-1} d\mu_i/dy + z_{Cl^-}^{-1} d\mu_{Cl^-}/dy)/\sum_i z_i^2 L_i. \qquad 8.18.11$$

Formulae (10) and (11) in contrast to the deceptively simpler formula (7) contain only such linear combinations of $d\mu_i$'s as satisfy the condition for unambiguity (7.04.18).

Formula (11) is a complete and unambiguous formula for the electromotive force in terms of the quantities L_i defined in our assumption (3). We can however transform the expression on the right of (11) into a more perspicuous form by considering the different condition where the two electrode solutions are identical and an external potential difference dE^e is applied across the electrodes. Under these conditions we have

$$z_i^{-1} d\mu_i = F dE^e \qquad \text{(all } i\text{)}. \qquad 8.18.12$$

Hence according to (3)

$$J_i = -z_i L_i F dE^e/dy \qquad \text{(all } i\text{)} \qquad 8.18.13$$

and the electrical current per unit cross-section carried by the ionic species i will be

$$z_i F J_i = -z_i^2 L_i F^2 \, dE^e/dy. \qquad 8.18.14$$

The fraction of the total current carried by the ionic species i, called the *transport number* t_i of the species i is then given by

$$t_i = z_i^2 L_i / \sum_i z_i^2 L_i. \qquad 8.18.15$$

Comparing (11) with (15) we deduce

$$F \, dE = \sum_i t_i(-z_i^{-1} \, d\mu_i + z_{Cl^-}^{-1} \, d\mu_{Cl^-}). \qquad 8.18.16$$

Returning now to cell (1) we see that this may always be regarded as several cells of type (2) in series, all electrodes other than the two extreme ones cancelling in pairs. We accordingly deduce from (16) for the electromotive force E of cell (1)

$$FE = \int \sum_i t_i(-z_i^{-1} \, d\mu_i + z_{Cl^-}^{-1} \, d\mu_{Cl^-}) \qquad 8.18.17$$

where the integration extends through all the bridge solutions from the left electrode solution I to the right electrode solution II.

We can rewrite (17) in terms of absolute activities as

$$FE/RT = \int \sum_i t_i(-z_i^{-1} \, d \ln \lambda_i + z_{Cl^-}^{-1} \, d \ln \lambda_{Cl^-}) \qquad 8.18.18$$

or in terms of molalities and activity coefficients as

$$FE/RT = \int \sum_i t_i \{-z_i^{-1} \, d \ln(m_i \gamma_i) + z_{Cl^-}^{-1} \, d \ln(m_{Cl^-} \gamma_{Cl^-})\}. \qquad 8.18.19$$

We again stress that only such combinations of activity coefficients occur in (19) as are, in accordance with the condition (7.04.18), unambiguously defined.

§8.19 *Cells containing single electrolytes*

Formula (8.18.19) gives an explicit value of the electromotive force E, but to apply it or test it we require to know the values of the transport numbers of all cations, and all anions throughout the bridge solutions. This in turn involves a knowledge of the compositions of all the continuous series of solutions forming the bridge. Since this knowledge is usually not available, formula (8.18.19) though exact is not of much use except in specially simple cases.

The simplest and the most useful example of a cell with transference is that in which there is only one kind of cation and one kind of anion in the whole cell. Let us consider for example the cell

$$\text{Ag} \,\Big|\, \text{AgCl} \,\Big|\, \begin{matrix} \text{Aqueous MgCl}_2 \\ \text{at molality } m_1 \end{matrix} \,\Big|\, \begin{matrix} \text{Transition} \\ \text{layer} \end{matrix} \,\Big|\, \begin{matrix} \text{Aqueous MgCl}_2 \\ \text{at molality } m_2 \end{matrix} \,\Big|\, \text{AgCl} \,\Big|\, \text{Ag}$$

8.19.1

We use the name *transition layer* to denote the naturally formed bridge between the two electrode solutions consisting entirely of solutions of $MgCl_2$ of intermediate compositions. For the cell (1) formula (8.18.19) reduces to

$$FE/RT = \int_{m=m_1}^{m=m_2} -t_{Mg^{2+}}\{\tfrac{1}{2}\mathrm{d}\ln(m_{Mg^{2+}}\gamma_{Mg^{2+}}) + \mathrm{d}\ln(m_{Cl^-}\gamma_{Cl^-})\}$$

$$= -\tfrac{1}{2}\int_{m=m_1}^{m=m_2} t_{Mg^{2+}}\,\mathrm{d}\ln(m^3\gamma_{Mg,Cl}^3) \qquad 8.19.2$$

where m denotes the molality and $\gamma_{Mg,Cl}$ the mean activity coefficient of the electrolyte $MgCl_2$.

Since in solutions containing only the single electrolyte $MgCl_2$ the value of $t_{Mg^{2+}}$ depends only on the molality, the integral in (2) is completely defined and is independent of how the molality varies across the transition layer. In particular it is independent of whether the transition layer has been formed mainly by mixing of the two electrode solutions or mainly by interdiffusion between them.

If the molalities of the two electrode solutions do not differ greatly from each other, it may be legitimate to neglect the variation of $t_{Mg^{2+}}$ with composition. In this case (2) simplifies to

$$FE/RT = -\tfrac{3}{2}t_{Mg^{2+}}\ln(m_2\gamma_2/m_1\gamma_1) \qquad 8.19.3$$

where γ_1, γ_2 denote the mean activity coefficients of $MgCl_2$ in solutions of molality m_1, m_2 respectively.

If the values of $\gamma_{Mg,Cl}$ are known either from measurements of the electromotive force of cells without transference or by freezing-point measurements combined with use of the Gibbs–Duhem relation, then formula (2) can be used to give information concerning the transport number $t_{Mg^{2+}}$. Conversely if there are independent measurements of this transport number, then formula (2) may be used to give information about the dependence of the mean activity coefficient $\gamma_{Mg,Cl}$ on the molality.

§8.20 Cells with transference having two dissimilar electrodes

In §8.14 we discussed cells without transference and in §8.18 cells with transference having two similar electrodes. We have still to consider cells with transference having two dissimilar electrodes. These are most easily disposed of by regarding them as a combination of the two types of cell previously discussed. This will be made clear by a typical example. The cell

$$\text{Pt, H}_2 \left| \begin{array}{c} \text{Solution I} \\ \text{containing HCl} \end{array} \right| \begin{array}{c} \text{Bridge} \\ \text{solutions} \end{array} \left| \begin{array}{c} \text{Solution II} \\ \text{containing HCl} \end{array} \right| \text{AgCl} \left| \text{Ag} \right| \text{Pt} \qquad 8.20.1$$

may be regarded as a combination of the two cells

$$\text{Pt, H}_2 \left| \begin{array}{c} \text{Solution I} \\ \text{containing HCl} \end{array} \right| \text{AgCl} \left| \text{Ag} \right| \text{Pt} \qquad 8.20.2$$

$$\text{Pt}\left|\text{Ag}\right|\text{AgCl} \left| \begin{array}{c} \text{Solution I} \\ \text{containing HCl} \end{array} \right| \begin{array}{c} \text{Bridge} \\ \text{solutions} \end{array} \left| \begin{array}{c} \text{Solution II} \\ \text{containing HCl} \end{array} \right| \text{AgCl}\left|\text{Ag}\right|\text{Pt} \qquad 8.20.3$$

Consequently the electromotive force of the cell (1) is the sum of those of the cells (2) and (3). But cell (2) is without transference and, as shown in §8.14, its electromotive force E_2 is given by

$$FE_2 = -\mu_{Ag}^{Ag} + \tfrac{1}{2}\mu_{H_2}^{G} + \mu_{AgCl}^{AgCl} - \mu_{HCl}^{I} \qquad 8.20.4$$

where the superscript $^\text{I}$ refers to the solution I. Cell (3) on the other hand has two similar electrodes and its electromotive force E_3 is given by (8.18.17)

$$FE_3 = -\int_I^{II} \sum_R t_R(z_R^{-1}\,d\mu_R - z_{Cl^-}^{-1}\,d\mu_{Cl^-})$$

$$- \int_I^{II} \sum_X t_X(z_X^{-1}\,d\mu_X - z_{Cl^-}^{-1}\,d\mu_{Cl^-}) \qquad 8.20.5$$

wherein we recall that $z_{Cl^-} = -1$ and all the z_X are negative integers. The electromotive force E_1 of the cell 1 is then given by

$$E_1 = E_2 + E_3. \qquad 8.20.6$$

The accurate expressions for the electromotive force of the most general type of cell with transference were formulated by P. B. Taylor.*

---

* Taylor, J. Phys. Chem. 1927 **31** 1478; Cf. Guggenheim, J. Phys. Chem. 1930 **34** 1758.

CHAPTER 9

GRAVITATIONAL FIELD

§9.01 Nature of gravitational field

The formulae of chapter 1 are easily extended so as to take account of the presence of a gravitational field. Such a field is characterized by a gravitational potential Φ with a definite value at each place. The modification of the gravitational field by the presence of matter in amounts dealt with in ordinary chemical and physical processes is completely negligible compared with the earth's field or any other field of comparable importance. We may therefore regard the gravitational field as completely independent of the state of the thermodynamic system considered. In this sense, we call the gravitational field an *external field*, and regard the gravitational potential at each point as independent of the presence or state of any matter there. It is owing to this fact that, although the abstract theories of gravitational potential and electrostatic potential are in some ways parallel, yet their significance for thermodynamic systems is different.

§9.02 Phases in gravitational field

Since a phase was defined as completely homogeneous in its properties and *state*, two portions of matter of identical temperature and composition will be considered as different phases if they are differently situated with respect to a gravitational field. It follows that the mere presence of a gravitational field excludes the possibility of a phase of finite depth in the direction of the field. In the presence of a gravitational field even the simplest possible kind of system must be considered as composed of a continuous sequence of phases each differing infinitesimally from its neighbours.

§9.03 Thermodynamic functions in gravitational field

The characteristic property of the gravitational potential is that the work w required to bring a quantity of matter of mass m from a place where the potential is Φ^α to a place where it is Φ^β is given by

$$w = m(\Phi^\beta - \Phi^\alpha) \qquad 9.03.1$$

thus depending on the mass but not on the chemical nature of the matter. In transferring an amount dn_i of the species i from the phase α to the phase β, the gravitational work is

$$(\Phi^\beta - \Phi^\alpha)M_i dn_i \qquad 9.03.2$$

where M_i is the proper mass of the species i. Thus formula (1.27.2) for dU^α must for each phase α contain the extra terms $\Sigma_i \Phi^\alpha M_i dn_i^\alpha$. That is to say

$$dU^\alpha = T^\alpha dS^\alpha - P^\alpha dV^\alpha + \sum_i (\mu_i^\alpha + M_i \Phi^\alpha) dn_i^\alpha \qquad 9.03.3$$

whence follows directly

$$dG^\alpha = -S^\alpha dT^\alpha + V^\alpha dP^\alpha + \sum_i (\mu_i^\alpha + M_i \Phi^\alpha) dn_i^\alpha. \qquad 9.03.4$$

It follows that to take account of the effect of a gravitational field one has merely to replace μ_i^α throughout by $\mu_i^\alpha + M_i \Phi^\alpha$.

Although in all thermodynamic formulae the quantity Φ^α occurs only in combinations of the form $\mu_i^\alpha + M_i \Phi^\alpha$, yet the gravitational potential difference $\Phi^\beta - \Phi^\alpha$ between two phases α and β, in contrast to the electric potential difference $\psi^\beta - \psi^\alpha$, is thermodynamically determinate owing to the fact that its value is independent of the presence and nature of the phase there. The phase may therefore be removed without altering Φ and then $\Phi^\beta - \Phi^\alpha$ can be determined in empty space by direct mechanical measurements.

§9.04 Equilibrium in gravitational field

For the equilibrium as regards the species i between two phases α and β defined not merely by their temperature, pressure, and composition, but also by their gravitational potentials, we have in analogy with (1.39.5) the general condition

$$\mu_i^\alpha + M_i \Phi^\alpha = \mu_i^\beta + M_i \Phi^\beta \qquad \text{(equilibrium)}. \qquad 9.04.1$$

§9.05 Dependence of μ_i on T and P

Observing that M_i and Φ^α are independent of T^α and P^α, we obtain, dropping the superscript $^\alpha$ throughout,

$$\partial\mu_i/\partial T = \partial^2 G/\partial n_i \partial T = -\partial S/\partial n_i = -S_i \qquad 9.05.1$$

$$\partial\mu_i/\partial P = \partial^2 G/\partial n_i \partial P = \partial V/\partial n_i = V_i \qquad 9.05.2$$

precisely the same as in the absence of a gravitational field.

§9.06 Single component in gravitational field

For the equilibrium of a single component i in a gravitational field we have according to (9.04.1)

$$d\mu_i + M_i d\Phi = 0 \qquad 9.06.1$$

or at constant temperature using (9.05.2)

$$V_i dP + M_i d\Phi = 0 \qquad (T \text{ const.}). \qquad 9.06.2$$

If ϱ denotes the density, then

$$\varrho_i = M_i/V_i. \qquad 9.06.3$$

Substituting (3) into (2) we obtain

$$dP = -\varrho_i d\Phi \qquad 9.06.4$$

in agreement with the general condition of hydrostatic equilibrium.

In the case of a single perfect gas we have

$$V_i = RT/P. \qquad 9.06.5$$

Substituting (5) into (2) we obtain

$$RT \, d \ln P + M_i d\Phi = 0 \qquad 9.06.6$$

and by integration

$$RT \ln(P^\beta/P^\alpha) = M_i(\Phi^\alpha - \Phi^\beta) \qquad 9.06.7$$

or

$$P^\beta/P^\alpha = \exp\{-M_i(\Phi^\beta - \Phi^\alpha)/RT\}. \qquad 9.06.8$$

For a liquid, on the other hand, neglecting compressibility and so treating V_i as independent of P, we can integrate (2) immediately obtaining

$$V_i(P^\beta - P^\alpha) = M_i(\Phi^\alpha - \Phi^\beta). \qquad 9.06.9$$

Alternatively integrating (4) we obtain

$$P^\beta - P^\alpha = \varrho_i(\Phi^\alpha - \Phi^\beta). \qquad 9.06.10$$

§9.07 Mixture in gravitational field

For the equilibrium of each species i of a mixture in a gravitational field we have according to (9.04.1)

$$d\mu_i + M_i d\Phi = 0. \qquad 9.07.1$$

Using (9.05.2) we obtain at constant temperature

$$D\mu_i + V_i dP + M_i d\Phi = 0 \qquad (T \text{ const.}) \qquad 9.07.2$$

where D denotes $\Sigma_i dx_i(\partial/\partial x_i)_{T,P}$. But according to the Gibbs–Duhem relation we have

$$\sum_i x_i D\mu_i = 0. \qquad 9.07.3$$

Multiplying (2) by x_i, summing over all species i, and using (3) we obtain

$$\sum_i x_i V_i dP + \sum_i x_i M_i d\Phi = 0. \qquad 9.07.4$$

Introducing the proper volume V_m and the proper mass M_m given respectively by

$$V_m = \sum_i x_i V_i \qquad 9.07.5$$

$$M_m = \sum_i x_i M_i \qquad 9.07.6$$

we can write (4) as

$$V_m dP + M_m d\Phi = 0. \qquad 9.07.7$$

But the density ϱ is related to V_m, M_m, by

$$\varrho = M_m / V_m. \qquad 9.07.8$$

Substituting (8) into (7), we recover the usual condition of hydrostatic equilibrium

$$dP = -\varrho d\Phi. \qquad 9.07.9$$

If we substitute for dP from (7) into (2), we obtain

$$D\mu_i + (M_i - V_i M_m/V_m) d\Phi \qquad 9.07.10$$

or, according to (8),

$$D\mu_i + (M_i - \varrho V_i) d\Phi = 0. \qquad 9.07.11$$

The differential equations of this section can be integrated only in certain exceptionally simple cases which we shall consider in turn.

§9.08 *Mixture of gases*

For a mixture of perfect gases it is possible to integrate (9.07.2), but the same result can be obtained as follows. For any component i in two gaseous mixtures α, β at the same temperature T, we have

$$\mu_i^\beta - \mu_i^\alpha = RT \ln(p_i^\beta/p_i^\alpha) \qquad 9.08.1$$

where p_i^α and p_i^β are the fugacities in the two phases.

Substituting (1) into (9.04.1) we obtain as the equilibrium condition for the species i in a gravitational field

$$RT \ln(p_i^\beta/p_i^\alpha) = M_i(\Phi^\alpha - \Phi^\beta) \qquad 9.08.2$$

or

$$p_i^\beta/p_i^\alpha = \exp\{-M_i(\Phi^\beta - \Phi^\alpha)/RT\}. \qquad 9.08.3$$

If we differentiate (2) we obtain

$$dp_i/p_i = -(M_i/RT)d\Phi. \qquad 9.08.4$$

If the gas mixture is perfect then using

$$p_i = y_i RT/V_m \qquad 9.08.5$$

we can rewrite (4) as

$$dp_i = -(y_i M_i/V_m)d\Phi. \qquad 9.08.6$$

Summing (6) over all species i, we obtain

$$dP = -(M_m/V_m)d\Phi = -\varrho\, d\Phi \qquad 9.08.7$$

thus verifying that (2) and (3) are consistent with hydrostatic equilibrium.

§9.09 *Ideal dilute solutions*

In the case of an ideal dilute solution we may replace (9.07.10) for each solute species s by

$$RT\, d\ln m_s + (M_s - V_s M_m/V_m)d\Phi = 0. \qquad 9.09.1$$

In the limit of *extreme dilution* we may replace M_m/V_m by M_1/V_1^0, where the superscript ^{0} relates to the pure solvent, and obtain

$$RT\, d\ln m_s + (M_s - V_s M_1/V_1^0)d\Phi = 0. \qquad 9.09.2$$

Neglecting compressibility, this can be integrated directly, giving

$$m_s^\beta/m_s^\alpha = \exp\{-(M_s - V_s M_1/V_1^0)(\Phi^\beta - \Phi^\alpha)/RT\}. \qquad 9.09.3$$

§9.10 Chemical reaction in gravitational field

For the chemical reaction

$$0 = \sum_B \nu_B B \qquad 9.10.1$$

the most general form for the condition of equilibrium in the absence of a gravitational field is

$$\sum_B \nu_B \mu_B = 0. \qquad 9.10.2$$

In the presence of a gravitational field the corresponding equilibrium condition is evidently

$$\sum_B \nu_B (\mu_B + M_B \Phi) = 0. \qquad 9.10.3$$

But owing to the conservation of mass we have

$$\sum_B \nu_B M_B = 0. \qquad 9.10.4$$

Multiplying (4) by Φ and subtracting from (3) we recover (2). It follows that any chemical equilibrium constant is independent of the gravitational potential or in other words is unaffected by the presence of a gravitational field.

CHAPTER 10

ELECTROSTATIC SYSTEMS

§10.01 Introduction

We now propose to study the thermodynamic properties of substances in an electrostatic field. For this purpose it will suffice to consider the field in a parallel-plate capacitor neglecting any edge effect. Thus when we refer to the extensive properties of a parallel-plate capacitor of area A, we really mean the difference between those of a capacitor of area $\mathscr{A}+A$ and those of a similar capacitor of area $\mathscr{A}$, where $\mathscr{A} \gg A$.

§10.02 Parallel-plate capacitor in vacuo

Consider a parallel-plate capacitor of area A, the distance between the plates being d. Let the charges on the two plates be $+Q$ and $-Q$. The capacitor being in vacuo let the work required to transfer an elementary charge $\mathrm{d}Q$ from the negative plate α to the positive plate β be $(\psi^\beta - \psi^\alpha)\mathrm{d}Q$. Then $\psi^\beta - \psi^\alpha$ is called the *potential difference* between the two plates and $E = -(\psi^\beta - \psi^\alpha)/d$ is called the *electric field strength* between the plates. Then the ratio ε_0 defined by

$$Qd/A(\psi^\beta - \psi^\alpha) = Q/AE = \varepsilon_0 \qquad 10.02.1$$

is a universal constant called the *rationalized permittivity of a vacuum*. The value of ε_0 is given by

$$\varepsilon_0 = 8.854 \times 10^{-12} \text{ C}^2 \text{ J}^{-1} \text{ m}^{-1}$$
$$4\pi\varepsilon_0 = 1.113 \times 10^{-10} \text{ C}^2 \text{ J}^{-1} \text{ m}^{-1}$$
$$= 1.113 \times 10^{-10} \text{ A s V}^{-1} \text{ m}^{-1}$$

§10.03 Parallel-plate capacitor in fluid

Now consider the same parallel-plate capacitor completely immersed in a homogeneous fluid. If the charges on the plates are again $+Q$ and $-Q$, and if the potential difference between the plates, defined as before, is again denoted by $\psi^\beta - \psi^\alpha$ then the ratio ε, defined by

$$Qd/A(\psi^\beta - \psi^\alpha) = Q/AE = \varepsilon \qquad 10.03.1$$

is called the *rationalized permittivity of the fluid*. The value of ε depends on the nature of the fluid, on its temperature, and possibly also on E, but is independent of the size and shape of the capacitor. ε has of course the same dimensions as ε_0. The ratio $\varepsilon_r = \varepsilon/\varepsilon_0$ is called the *relative permittivity* or the *dielectric coefficient* of the fluid.

§10.04 Work of charging a capacitor

According to (10.03.1) we have

$$\psi^\beta - \psi^\alpha = Qd/\varepsilon A \qquad 10.04.1$$

and so the work required to bring an element of charge dQ from the negative plate α to the positive plate β is

$$(Qd/\varepsilon A)dQ. \qquad 10.04.2$$

From (10.03.1) we have also

$$Q = A\varepsilon E \qquad 10.04.3$$

$$dQ = A\,d(\varepsilon E). \qquad 10.04.4$$

Substituting (3) and (4) into (2) we obtain for the work w required to increase the field strength from E to $E + dE$

$$w = AdE\,d(\varepsilon E) = V_c E\,d(\varepsilon E) \qquad 10.04.5$$

where V_c denotes the volume between the plates of the capacitor and is assumed independent of temperature and pressure.

Formula (5) is valid for any infinitesimal change, including in particular an adiabatic change and an isothermal change, but the dependence of εE on E will in general not be the same in these two cases. The quantity εE is called the *electric displacement*.

§10.05 Characteristic functions

If we now consider the system consisting of the whole fluid of volume V surrounding and including the capacitor, we obtain by using (10.04.5) the relations

$$dU = T\,dS - P\,dV + V_c E\,d(\varepsilon E) + \sum_i \mu_i\,dn_i \qquad 10.05.1$$

$$d\mathscr{F} = -S\,dT - P\,dV + V_c E\,d(\varepsilon E) + \sum_i \mu_i\,dn_i. \qquad 10.05.2$$

Formulae (1) and (2) are the extensions of (1.28.6), (1.28.7) respectively including the extra term (10.04.5) representing the work required to change the field E between the plates of the capacitor.

We now define the characteristic function G by

$$G = U - TS + PV - V_c \varepsilon E^2. \qquad 10.05.3$$

Differentiating (3) and substituting into (1), we obtain

$$dG = -S\,dT + V\,dP - V_c \varepsilon E\,dE + \sum_i \mu_i\,dn_i. \qquad 10.05.4$$

In all the above formulae V denotes the total volume of fluid in which the capacitor is immersed and P denotes the pressure acting on the outside boundary of the fluid in which the capacitor is completely immersed. We have carefully avoided reference to any pressure within the fluid between the plates of the capacitor, for the definition of such a pressure would require special caution and its use as an independent variable would lead to more complicated formulae.

§10.06 *Analogues of Maxwell's relations*

By forming the second differential coefficients of the characteristic functions we can obtain several relations analogous to Maxwell's relations obtained in §1.47. In particular from (10.05.4) we derive

$$(\partial S/\partial E)_{T,P} = V_c(\partial[\varepsilon E]/\partial T)_{P,E} = V_c E(\partial \varepsilon/\partial T)_{P,E} \qquad 10.06.1$$

$$(\partial V/\partial E)_{T,P} = -V_c(\partial[\varepsilon E]/\partial P)_{T,E} = -V_c E(\partial \varepsilon/\partial P)_{T,E}. \qquad 10.06.2$$

This change in volume accompanying change in field strength at constant temperature and pressure is called *electrostriction*.

§10.07 *Constant permittivity. Dielectric constant*

For the sake of generality we have hitherto made no assumption concerning the dependence of the permittivity ε on the field strength E. For almost all substances at field strengths met in an ordinary laboratory the permittivity ε is for a given temperature and pressure independent of the field strength. We shall from here onwards assume this to be the case. The *relative permittivity* or *dielectric coefficient* $\varepsilon_r = \varepsilon/\varepsilon_0$ is then called the *dielectric constant*.

Formula (10.06.1) may now be written more simply as

$$(\partial S/\partial E)_{T,P} = V_c E(\partial \varepsilon/\partial T)_P \qquad 10.07.1$$

and the electrostriction formula (10.06.2) as

$$(\partial V/\partial E)_{T,P} = -V_c E(\partial \varepsilon/\partial P)_T. \qquad 10.07.2$$

§10.08 *Integrated formulae*

When we assume that ε is independent of E we can integrate (10.05.4) at constant T, P, n_i obtaining

$$G = G^0 - \tfrac{1}{2}\varepsilon E^2 V_c \qquad 10.08.1$$

where the superscript ^{0} denotes the value at zero field at the given temperature, pressure, and composition.

By differentiation of (1) we obtain

$$S = S^0 + \tfrac{1}{2}(\partial \varepsilon/\partial T)E^2 V_c \qquad 10.08.2$$

$$V = V^0 - \tfrac{1}{2}(\partial \varepsilon/\partial P)E^2 V_c \qquad 10.08.3$$

$$\mu_i = \mu_i^0 - \tfrac{1}{2}(\partial \varepsilon/\partial n_i)E^2 V_c \qquad 10.08.4$$

from which we deduce

$$H = H^0 + \tfrac{1}{2}\{\varepsilon + T(\partial \varepsilon/\partial T)\}E^2 V_c \qquad 10.08.5$$

$$U = U^0 + \tfrac{1}{2}\{\varepsilon + T(\partial \varepsilon/\partial T) + P(\partial \varepsilon/\partial P)\}E^2 V_c. \qquad 10.08.6$$

We must point out that the statement occurring in text-books on electricity that the energy density due to the field is $\tfrac{1}{2}\varepsilon E^2$ is false.

§10.09 *Application to perfect gas*

We shall illustrate the use of the relation (10.08.4) by its application to the simplest case of a single perfect gas.

The rationalized permittivity ε of a perfect gas is related to the rationalized permittivity ε_0 of a vacuum by

$$\varepsilon - \varepsilon_0 = (Ln/V)(\alpha + \beta/T) \qquad 10.09.1$$

where α is equal to the molecular polarizability and β is given by

$$\beta = \mu^2/3k \qquad 10.09.2$$

where μ is the electric moment of the molecule and k the Boltzmann constant.

Substituting (1) into (10.08.4) we obtain

$$\mu = \mu^0 - \tfrac{1}{2}E^2 L(\alpha + \beta/T)$$
$$= \mu^\ominus + RT \ln(nRT/P^\ominus V) - \tfrac{1}{2}E^2 L(\alpha + \beta/T). \qquad 10.09.3$$

Let us now consider the equilibrium distribution of a gas between the region, denoted by the superscript ^{i}, inside a capacitor where the field strength is E and the region, denoted by the superscript ^{e}, exterior to this field. We have then

$$\mu^i = \mu^\ominus + RT \ln(n^i RT/V^i P^\ominus) - \tfrac{1}{2}E^2 L(\alpha + \beta/T) \qquad 10.09.4$$
$$\mu^e = \mu^\ominus + RT \ln(n^e RT/V^e P^\ominus). \qquad 10.09.5$$

The equilibrium distribution is determined by

$$\mu^i = \mu^e. \qquad 10.09.6$$

Substituting (4) and (5) into (6), we obtain, writing c for n/V,

$$RT \ln(c^i/c^e) - \tfrac{1}{2}E^2 L(\alpha + \beta/T) = 0 \qquad 10.09.7$$

or

$$c^i/c^e = \exp\{(\tfrac{1}{2}E^2/RT)(\alpha + \beta/T)\}. \qquad 10.09.8$$

Since α is always positive and β is either positive or zero, it follows that c is always greater inside the field than outside it. Thus every perfect gas is attracted into an electric field.

CHAPTER 11

MAGNETIC SYSTEMS

§11.01 *Introduction*

In order to apply thermodynamics to magnetic systems we have merely to extend our previous formulae by including extra terms for the magnetic work. In principle the procedure is straightforward and should cause no difficulty. There is however a serious incidental difficulty, namely that of finding the correct general expression for magnetic work. We should expect to be able to discover such an expression by consulting any reputable text-book on electromagnetism. Unfortunately this is far from the case. The treatment given in most text-books is altogether inadequate. In most cases the derivations of formulae for magnetic work assume either explicitly or implicitly that the permeability of each piece of matter is a constant, whereas from a thermodynamic viewpoint one of the questions of greatest interest is how the permeability varies with the temperature. It is therefore desirable, if not essential, to start from formulae which are not restricted to the assumption that the permeability of each piece of matter is invariant. In many, if not most, text-books on electromagnetism the treatment of magnetic work suffers from other even more serious defects. In some text-books the treatment is based on a discussion of permanent magnets imagined to be constructed by bringing together (reversibly?!) from infinity an infinite number of infinitesimal magnetic elements. Actually a permanent magnet is an idealization far from reality. It is true that magnets can be made which are nearly permanent with respect to changes in position, but they are never permanent with respect to changes of temperature. Increase of temperature is usually accompanied by an irreversible loss of magnetization. Whatever may be the use of the conception of a permanent magnet in the theory of such instruments as compasses, galvanometers, and dynamos, it is not a useful conception as a basis for the analysis of magnetic work when changes of temperature may be important. The worst text-books give formulae for magnetic work which not only are of restricted applicability, but even

MAGNETIC SYSTEMS 339

contain wrong signs. Others confuse the external magnetic field B_e with the local internal field B. Fortunately there are a few text-books* on electromagnetism which give a clear correct treatment of magnetic work. Here we shall assume the correct formula for magnetic work after first recalling the physical meaning of the several electromagnetic quantities involved and how they are related to one another.

§11.02 *Fundamental electromagnetic vectors*

As elsewhere in this book we use the rationalized system of electromagnetic quantities. We recall that the strength and direction of an electrostatic field is described at each point by a vector E such that the force acting on a small stationary test charge Q placed at this point is QE. This vector E is called the *electric field strength*. The analogous magnetic vector describing the force acting on a small test element of current is denoted by B and has the property[†] that the force on each element ds of a linear conductor of current i is given by the vector product $i\mathrm{d}s \times B$. This vector B is called the *magnetic induction*.

§11.03 *Permittivity and permeability in a vacuum*

In a vacuum the value of E at each point is determined by the distribution of electric charges and is the sum of independent contributions from each charge. The contribution to E of a charge Q at a distance r is directed along r and is of magnitude

$$Q/4\pi\varepsilon_0 r^2 \qquad 11.03.1$$

where ε_0 is a universal constant called the *rationalized permittivity of a vacuum*. Alternatively we may say that each charge Q makes an additive contribution

$$Q/4\pi\varepsilon_0 r \qquad 11.03.2$$

to the *electrostatic potential* ψ and that E is then determined by

$$E = -\operatorname{grad}\psi. \qquad 11.03.3$$

We turn now to the analogous magnetic formulae. Each element ds of a linear conductor carrying a current i makes an additive contribution

$$\mu_0\, i\mathrm{d}s/4\pi r \qquad 11.03.4$$

---
* In particular Stratton, Electromagnetic Theory, McGraw-Hill 1941, hereafter referred to as S., E.T.
[†] S., E.T. p. 96.

340   MAGNETIC SYSTEMS

to A, called the *magnetic vector potential*, and B is then determined by

$$B = \operatorname{curl} A. \qquad 11.03.5$$

The quantity μ_0 occurring in (4) is a universal constant called the *rationalized permeability of a vacuum*.

Before proceeding further it is instructive to consider the physical dimensions of the quantities occurring above in terms of the four independent dimensions length L, time T, energy U, and electric charge Q. For the present purpose it is more convenient to choose energy than mass as one of the four independent dimensions. The dimensions of the most important quantities are given in table 11.1.

TABLE 11.1

Dimensions of electromagnetic quantities

L denotes length, T time, U energy, and Q electric charge

Symbol	Name	Dimensions
Q	Electric charge	Q
i	Current	QT^{-1}
ds	Element of length	L
ids	Element of current	QLT^{-1}
ψ	Electrostatic potential	UQ^{-1}
A	Magnetic vector potential	$UL^{-1}TQ^{-1}$
E	Electric field strength	$UL^{-1}Q^{-1}$
B	Magnetic induction	$UL^{-2}TQ^{-1}$
$\varepsilon_0 E$		QL^{-2}
$\mu_0^{-1} B$		$QL^{-1}T^{-1}$
ε_0		$Q^2 L^{-1} U^{-1}$
μ_0		$UQ^{-2}L^{-1}T^2$
μ_0^{-1}		$Q^2 LT^{-2} U^{-1}$
$\varepsilon_0 \mu_0$		$L^{-2}T^2$
$\varepsilon_0 E^2$		UL^{-3}
$\mu_0^{-1} B^2$		UL^{-3}

The following points are worthy of note.

1. Inasmuch as an element of current is the analogue in a magnetic system of an element of electric charge in an electrostatic system, it is clear* that μ_0^{-1}, not μ_0, is the analogue of ε_0.
2. $(\varepsilon_0 \mu_0)^{-\frac{1}{2}}$ has the dimensions of a velocity; it is well known that this quantity is equal to the speed of propagation of electromagnetic waves in a vacuum.

* Sommerfeld, 'Electrodynamics', translated by Ramberg, Academic Press 1952 p. 21.

3. The quantities $\varepsilon_0 E^2$ and $\mu_0^{-1} B^2$ both have the dimensions of energy density or pressure.

The values of ε_0 and μ_0 and related quantities in the rational system are as follows:

$$\varepsilon_0 = 8.854 \times 10^{-12} \text{ C}^2 \text{ J}^{-1} \text{ m}^{-1}$$
$$\mu_0 = 4\pi \times 10^{-7} \text{ J s}^2 \text{ C}^{-2} \text{ m}^{-1} = 1.2566 \times 10^{-6} \text{ J s}^2 \text{ C}^{-2} \text{ m}^{-1}$$
$$\varepsilon_0 \mu_0 = 1.1126 \times 10^{-17} \text{ s}^2 \text{ m}^{-2} = (2.9979 \times 10^8 \text{ m s}^{-1})^{-2}.$$

§11.04 Simplest examples of fields in a vacuum

The formulae of the previous section are sufficient to specify completely the E field due to any given distribution of charges in a vacuum or the B field due to any given distribution of currents in a vacuum. The quantitative application of these formulae is however complicated and tedious except for systems having a high degree of symmetry. We shall consider briefly one such electrostatic system and one such magnetic system.

As the electrostatic system we choose the parallel-plate capacitor, already discussed in the previous chapter, neglecting edge effects. If charges Q and $-Q$ are distributed uniformly over the two plates each of area A at a distance d apart, then in the absence of any matter between the plates the electric field is uniform, normal to the plates, and has the value

$$|E| = Q/\varepsilon_0 A. \qquad 11.04.1$$

As an example of a magnetic system having simple symmetry we choose a long uniform solenoid and we ignore end effects. The magnetic induction inside the empty solenoid is then uniform, parallel to the axis, and has the value

$$|B| = \mu_0 i/l \qquad 11.04.2$$

when the current is i and there is one turn per length l.

For reasons which will appear later it is instructive to rewrite (1) and (2) in somewhat different forms. We rewrite (1) as

$$\varepsilon_0 |E| = Qd/V_c \qquad 11.04.3$$

where d is the distance between the plates so that $V_c = dA$ is the volume included between the plates of the capacitor. The product Qd of the charge on a plate and the distance between the plates may be called the *electric moment* of the charged capacitor. Thus according to (3) we observe that in this system with simple symmetry $\varepsilon_0 |E|$ is equal to the *electric moment per unit volume*.

We likewise rewrite (2) in the form

$$\mu_0^{-1}|\boldsymbol{B}| = niA/V_s \qquad 11.04.4$$

where n denotes the total number of turns, A denotes the cross-section of the solenoid, and $V_s = nlA$ denotes the volume contained by the solenoid. We may regard the solenoid as an *electromagnet* and we call the product niA its *magnetic moment*. We see then according to (4) that $\mu_0^{-1}|\boldsymbol{B}|$ is equal to the *magnetic moment per unit volume* of the solenoid.

From these relations we again perceive that μ_0^{-1}, not μ_0, is the analogue of ε_0.

§11.05 *Presence of matter*

We shall now discuss briefly the effect of filling the parallel-plate capacitor and the solenoid respectively with uniform matter.

When the space between the plates of the capacitor is filled with uniform matter, this matter becomes electrically polarized as a result of the field due to the charges on the plates. The *electric polarization* $\boldsymbol{P}$ is defined as the electric moment per unit volume induced in the matter. Owing to the symmetry of the system under consideration $\boldsymbol{P}$ is in this case uniform and normal to the plates. It is not difficult to see what will be the resultant effect on the field $\boldsymbol{E}$. We interpreted formula (11.04.3) to mean that $\varepsilon_0|\boldsymbol{E}|$ is equal to the *electric moment per unit volume* of the charged capacitor. It is evident that $\varepsilon_0|\boldsymbol{E}|$ will now be equal to the resultant electric moment per unit volume due partly to the charges $\pm Q$ on the plates and partly to the polarization of the matter between the plates. Thus in place of (11.04.3) we shall have

$$\varepsilon_0|\boldsymbol{E}| = Qd/V_c - |\boldsymbol{P}| \qquad 11.05.1$$

or

$$\varepsilon_0|\boldsymbol{E}| + |\boldsymbol{P}| = Qd/V_c = Q/A. \qquad 11.05.2$$

Thus $\varepsilon_0 \boldsymbol{E} + \boldsymbol{P}$ is now related to the charge on the capacitor plates in precisely the same manner as $\varepsilon_0 \boldsymbol{E}$ was related to it when the capacitor was empty. In other systems having lower symmetry the situation is less simple because $\boldsymbol{E}$ and $\boldsymbol{P}$ vary from place to place. The composite vector $\varepsilon_0 \boldsymbol{E} + \boldsymbol{P}$ still however plays an important role. It is called by the curious name *electric displacement* and is denoted by $\boldsymbol{D}$. Thus by definition

$$\boldsymbol{D} = \varepsilon_0 \boldsymbol{E} + \boldsymbol{P}. \qquad 11.05.3$$

From the identity (3) it is evident that any two of the vectors $\boldsymbol{E}, \boldsymbol{P}, \boldsymbol{D}$

completely determine the remaining one. It is however a fundamental assumption of electrostatics, borne out by experiment, that at any point in a piece of matter of given composition, given temperature, and given pressure any one of the vectors E, P, D completely determines the other two. If moreover the matter is isotropic, then E, P, and D have the same direction. If then we write

$$D = \varepsilon E \qquad 11.05.4$$

the coefficient ε is a scalar quantity, provided the matter is isotropic. (Otherwise ε would be a tensor of rank two.) The quantity ε defined by (4) is called the *permittivity* of the matter. Its value in general depends on the composition of the matter, the temperature, the pressure, and the field strength. The ratio

$$D/\varepsilon_0 E = \varepsilon/\varepsilon_0 \qquad 11.05.5$$

is called the *relative permittivity* or the *dielectric coefficient* or, when its value is independent of E, the *dielectric constant*. It is evident from (3) that P and D have the same dimensions as $\varepsilon_0 E$, namely that of charge/area. It is likewise evident from (3) and (4) that ε has the same dimensions as ε_0, so that the *dielectric coefficient* $\varepsilon/\varepsilon_0$ is a dimensionless number.

Much of the above was implicitly assumed in the previous chapter, is moreover well known, and is seemingly irrelevant to magnetic systems. It is however convenient to have these relations before us for comparison with analogous but less understood magnetic relations.

We turn now to consider the effect of filling the uniform solenoid with uniform matter. As a result of the current in the solenoid the matter will behave as if it contained induced microscopic molecular current circuits or elementary magnets. According to (11.04.4) their contribution to $\mu_0^{-1} B$ will be equal to the magnetic moment per unit volume; this quantity is called the *magnetization* and is denoted by M. Owing to the symmetry of the solenoid, M will be parallel to the axis and so (11.04.4) has to be replaced by

$$\mu_0^{-1}|B| = niA/V_s + |M| \qquad 11.05.6$$

or

$$\mu_0^{-1}|B| - |M| = niA/V_s = i/l. \qquad 11.05.7$$

Thus the composite vector $\mu_0^{-1} B - M$ is now related to the current through the solenoid in precisely the same manner as $\mu_0^{-1} B$ was related to it when the inside of the solenoid was empty. In other systems having lower symmetry the situation is less simple because B and M vary from place to place. The composite vector $\mu_0^{-1} B - M$ still however plays an important role. It is denoted by H and is called by the misleading name *magnetic field intensity*.

Thus
$$H = \mu_0^{-1} B - M. \qquad 11.05.8$$

The names generally used for E, B, D, and H are extremely confusing. A few of the better authorities use better names. In particular Sommerfeld* uses the names

> E electric field strength $\qquad$ B magnetic field strength
> D electric excitation $\qquad$ H magnetic excitation

while Stratton[†] uses the names

> E electric force vector $\qquad$ B magnetic force vector
> D electric derived vector $\qquad$ H magnetic derived vector.

From the identity (8) it is evident that any two of the vectors B, M, H completely determine the remaining one. It is however a fundamental assumption of electromagnetic theory that at any point in a piece of matter of given composition, given temperature, and given pressure any one of the vectors B, M, H completely determines the other two. The phenomenon known as *hysteresis* contradicts the assumption; such phenomena are here expressly excluded from consideration. With this proviso we write

$$H = \mu^{-1} B \qquad 11.05.9$$

and the coefficient μ is called the *permeability* of the matter. Provided the matter is isotropic μ is a scalar. (Otherwise μ would be a tensor of rank two.) The value of μ in general depends on the composition, the temperature, the pressure, and the field strength. The ratio μ/μ_0 is called the *relative permeability* of the substance and is denoted by μ_r.

§11.06 *Electric and magnetic work*

Having completed our elementary review of the physical significance of the vectors E, D and B, H we shall quote without proof general formulae for electric and magnetic work.

We first consider an electrostatic system consisting of charged conductors and dielectrics. For any infinitesimal change in the system, produced by moving either an electric charge or a conductor or a dielectric, the electric work w done on the system is given by[‡]

$$w = \int dV\, E\, dD \qquad 11.06.1$$

---
* Sommerfeld, 'Electrodynamics', translated by Ramberg, Academic Press 1952 Part 1 § 2.
† S., E.T. p. 12.
‡ S., E.T. p. 108.

where dD denotes the increment of D in the element of volume dV and the integration extends over all space, or that part of space where the electric field does not vanish.

In the simplest case of a parallel-plate capacitor containing a uniform dielectric, if we neglect edge effects, E and D vanish outside the capacitor, while between the plates they are uniform having the values

$$|D| = Q/A \qquad 11.06.2$$

$$|E| = Q/\varepsilon A \qquad 11.06.3$$

where $\pm Q$ denotes the charge on either plate of area A. If then d denotes the distance between the plates and V_c the volume contained between them, formula (1) reduces to

$$w = V_c Q \, dQ/\varepsilon A^2 = (Qd/\varepsilon A) \, dQ \qquad 11.06.4$$

in agreement with formula (10.04.2).

We turn now to a magnetic system consisting of current circuits and magnetic matter, concerning which our only restrictive assumption is the absence of hysteresis. For any infinitesimal change in the system either by changing the current in any circuit or by moving any conductor carrying a current, the magnetic work done on the system is*

$$w = \int dV \, H \, dB \qquad 11.06.5$$

where dB is the increment of B in the element of volume dV and the integration extends over all space, or that part of space where the magnetic field does not vanish.

Since we have been at pains to emphasize that B is the analogue of the force vector E, while H is the analogue of the derived vector D, the reader may justifiably express surprise that formula (5) contains as integrand HdB, not BdH. The explanation of this paradox is that the analogy must not be pushed too far, because, whereas the electrostatic energy due to fixed charges is potential energy, the magnetic energy due to electric currents is kinetic energy. More precisely* while the Hamiltonian contains as integrands EdD and HdB, the Lagrangian contains as integrands $-E$dD and BdH.

In the simplest case of a long solenoid filled with a uniform isotropic substance, if we neglect end effects, B and H vanish outside the solenoid,

---

* Guggenheim, Proc. Roy. Soc. A 1936 **155** 63; Broer, Physica 1946 **12** 49.

while inside they are uniform having the values

$$|H| = i/l \qquad 11.06.6$$

$$|B| = \mu i/l \qquad 11.06.7$$

where i denotes the current and l the length per turn. If then V_s denotes the internal volume of the solenoid, L its length, A its cross-section, and n the total number of turns, formula (5) becomes

$$w = V_s(i/l)\mathrm{d}(\mu i)/l = (A/L)n^2 i\mathrm{d}(\mu i). \qquad 11.06.8$$

§11.07 Formula for Helmholtz function

Once we know the general formula for magnetic work it is, as already mentioned in §11.01, a straightforward matter to write down thermodynamic formulae of general validity. For the sake of brevity and simplicity we shall neglect changes of volume whether due to change of temperature (expansivity) or to change of pressure (compressibility) or change of magnetic field (*magnetostriction*). The formulae may be applied to solid and liquid phases at constant pressure as an approximation.

Consider now a system consisting of linear conductors and magnetic matter and suppose the currents gradually increased from zero to final values corresponding to final values of B and H at each point of the system. Then the magnetic work w done on the system when the field is thus built up is

$$w = \int \mathrm{d}V \int_0^B H \mathrm{d}B \qquad 11.07.1$$

where the first integration extends over all space. The second integral will depend on the relation between B and H which in turn depends on the temperature at each stage. Let us now specify that the path of integration shall be isothermal. Then the work w is equal to the increase in the Helmholtz function F of the system. We accordingly have

$$F = F^0 + \int \mathrm{d}V \int_0^B H \mathrm{d}B \qquad 11.07.2$$
$$(T \text{ const.})$$

where the superscript ^{0} denotes the value when B is everywhere zero, that is to say when no currents are flowing.

In the simplest case of a uniform field, as when a long solenoid of volume V is filled with a uniform substance, (2) can be written as

$$(F - F^0)/V = \int_0^B H \mathrm{d}B = \int_0^B \mu^{-1} B \mathrm{d}B. \qquad 11.07.3$$
$$(T \text{ const.}) \qquad (T \text{ const.})$$

§11.08 Other thermodynamic functions

From the formula for the Helmholtz function we can immediately derive formulae for the entropy S and the total energy U by differentation with respect to T. For the sake of brevity and simplicity we shall confine ourselves to the formulae valid in a region of volume V where composition and field are uniform. Using the superscript ^{0} to denote values of a function when $\boldsymbol{B}$ is zero, we derive from (11.07.3)

$$\frac{S-S^0}{V} = -\frac{\partial}{\partial T}\int_0^B H\,\mathrm{d}\boldsymbol{B} = -\int_0^B \frac{\partial H}{\partial T}\,\mathrm{d}\boldsymbol{B}$$
$$\text{(}T\text{ const.)} \qquad \text{(}T\text{ const.)}$$
$$= -\int_0^B \frac{\partial(1/\mu)}{\partial T} B\,\mathrm{d}\boldsymbol{B} \qquad 11.08.1$$
$$\text{(}T\text{ const.)}$$

$$\frac{U-U^0}{V} = \int_0^B \left\{\frac{1}{\mu} - T\frac{\partial(1/\mu)}{\partial T}\right\} B\,\mathrm{d}\boldsymbol{B}. \qquad 11.08.2$$
$$\text{(}T\text{ const.)}$$

We can introduce other characteristic functions in particular $\mathcal{J}$ defined by

$$\mathcal{J}(T, H) = \mathcal{F} - \int \mathrm{d}V\, HB = \mathcal{F}^0 - \int \mathrm{d}V \int_0^H B\,\mathrm{d}\boldsymbol{H}. \qquad 11.08.3$$

In the absence of permanent magnets, when $\boldsymbol{B}=0$ implies $\boldsymbol{H}=0$ throughout space, $\mathcal{J}$ has the property

$$\mathrm{d}\mathcal{J} = -S\,\mathrm{d}T - \int_0^H \mathrm{d}V\, B\,\mathrm{d}\boldsymbol{H}. \qquad 11.08.4$$

From (4) we derive

$$(\partial S/\partial H)_T = \int \mathrm{d}V (\partial B/\partial T)_H \qquad 11.08.5$$

and consequently

$$S = S^0 + \int \mathrm{d}V \int_0^H (\partial B/\partial T)_H \mathrm{d}H \qquad 11.08.6$$
$$\text{(}T\text{ const.)}$$

or in the case of a uniform field

$$(S-S^0)/V = \int_0^H (\partial B/\partial T)_H \mathrm{d}H = \int_0^H (\partial \mu/\partial T) H\,\mathrm{d}H. \qquad 11.08.7$$
$$\text{(}T\text{ const.)} \qquad \text{(}T\text{ const.)}$$

At first sight formulae (1) and (7) may seem to disagree but in fact their equivalence follows from

$$-\int_0^H (\partial B/\partial T)_H \, dH = -\int_0^B (\partial B/\partial T)_H (\partial H/\partial B)_T \, dB = \int_0^B (\partial H/\partial T)_B \, dB. \quad 11.08.8$$
(T const.) \qquad (T const.) \qquad (T const.)

§11.09 *Case of linear induction*

Hitherto we have imposed no restriction on the relation between H and B. The permeability μ was defined by

$$\mu = B/H \qquad 11.09.1$$

and in general μ depends on B (or H) as well as on the temperature. For most materials, other than those exhibiting hysteresis, at the field strengths ordinarily used in the laboratory and at ordinary temperatures, it is found that μ is, at a given temperature, independent of B. Under these conditions the integrations in the formulae of the previous two sections can be performed explicitly. Thus formulae (11.07.3), (11.08.1), and (11.08.2) reduce respectively to

$$(\mathcal{F} - \mathcal{F}^0)/V = \tfrac{1}{2}B^2/\mu = \tfrac{1}{2}HB = \tfrac{1}{2}\mu H^2 \qquad 11.09.2$$

$$(S - S^0)/V = \tfrac{1}{2}B^2(d\mu/dT)/\mu^2 = \tfrac{1}{2}H^2 \, d\mu/dT \qquad 11.09.3$$

$$(U - U^0)/V = \tfrac{1}{2}H^2(\mu + T \, d\mu/dT). \qquad 11.09.4$$

Although a variation of μ with B at constant temperature is the exception, it does occur especially at low temperatures. In particular this phenomenon of *magnetic saturation* has been observed for hydrated gadolinium sulphate.* The formulae of the present section are then not applicable.

§11.10 *Specimen in uniform external field*

The relations developed so far involve integration over all space or that part of space where the field does not vanish. These integrations are usually too complicated to be practicable except in the case of a long solenoid completely filled with a uniform material. Unfortunately this example is of little practical interest. The experimenter is more interested in the behaviour of a specimen of matter introduced into a magnetic field which was uniform before the introduction of the specimen. We shall therefore transform our formulae to describe the behaviour of a specimen of magnetic material in a magnetic field which before the introduction of the specimen was uniform and of magnitude defined by $B = B_e$. We call this the external

---

* Woltjer and Onnes, Comm. Phys. Lab. Leiden 1923 no. 167c.

field. In contrast to B_e the force vector (induction) of the uniform field before the specimen was introduced, we continue to use B, H, M to refer to the state with the specimen present. M of course vanishes outside the specimen. We define H_e, B_i, H_i, respectively by

$$B_e = \mu_0 H_e \qquad 11.10.1$$

$$B = B_e + B_i = \mu_0(H_e + H_i + M). \qquad 11.10.2$$

By virtue of Maxwell's electromagnetic equations the following conditions are obeyed

$$\text{div } B = 0 \qquad \text{div } B_e = 0 \qquad \text{div } B_i = 0 \qquad 11.10.3$$

$$\text{curl } H_i = 0 \qquad \text{curl } (dH_i) = 0. \qquad 11.10.4$$

We now have

$$\int dV H dB - \int dV H_e dB_e$$

$$= \int dV H_i dB + \int dV H_e dB_i$$

$$= \int dV H_i dB + \int dV H_e \mu_0 dH_i + \int dV H_e \mu_0 dM$$

$$= \int dV H_i dB + \int dV B_e dH_i + \int dV B_e dM. \qquad 11.10.5$$

But as a consequence of (3) and (4)

$$\int dV H_i dB = 0 \qquad \int dV B_e dH_i = 0. \qquad 11.10.6$$

Using (6) in (5) we obtain

$$\int dV H dB - \int dV H_e dB_e = \int dV B_e dM. \qquad 11.10.7$$

This relation and its elegant derivation are due to Casimir.*

If we integrate (7) at constant temperature the left side is the Helmholtz function with the specimen present less the Helmholtz function with the specimen absent. We call this the Helmholtz function of the interaction between the external field and the specimen and we use the superscript [i] to denote this. We have then

$$\mathscr{F}^i = \int dV \int B_e dM = \int B_e dm \qquad 11.10.8$$
$$\quad\quad\;\;\; (T \text{ const.}) \quad\;\; (T \text{ const.})$$

where $m = \int dV M$ is the magnetic moment of the specimen.

* Casimir, private communication, 1951; Heine, Proc. Cambridge Phil. Soc. 1955 **52** 546.

§11.11 Other thermodynamic functions

It is clear from (11.10.8) that $\mathcal{F}^i$ is the characteristic function of interaction for the independent variables T, $\mathbf{m}$. For most purposes a more useful function is the characteristic function of interaction for the independent variables T, $\mathbf{B}_e$ denoted by $\mathcal{J}^i$ and defined by

$$\mathcal{J}^i = \mathcal{F} - \mathbf{B}_e \mathbf{m} \qquad 11.11.1$$

and obeying the relation

$$d\mathcal{J}^i = -S^i dT - \mathbf{m} d\mathbf{B}_e. \qquad 11.11.2$$

From (2) we derive the Maxwell-type relation

$$(\partial S^i / \partial \mathbf{B}_e)_T = (\partial \mathbf{m} / \partial T)_{\mathbf{B}_e} \qquad 11.11.3$$

and consequently

$$S^i = \int_0^{\mathbf{B}_e} (\partial \mathbf{m} / \partial T)_{\mathbf{B}_e} d\mathbf{B}_e. \qquad 11.11.4$$
$(T \text{ const.})$

§11.12 Specimens of simple shape

The relations containing $\mathbf{B}_e$ and $\mathbf{M}$, while formally correct, are not of much use unless we know the relationship between $\mathbf{B}_e$ and $\mathbf{M}$. This relationship is complicated unless the magnetic specimen has the shape of a spheroid having its axis of symmetry parallel to the external field $\mathbf{B}_e$. For a spheroidal specimen with semi-axes a, b, b the vectors $\mathbf{B}_e$, $\mathbf{B}$, $\mathbf{M}$ are parallel throughout the specimen and obey the linear relation

$$\mathbf{B}_e / \mathbf{B} = \{\mu_0 + D(\mu - \mu_0)\}/\mu \qquad 11.12.1$$

where D is a constant determined by the ratio b/a. It has the curious name *demagnetizing coefficient*.

When $a/b \to 0$ so that the specimen becomes a circular disc with its plane normal to the field

$$D = 1 \qquad \mathbf{B}/\mathbf{B}_e = 1 \qquad \text{(circular disc)}. \qquad 11.12.2$$

When $b/a \to 0$ so that the specimen has the shape of a needle parallel to the field

$$D = 0 \qquad \mathbf{B}/\mathbf{B}_e = \mu/\mu_0 \qquad \text{(needle)}. \qquad 11.12.3$$

When $b = a$ so that the specimen is spherical

$$D = \tfrac{1}{3} \mathbf{B}/\mathbf{B}_e = 3\mu/(2\mu_0 + \mu) \qquad \text{(sphere)}. \qquad 11.12.4$$

From (11.05.8) and (11.05.9) we deduce

$$M = B(1/\mu_0 - 1/\mu).\qquad 11.12.5$$

Combining (5) with (1) we obtain

$$B_e/\mu_0 M = \mu_0/(\mu - \mu_0) + D.\qquad 11.12.6$$

§11.13 Diamagnetic, paramagnetic, and ferromagnetic substances

Substances are divided into three classes according to their magnetic properties. These have the names *diamagnetic*, *paramagnetic*, and *ferromagnetic*.

In a *diamagnetic* substance μ has a constant value less than μ_0, independent of the field strength and of the temperature. For such a substance there is no magnetic term in the entropy and consequently there is no distinction between the energy and the Helmholtz function. Thus the thermodynamics of diamagnetic substances is trivial.

In a *paramagnetic* substance μ has a value greater than μ_0 and varying with the temperature. The value of μ also depends on the field, but usually varies but little with the field except in high fields or at low temperatures. Paramagnetic substances form the class to which the application of thermodynamics is most interesting and useful. The remaining sections of this chapter will be devoted almost entirely to paramagnetic substances.

A characteristic of *ferromagnetic* substances is the occurrence of *hysteresis*. This means that M is not a single-valued function of the field. When the field is varied the changes in magnetization are usually not reversible. The application of thermodynamics is accordingly difficult. Such attempts as have been made to apply thermodynamics to ferromagnetic substances are still controversial and nothing further will be said of them. Our only remarks concerning ferromagnetic substances will be of a general qualitative nature.

In ferromagnetic substances μ is greater than μ_0 and usually considerably greater than in paramagnetic substances. There can even be magnetization in the absence of any external field. This is called *permanent magnetization* or *remanent magnetization*.

When the temperature of a ferromagnetic substance is raised, the substance eventually becomes paramagnetic. The temperature at which this change occurs is called the *Curie temperature*. The change is a transition of higher order as defined in chapter 6. Thus the Curie temperature is a lambda point, in fact the first example of a lambda point to be discovered.

§11.14 Simple paramagnetic behaviour

We shall describe in some detail the behaviour of those paramagnetic substances whose magnetic properties are entirely due to electron spin. The behaviour of the larger class whose magnetic properties are due, partly or entirely, to orbital angular momentum is qualitatively similar but quantitatively more complicated. A description of these will not be attempted here as it would require too much space. The reader interested will have to turn to a more specialized source of information.*

The fundamental unit of magnetic moment in electron theory is *Bohr's magneton* and all magnetic moments will be expressed in terms of this unit. Bohr's magneton is denoted by β and is defined by

$$\beta = eh/4\pi m_e \qquad 11.14.1$$

where $-e$ denotes the charge and m_e the mass of an electron while h, as usual, denotes the Planck constant. If we multiply (1) by the Avogadro constant L, we obtain the corresponding proper unit

$$L\beta = Fh/4\pi m_e \qquad 11.14.2$$

where F denotes the Faraday constant. Inserting the numerical values

$$F = 9.649 \times 10^4 \text{ C mole}^{-1}$$
$$m_e = 9.109 \times 10^{-31} \text{ kg}$$
$$h = 6.626 \times 10^{-34} \text{ kg m}^2 \text{ s}^{-1}$$

we obtain

$$L\beta = 5.586 \text{ A m}^2 \text{ mole}^{-1}. \qquad 11.14.3$$

Correspondingly for β we have

$$\beta = 5.586 \times 1.6601 \times 10^{-24} \text{ A m}^2$$
$$= 9.272 \times 10^{-24} \text{ A m}^2. \qquad 11.14.4$$

Following standard spectroscopic notation we shall denote the resultant spin quantum number by S, so that the multiplicity is $2S+1$. Examples of values of S for some typical paramagnetic ions of transition elements are given in table 11.2. The first and last ions in the table, having $S=0$, are diamagnetic.

We now consider a substance such as ammonium ferric alum $NH_4Fe(SO_4)_2 \cdot 12H_2O$ each molecule of which contains a considerable number of atoms, in this example 52, only one of which, in this case Fe, is paramagnetic. In such a substance the paramagnetic ions, in this case

---

* Van Vleck, Electric and Magnetic Susceptibilities, Clarendon Press 1932 p. 259.

Fe^{3+}, may usually be considered as mutually independent, each making its own contribution to the paramagnetism of the substance. We shall denote the proper volume as usual by V_m, this being the volume which contains L paramagnetic ions.

TABLE 11.2

Multiplicities of typical paramagnetic ions of transition elements

Ions	Number of 3d electrons	S	$2S+1$
Sc^{3+}	0	0	1
Sc^{2+}, Ti^{3+}, V^{4+}	1	$\frac{1}{2}$	2
Ti^{2+}, V^{3+}	2	1	3
V^{2+}, Cr^{3+}	3	$1\frac{1}{2}$	4
Cr^{2+}, Mn^{3+}	4	2	5
Mn^{2+}, Fe^{3+}	5	$2\frac{1}{2}$	6
Fe^{2+}	6	2	5
Co^{2+}	7	$1\frac{1}{2}$	4
Ni^{2+}	8	1	3
Cu^{2+}	9	$\frac{1}{2}$	2
Cu^{+}, Zn^{2+}	10	0	1

We consider a small spherical specimen of such a substance placed in a uniform external magnetic field with induction B_e. Then for the independent variables T, B_e the characteristic function $\mathcal{J}^i$ of the interaction between the specimen and the field is given by

$$\mathcal{J}^i \frac{V_m}{V_s} = -RT \ln \frac{\sinh\{(2S+1)L\beta B_e/RT\}}{\sinh\{L\beta B_e/RT\}} \qquad 11.14.5$$

where V_s denotes the volume of the specimen. Formula (5) is essentially due* to Brillouin.

From formula (5) we can derive all the thermodynamic formulae relating to the magnetic properties of the specimen. The magnetic moment m of the specimen is determined by

$$m = -\partial \mathcal{J}^i/\partial B_e \qquad 11.14.6$$

and the magnetization M by

$$MV_s = -\partial \mathcal{J}^i/\partial B_e. \qquad 11.14.7$$

* Van Vleck, Electric and Magnetic Susceptibilities, Clarendon Press 1932; Stoner, Magnetism and Matter 1934.

From (5) and (7) we derive

$$MV_m = (2S+1)L\beta \coth\{(2S+1)L\beta B_e/RT\} - L\beta \coth\{L\beta B_e/RT\}. \quad 11.14.8$$

We shall study the particular case $S=\tfrac{1}{2}$ before continuing with the general case. When $S=\tfrac{1}{2}$, formula (8) reduces to the simple form

$$MV_m = L\beta \tanh\{L\beta B_e/RT\}. \quad 11.14.9$$

We see at once that for sufficiently small field strengths we may replace (9) by the approximation

$$MV_m = (L\beta)^2 B_e/RT \quad (L\beta B_e \ll RT) \quad 11.14.10$$

so that M is directly proportional to B_e and inversely proportional to T. This behaviour is known as *Curie's law*. At the opposite extreme of sufficiently high values of B_e we may replace (9) by the approximation

$$MV_m = L\beta \quad (L\beta B_e \gg RT) \quad 11.14.11$$

so that M is independent of B_e and of T. This behaviour is called *magnetic saturation*. We shall soon see that for all values of S Curie's law holds in sufficiently low fields and saturation occurs in sufficiently high fields.

We now return to the general formula (5) and consider its simplification in the two extremes of large and of small B_e. Considering first large values of B_e we replace each sinh by $\tfrac{1}{2}$ exp and obtain immediately

$$\varDelta^i V_m/V_s = -2SL\beta B_e \quad (L\beta B_e \gg RT). \quad 11.14.12$$

From (7) and (12) we derive

$$MV_m = 2SL\beta \quad (L\beta B_e \gg RT) \quad 11.14.13$$

representing saturation.

We turn now to the opposite extreme of small B_e. We expand each sinh as a power series retaining the first two terms. We then expand the logarithm, again retaining the first two terms. We thus obtain

$$\varDelta^i V_m/V_s = -RT \ln(2S+1) - 4S(S+1)(L\beta B_e)^2/6RT \quad (L\beta B_e \ll RT). \quad 11.14.14$$

From (7) and (14) we derive

$$MV_m = 4S(S+1)(L\beta)^2 B_e/3RT \quad (L\beta B_e \ll RT) \quad 11.14.15$$

so that M is directly proportional to B_e and inversely proportional to T in accordance with *Curie's law*.

Formula (15) has been verified experimentally for numerous substances. The more general theoretical relation (8) between M and B_e extending from the extreme of Curie's law to the opposite extreme of saturation has been quantitatively verified* for hydrated gadolinium sulphate, in which the paramagnetic Gd^{3+} ion is in an ^8S state with $S=3\frac{1}{2}$.

§11.15 *Entropy of simple paramagnetic substances*

We continue to restrict our discussion to substances whose paramagnetism is due entirely to electron spin. The behaviour of other paramagnetic substances is qualitatively similar but more complicated.

By differentiating (11.14.5) with respect to T we can obtain a general formula for S^i, the entropy of interaction between the field and the specimen. For the sake of brevity we shall however confine ourselves to the two extreme cases of B_e large and of B_e small.

At magnetic saturation according to (11.14.12) the function A^i is independent of the temperature and the entropy S^i vanishes.

Under the opposite conditions of small field we derive from (11.14.14)

$$\frac{S^i}{R}\frac{V_m}{V_s} = \ln(2S+1) - \frac{4S(S+1)}{6}\left(\frac{L\beta B_e}{RT}\right)^2 \qquad (L\beta B_e \ll RT). \qquad 11.15.1$$

§11.16 *Adiabatic demagnetization*

In a system whose state can be completely defined by the temperature T and the external magnetic field B_e (all other degrees of freedom such as pressure and composition being either irrelevant or held constant), the equation for a reversible adiabatic process is

$$S(T, B_e) = \text{const.} \qquad \text{(adiabatic).} \qquad 11.16.1$$

In a sample of a paramagnetic substance, such as ferric alum, in the temperature range 2 K to 4 K all contributions to the entropy from translational, rotational, intramolecular, and vibrational degrees of freedom are effectively zero, while any contributions from intranuclear degrees of freedom remain constant. Hence for adiabatic variations of the field B_e we have

$$S^i(T, B_e) = \text{const.} \qquad \text{(adiabatic).} \qquad 11.16.2$$

Provided B_e is not too great, we may use formula (11.15.1) for S^i, so that (2) leads to

$$B_e/T = \text{const.} \qquad \text{(adiabatic).} \qquad 11.16.3$$

* Woltjer and Onnes, Comm. Phys. Lab. Leiden 1923 no. 167c.

Thus when the field is reduced the temperature drops proportionally. This is the principle of cooling by *adiabatic demagnetization*.

§11.17 Unattainability of zero temperature

By means of adiabatic demagnetization temperatures as low as 10^{-3} K have been reached. It would appear from formula (11.16.3) that by reducing the external field to zero, we should reach $T=0$ in contradiction of Nernst's heat theorem. The resolution of this paradox is that before $T=0$ is reached, usually in the region $T \approx 10^{-2}$ K, the formulae of §11.14 and §11.15 cease to be applicable. In other words, at some such temperature the substance ceases to be paramagnetic but becomes eventually either diamagnetic or ferromagnetic.

In the change from the paramagnetic to the diamagnetic or ferromagnetic state, the proper entropy in zero magnetic field is reduced by an amount

$$R \ln(2S+1). \qquad 11.17.1$$

Hence by comparison with (11.15.1), we see that the value of S^i for zero field falls to zero. This is in agreement with the third principle of thermodynamics as expounded in chapter 3. The reader must turn elsewhere* for details of such changes.

* For example Debye, Ann. Phys. Lpz. 1938 **32** 85. An excellent elementary account is given by Simon, Very Low Temperatures, Science Museum Handbook 1937 No. 3 p. 58.

CHAPTER 12

RADIATION

§12.01 General considerations

There are several alternative ways of approach to the thermodynamics of radiation. We shall choose the one according to which the radiation is regarded as a collection of photons. Each photon is characterized by a frequency, a direction of propagation, and a plane of polarization. In empty space all photons have equal speeds c. Each photon has an energy U_i related to its frequency v_i by Planck's relation

$$U_i = hv_i \qquad 12.01.1$$

and a momentum of magnitude hv_i/c. It is convenient to group together all the species of photons having equal frequencies, and so equal energies, but different directions of propagation and planes of polarization. We denote by g_i the number of distinguishable kinds of photons having frequencies v_i and energies U_i. More precisely $g_i dv_i$ denotes the number of distinguishable kinds of photons having frequencies between v_i and $v_i + dv_i$ and energies between U_i and $U_i + dU_i$. By purely geometrical considerations it can be shown[*] that in an enclosure of volume V

$$g_i dv_i = 2 \times 4\pi V c^{-3} v_i^2 dv_i \qquad 12.01.2$$

the factor 2 being due to the two independent planes of polarization.

§12.02 Energy and entropy in terms of g_i's

We denote by N_i the number of photons having energy U_i and frequency v_i interrelated by (12.01.1). Then the total energy U is given by

$$U = \sum_i N_i U_i. \qquad 12.02.1$$

---

[*] Brillouin, Die Quantenstatistik, Springer 1931 ch. 2; Fowler and Guggenheim, Statistical Thermodynamics, Cambridge University Press 1939 §§ 401–403.

From the fact that photons obey Bose–Einstein statistics it can be shown*
that the entropy S of the system is given by

$$S/k = \sum_i \ln\{(g_i+N_i)!/g_i!N_i!\}. \qquad 12.02.2$$

Differentiating (1) and (2) at constant g_i, that is to say constant V, we have

$$dU = \sum_i U_i dN_i \qquad 12.02.3$$

$$dS/k = \sum_i \ln\{(g_i+N_i)/N_i\}dN_i. \qquad 12.02.4$$

The condition for equilibrium is according to (1.35.1) that S should be a maximum for given U, V. Hence for the most general possible variation, the expressions (3) and (4) must vanish simultaneously. It follows that

$$U_i/\ln\{(g_i+N_i)/N_i\} = U_k/\ln\{(g_k+N_k)/N_k\} \quad \text{(all } i, k\text{)} \qquad 12.02.5$$

and consequently using (3) and (4)

$$U_i/\ln\{(g_i+N_i)/N_i\} = \sum_i U_i dN_i / \sum_i \ln\{(g_i+N_i)/N_i\}dN_i$$

$$= k\,dU/dS = kT \qquad 12.02.6$$

since at constant volume

$$dU = T\,dS \quad (V \text{ const.}). \qquad 12.02.7$$

From (6) we have

$$N_i/(g_i+N_i) = \exp(-U_i/kT) \qquad 12.02.8$$

and so

$$N_i = g_i/\{\exp(U_i/kT)-1\}. \qquad 12.02.9$$

Substituting (9) into (1), we obtain

$$U = \sum_i g_i U_i/\{\exp(U_i/kT)-1\}. \qquad 12.02.10$$

For the entropy we obtain from (2), using Stirling's approximation for the factorials, and by use of (8)

$$S = \sum_i N_i \ln\{(g_i+N_i)/N_i\} + \sum_i g_i \ln\{(g_i+N_i)/g_i\}$$

$$= \sum_i N_i U_i/kT - \sum_i g_i \ln\{1-\exp(-U_i/kT)\}. \qquad 12.02.11$$

For the Helmholtz function $\mathscr{F}$ we deduce from (1) and (11)

$$\mathscr{F} = kT \sum_i g_i \ln\{1-\exp(-U_i/kT)\}. \qquad 12.02.12$$

* Brillouin, Die Quantenstatistik, Springer 1931 ch. 6.

§12.03 Thermodynamic functions

In the previous section we obtained formulae for the energy, the entropy, and the Helmholtz function in terms of the U_i's and g_i's without making any use of (12.01.1) or (12.01.2). If we now substitute the values of U_i and g_i, given by these formulae, into the relations of the previous section we obtain

$$F = 8\pi V c^{-3} kT \int_0^\infty \ln\{1 - \exp(-hv/kT)\} v^2 \, dv \qquad 12.03.1$$

$$U = 8\pi V c^{-3} \int_0^\infty hv^3 \{\exp(hv/kT) - 1\}^{-1} \, dv. \qquad 12.03.2$$

We can write (2) in the form

$$U = \int_0^\infty U_v \, dv \qquad 12.03.3$$

$$U_v = 8\pi V c^{-3} hv^3 \{\exp(hv/kT) - 1\}^{-1} \qquad 12.03.4$$

which is *Planck's formula* from which quantum theory originated.

§12.04 Evaluation of integrals

We can rewrite (12.03.1) as

$$F = -8\pi V k^4 T^4 h^{-3} c^{-3} I \qquad 12.04.1$$

where I is the integral defined by

$$I \equiv -\int_0^\infty \xi^2 \ln\{1 - \exp(-\xi)\} \, d\xi. \qquad 12.04.2$$

Using the power series for the logarithm and then integrating term by term, we obtain

$$I = \int_0^\infty \sum_{n=1}^\infty n^{-1} \xi^2 \exp(-n\xi) \, d\xi = \sum_{n=1}^\infty n^{-4} \int_0^\infty \eta^2 \exp(-\eta) \, d\eta$$

$$= 2 \sum_{n=1}^\infty n^{-4} = \pi^4/45. \qquad 12.04.3$$

Substituting (3) into (1) we obtain finally

$$F = -(8\pi^5 k^4/45 c^3 h^3) T^4 V. \qquad 12.04.4$$

§12.05 *Stefan–Boltzmann law*

We could obtain formulae analogous to (12.04.4) for U and S by evaluation of the relevant integrals, but it is more convenient to obtain these formulae by differentiation of (12.04.4).

We first abbreviate (12.04.4) to

$$F = -\tfrac{1}{3}aT^4 V \qquad\qquad 12.05.1$$

where a is a universal constant defined by

$$a = 8\pi^5 k^4 / 15 c^3 h^3 = 7.5646 \times 10^{-16} \text{ J m}^{-3} \text{ K}^{-4}. \qquad 12.05.2$$

From (1) we deduce immediately

$$S = \tfrac{4}{3} a T^3 V \qquad\qquad 12.05.3$$

$$U = a T^4 V \qquad\qquad 12.05.4$$

$$P = \tfrac{1}{3} a T^4 = \tfrac{1}{3} U/V \qquad\qquad 12.05.5$$

$$G = U - TS + PV = 0. \qquad\qquad 12.05.6$$

Formula (5) can be derived from classical electromagnetic theory. Formula (4) was discovered by Stefan and derived theoretically by Boltzmann. It is called the *Stefan–Boltzmann law*.

From (4) we see that aT^4 is the equilibrium value of the radiation per unit volume in an enclosure. If a small hole is made in such an enclosure then it can be shown by geometrical considerations that the radiation emitted through the hole per unit area and per unit time is σT^4, where σ is given by

$$\sigma = \tfrac{1}{4} ac = 5.670 \times 10^{-8} \text{ J m}^{-2} \text{ s}^{-1} \text{ K}^{-4} \qquad 12.05.7$$

in which c denotes the speed of light. This constant σ is called the *Stefan–Boltzmann constant*.

§12.06 *Adiabatic changes*

Suppose that radiation is confined by perfectly reflecting walls and that the volume of the container is altered by moving a piston. If the radiation remains in thermal equilibrium its temperature will change. For such a reversible adiabatic change, we have

$$S = \text{const.} \qquad\qquad 12.06.1$$

From (12.05.3) and (1) it follows that

$$VT^3 = \text{const.} \quad \text{(adiabatic).} \qquad\qquad 12.06.2$$

From (12.05.4) and (12.05.5) we have

$$P/T^4 = \text{const.} \qquad 12.06.3$$

so that

$$PV/T = \text{const.} \quad \text{(adiabatic)} \qquad 12.06.4$$

and

$$PV^{\frac{4}{3}} = \text{const.} \quad \text{(adiabatic).} \qquad 12.06.5$$

From (2), (3), (4), (5) it appears that the relations for a reversible adiabatic change in radiation are formally similar to those for a perfect gas such that the ratio C_P/C_V has the constant value $\frac{4}{3}$. This apparent resemblance is however accidental, for the ratio C_P/C_V of radiation is not $\frac{4}{3}$. In fact for radiation

$$C_V = (\partial U/\partial T)_V = T(\partial S/\partial T)_V = 4aT^3V \qquad 12.06.6$$

while

$$C_P = T(\partial S/\partial T)_P \to \infty \qquad 12.06.7$$

since no increase in S, however great, can increase T without increasing P.

CHAPTER 13

ONSAGER'S RECIPROCAL RELATIONS

§13.01 *Introduction*

We recall the fundamental properties of entropy stated in §1.19. The entropy of a system can change in two distinct ways namely by external and internal changes. This is expressed symbolically by

$$dS = d_e S + d_i S \qquad 13.01.1$$

where $d_e S$ denotes the part of dS due to interaction with the surroundings and $d_i S$ denotes the part due to changes taking place in the system. We have the now familiar equality

$$d_e S = q/T \qquad 13.01.2$$

where T is the thermodynamic temperature of the system. As regards $d_i S$ the only property hitherto stressed is the inequality

$$d_i S > 0. \qquad 13.01.3$$

We shall in this chapter consider more quantitatively the value of $d_i S$ or rather of $d_i S/dt$ which is the rate of internal production of entropy. Such considerations constitute a subject often called *thermodynamics of irreversible processes*. It is a subject on which whole books* have been written and it is not practicable to devote sufficient space here for an exhaustive discussion. For the sake of brevity it has been decided to exclude from the present discussion the interesting field of thermal diffusion.

§13.02 *Electric insulators and conductors*

We introduce the subject of internal entropy production by considering electric conductors which we shall compare and contrast with electric

---

* De Groot, Thermodynamics of Irreversible Processes, North-Holland 1951; Denbigh, The Thermodynamics of the Steady State, Methuen 1951; Prigogine and Defay, Etude Thermodynamique des Phénomènes Irréversibles, Dunod 1947.

insulators. We begin with the simplest case of isotropic media and then pass on to the more interesting case of anisotropic media.

The Helmholtz function of an electric insulator is related to the *electric field strength* E and the *electric displacement* D, under isothermal conditions, by

$$d\mathcal{F} = VE\,d D. \qquad 13.02.1$$

This follows immediately from formula (11.06.1). We recall that D is related to E by

$$D = \varepsilon E \qquad 13.02.2$$

where ε is the permittivity. Under ordinary conditions ε is independent of E and this will be assumed here.

In an isotropic medium D and E have the same direction and ε is a scalar quantity. In an anisotropic medium D and E generally have different directions. A quantity such as ε relating two non-parallel vectors D and E according to (2) is called a tensor. Without any prior knowledge of tensors we can see what this means by considering the cartesian components of (2). The relations for these components have the form

$$D_x = \varepsilon_{xx} E_x + \varepsilon_{xy} E_y + \varepsilon_{xz} E_z \qquad 13.02.3$$

$$D_y = \varepsilon_{yx} E_x + \varepsilon_{yy} E_y + \varepsilon_{yz} E_z \qquad 13.02.4$$

$$D_z = \varepsilon_{zx} E_x + \varepsilon_{zy} E_y + \varepsilon_{zz} E_z \qquad 13.02.5$$

where all the quantities are scalars.

Since we are assuming that ε is independent of E, we can substitute (2) into (1) and integrate obtaining, apart from a trivial integration constant,

$$\mathcal{F}/V = \tfrac{1}{2}\varepsilon E^2. \qquad 13.02.6$$

In an isotropic medium ε is as we have already mentioned a scalar and there is no difficulty. In an anisotropic medium the meaning of (6) is by no means so simple and its interpretation requires at least an elementary knowledge of tensors. However all that we need to record here is that the existence of the Helmholtz function related to E and D by (1) requires the symmetry conditions

$$\varepsilon_{xy} = \varepsilon_{yx} \qquad \varepsilon_{yz} = \varepsilon_{zy} \qquad \varepsilon_{zx} = \varepsilon_{xz}. \qquad 13.02.7$$

In the terminology of tensors we say that ε must be a symmetrical tensor. When the relations (7) are assumed, the expression for the Helmholtz function becomes

$$\mathcal{F}/V = \tfrac{1}{2}\varepsilon_{xx} E_x^2 + \tfrac{1}{2}\varepsilon_{yy} E_y^2 + \tfrac{1}{2}\varepsilon_{zz} E_z^2 + \varepsilon_{xy} E_x E_y + \varepsilon_{yz} E_y E_z + \varepsilon_{zx} E_z E_x. \qquad 13.02.8$$

When we assume (8) and (7) we can immediately derive (3), (4), and (5) by means of (1). Without the relations (7) it is impossible to find any formula for the Helmholtz function consistent with (1).

We now turn from insulators to conductors. If we denote electric field by E and current density by J we may write

$$J = \sigma E \qquad 13.02.9$$

where σ denotes the electric conductivity. Under ordinary conditions σ is independent of E and we shall assume this. In an isotropic medium J and E have the same direction and σ is a scalar quantity. In an anisotropic medium J and E generally have different directions so that σ is a tensor. The relations between the cartesian components have the form

$$J_x = \sigma_{xx} E_x + \sigma_{xy} E_y + \sigma_{xz} E_z \qquad 13.02.10$$

$$J_y = \sigma_{yx} E_x + \sigma_{yy} E_y + \sigma_{yz} E_z \qquad 13.02.11$$

$$J_z = \sigma_{zx} E_x + \sigma_{zy} E_y + \sigma_{zz} E_z \qquad 13.02.12$$

where all the quantities are scalars.

Let us now determine the rate of internal production of entropy in the simple case of an isotropic medium so that J has the same direction as E. It is simplest to assume that the conductor is maintained at a constant temperature and that J and E are independent of the time. The conductor is then maintained in an unchanging state so that

$$dS/dt = 0. \qquad 13.02.13$$

Substituting from (13.01.1) and (13.01.2) into (13) we obtain

$$d_i S/dt = -d_e S/dt = -T^{-1} dq/dt \qquad 13.02.14$$

where $-q$ is the heat given up to the thermostat. From elementary electrical theory we have

$$-dq/dt = VJE. \qquad 13.02.15$$

Substituting (15) into (14) we obtain

$$T d_i S/dt = VJE. \qquad 13.02.16$$

Finally substituting (9) into (16) we obtain

$$T d_i S/dt = V\sigma E^2. \qquad 13.02.17$$

In the more complicated and more interesting case of an anisotropic conductor, we obtain by similar but more difficult reasoning

$$V^{-1} T d_i S/dt = J_x E_x + J_y E_y + J_z E_z. \qquad 13.02.18$$

Substituting (10), (11), and (12) into (18) we derive

$$V^{-1}T\,\mathrm{d}_i S/\mathrm{d}t = \sigma_{xx}E_x^2 + \sigma_{yy}E_y^2 + \sigma_{zz}E_z^2$$
$$+ (\sigma_{xy}+\sigma_{yx})E_xE_y + (\sigma_{yz}+\sigma_{zy})E_yE_z + (\sigma_{zx}+\sigma_{xz})E_zE_x. \quad 13.02.19$$

If we compare (19) with (8) we notice a superficial resemblance but also two differences which are interrelated. In the first place $\mathcal{F}$ occurring on the left of (8) is a function of the state of the medium while $\mathrm{d}_i S/\mathrm{d}t$ occurring on the left of (19) is not. In the second place, and as a consequence of the first difference, there is no compelling reason from classical thermodynamics why relations between the σ's analogous to the relations (7), namely

$$\sigma_{xy} = \sigma_{yx} \qquad \sigma_{yz} = \sigma_{zy} \qquad \sigma_{zx} = \sigma_{xz} \qquad 13.02.20$$

should always be true. They must in fact be true for reasons of geometrical symmetry except for crystals of low symmetry, namely those in which the only element of symmetry is an axis of rotation. It was however suggested by Clerk Maxwell* that (20) is always true. Even earlier Stokes[†] had surmised the truth of relations analogous to (20) for thermal conductivity.

There are moreover reasons based on the kinetic principle of detailed balancing[‡], which we shall not here go into, for assuming such relations and they are in good agreement with experiment. The equations (20) constitute one of the simplest examples of *Onsager's reciprocal relations*.

§13.03 *Onsager's reciprocal relations*

We are now ready for a more general statement of Onsager's reciprocal relations. We denote by J_i the flux in a certain direction of something such as electric charge, as in the previous section, or a particular molecular or ionic species, or a quantity of energy. We denote by X_i the *driving force* corresponding to J_i. We make our meaning more precise by the statement that the rate of internal production of entropy per unit volume is given by

$$V^{-1}T\,\mathrm{d}_i S/\mathrm{d}t = \sum_i J_i X_i. \qquad 13.03.1$$

For example when J_i is electric current density, then X_i is the electric field. When J_i is the flux of a molecular species i, then X_i is minus the gradient of its chemical potential. When J_i is the flux of an ionic species i, then X_i

---
* Maxwell, Electricity and Magnetism, Oxford University, 1st ed. 1873; 3rd ed. 1892 ch. 8.
[†] Stokes, Cambridge and Dublin Math. J. 1851 **6** 235.
[‡] Onsager, Phys. Rev. 1931 **37** 405; De Groot, Thermodynamics of Irreversible Processes, North-Holland 1951.

is minus the gradient of its electrochemical potential. When J_i is the flux of energy then X_i must be closely related to the temperature gradient.

Provided the gradients X_i are not too great the fluxes J_i will generally be linear functions of the driving forces. This may be expressed as

$$J_i = \sum_k L_{ik} X_k \quad (L_{ik} \text{ const.}). \qquad 13.03.2$$

In the simple case of only two kinds of flow (2) reduces to

$$J_1 = L_{11} X_1 + L_{12} X_2 \qquad 13.03.3$$
$$J_2 = L_{21} X_1 + L_{22} X_2. \qquad 13.03.4$$

We now state Onsager's reciprocal relations in the general form

$$L_{ik} = L_{ki} \quad (\text{all } i, k). \qquad 13.03.5$$

In the simple case of only two kinds of flow (5) reduces to

$$L_{12} = L_{21}. \qquad 13.03.6$$

§13.04 Electrokinetic effects

Electrokinetic phenomena occur when a liquid which is a poor electric conductor flows through a tube. Generally the tube walls and the liquid have opposite electric charges together constituting an electric double layer. There is a consequent interplay between the flow of matter and the flow of electric charge.

We consider a tube of length l and uniform cross-section A. We denote the electric current density by $\mathbf{J}$ and the electric field, which is equal and opposite to the electric potential gradient, by $\mathbf{E}$. We measure the rate of flow of liquid by the volume per unit time and we denote this by fA. We denote the pressure gradient by P_l. It can be verified that the rate of internal production of entropy per unit volume is given by

$$V^{-1} T d_i S/dt = \mathbf{J}\mathbf{E} + f P_l. \qquad 13.04.1$$

From the form of (1) we see that we may regard $\mathbf{E}$ and P_l as the driving forces corresponding to the fluxes $\mathbf{J}$ and f respectively. We assume the linear relations

$$\mathbf{J} = L_{11} \mathbf{E} + L_{12} P_l \qquad 13.04.2$$
$$f = L_{21} \mathbf{E} + L_{22} P_l. \qquad 13.04.3$$

We then have Onsager's reciprocal relation

$$L_{12} = L_{21}. \qquad 13.04.4$$

Let us now consider briefly the physical significance of the L's. In the absence of a pressure gradient (2) reduces to

$$J = L_{11} E \qquad (P_l = 0) \qquad 13.04.5$$

from which we see that L_{11} is just the *electric conductivity*. In the absence of an electric field (3) reduces to

$$f = L_{22} P_l \qquad 13.04.6$$

from which we have according to Poiseuille's law

$$L_{22} = r^2/8\eta \qquad 13.04.7$$

where η is the viscosity and r the radius of the tube (or the effective radius if the cross-section is not circular).

The essential consequence of Onsager's relation is this. All the electrokinetic effects require for their quantitative description a knowledge of the electric conductivity, the viscosity, and *one other coefficient*, not two. We shall now formulate briefly the relations for some of the most important electrokinetic effects.*

In the first place we have the *streaming potential*, defined as the electric potential difference per unit pressure difference in a stationary state with zero electric current. According to (2) it is given by

$$E/P_l = -L_{12}/L_{11} \qquad (J=0). \qquad 13.04.8$$

In the second place we have *electro-osmosis*, which is the flow of liquid per unit electric current in a state with zero pressure gradient. According to (2) and (3) it is given by

$$f/J = L_{21}/L_{11} \qquad (P_l = 0). \qquad 13.04.9$$

The third effect is the *electro-osmotic pressure*, which is the pressure difference per unit potential difference in the stationary state with zero material flow. According to (3) it is given by

$$P_l/E = -L_{21}/L_{22} \qquad (f=0). \qquad 13.04.10$$

The fourth effect is the *streaming current*, which is the electric current per unit material flow for the steady state of zero electric field. According to (2) and (3) it is given by

$$J/f = L_{12}/L_{22} \qquad (E=0). \qquad 13.04.11$$

* De Groot, Thermodynamics of Irreversible Processes, North-Holland 1951 p. 187.

As a consequence of Onsager's reciprocal relation (4) we deduce from (8) and (9)

$$(E/P_l)_{J=0} = -(f/J)_{P_l=0} \qquad 13.04.12$$

and from (10) and (11)

$$(P_l/E)_{f=0} = -(J/f)_{E=0}. \qquad 13.04.13$$

Formula (13), known as Saxén's relation, has been verified experimentally with an accuracy of about 2 per cent*. Formula (12) has also been confirmed experimentally but only with an accuracy of about 15 per cent.*

§13.05 *Electric double layer*

We have seen in the previous section how the several electrokinetic effects can be expressed quantitatively in terms of the conductivity, the viscosity, and one further parameter denoted by $L_{12} = L_{21}$. It was not necessary to consider the physical significance of L_{12}. We shall now show that this quantity is closely related to the strength of the electric double layer at the boundary between the wall of the tube and the liquid in the tube. It does not matter which of the electrokinetic effects we consider in order to establish the required relation. We choose to consider electro-osmosis.

We consider a thin strip of liquid near to and parallel to the wall. We denote by du the difference of velocity along the tube between the inner and outer surface of this strip. We denote by $d\tau$ the strength of the electric double layer in this strip, that is to say the electric moment per unit area of the strip. We consider a steady state of motion under an applied electric field E. We now equate the two opposing couples due to the viscous effect of the velocity gradient on the one hand and the effect of the electric field on the dipoles on the other. We thus obtain the condition

$$\eta \, du = E \, d\tau. \qquad 13.05.1$$

Integrating from the wall, where the liquid is stationary, to the interior we obtain for the velocity u in the interior of the liquid

$$\eta u = E\tau \qquad 13.05.2$$

where τ is the strength of the whole double layer, that is to say the total electric moment per unit area of the wall.

The flow fA expressed as volume of liquid per unit time is related to u by

$$fA = uA' \qquad 13.05.3$$

* Miller, Chem. Rev. 1960 **60** 20.

where A' is an area less than the total internal cross-section A of the tube but greater than the cross-section of liquid having a velocity inappreciably different from u. In practice the thickness of the double layer is small compared with the width of the tube and consequently the difference between the cross-sections A and A' is negligible. We may then regard A' as the internal cross-section of the tube and substituting (2) into (3) we have

$$f = \tau E/\eta \qquad 13.05.4$$

while from (13.04.3) we have

$$f = L_{21} E \qquad (P_l = 0). \qquad 13.05.5$$

Comparing (4) with (5) we deduce

$$L_{21} = \tau/\eta. \qquad 13.05.6$$

Formula (6) expresses a relation essentially due to Helmholtz* although he did not use the same notation. Most authors instead of using Helmholtz's strength of the double layer here denoted by τ prefer to consider another quantity introduced by Perrin[†] and subsequently denoted by ζ by Freundlich.[‡] This quantity has the dimensions of an electric potential and is called the ζ-potential. It is derived from τ by division by the rational permittivity. The introduction of this subsidiary quantity adds nothing except unnecessary complications.[§]

§13.06 *Electrochemical cells with transference*

We shall now use Onsager's reciprocal relations to obtain a stricter derivation of formula (8.18.16) for the electromotive force of the cell with transference described by (8.18.2). We shall not repeat the whole of the textual argument, but shall merely revise the formulae. The first change is that we replace (8.18.3) by the less restrictive assumption

$$J_i = -\sum_k L_{ik} \, d\mu_k/dy. \qquad 13.06.1$$

The condition for zero electric current then becomes instead of (8.18.9)

$$\sum_i \sum_k z_i L_{ik} \, d\mu_k/dy = 0. \qquad 13.06.2$$

---

* Helmholtz, Ann. Phys. Lpz. 1879 **7** 337.
† Perrin, J. Chim. Phys. 1904 **2** 601.
‡ Freundlich, Colloid and Capillary Chemistry, Methuen 1926 p. 242.
§ Guggenheim, Trans. Faraday Soc. 1940 **36** 139, 722.

We still have for the electromotive force formula (8.18.7)

$$F\,dE/dy = z_{Cl^-}^{-1}\,d\mu_{Cl^-}/dy. \qquad 13.06.3$$

Combining (2) with (3) we have

$$\sum_i \sum_k z_i z_k L_{ik}(-z_k^{-1}\,d\mu_k/dy + z_{Cl^-}^{-1}\,d\mu_{Cl^-}/dy)$$

$$= \sum_i \sum_k z_i z_k L_{ik} z_{Cl^-}^{-1}\,d\mu_{Cl^-}/dy$$

$$= \sum_i \sum_k z_i z_k L_{ik} F\,dE/dy. \qquad 13.06.4$$

Consequently, instead of (8.18.11) we have

$$F\,dE/dy = \sum_k \{(-z_k^{-1}\,d\mu_k/dy + z_{Cl^-}^{-1}\,d\mu_{Cl^-}/dy)(z_k \sum_i z_i L_{ik}/\sum_l \sum_m z_l z_m L_{lm})\}. \qquad 13.06.5$$

We now, as in §8.18, turn to the different condition where the two electrode solutions are identical and an external potential difference dE^e is applied across the electrodes. We have by (8.18.12)

$$z_k^{-1}\,d\mu_k = F\,dE^e \quad \text{(all } k\text{)} \qquad 13.06.6$$

so that by (1)

$$J_i = -\sum_k z_k L_{ik} F\,dE^e/dy. \qquad 13.06.7$$

The electric current per unit cross-section carried by the ionic species i will be

$$z_i F J_i = -z_i \sum_k z_k L_{ik} F^2\,dE^e/dy. \qquad 13.06.8$$

The transport number t_i, being the fraction of the total current carried by the ionic species i, is therefore

$$t_i = z_i \sum_k z_k L_{ik}/\sum_i \sum_k z_i z_k L_{ik}. \qquad 13.06.9$$

The transport number of the ionic species k is likewise

$$t_k = z_k \sum_i z_i L_{ki}/\sum_k \sum_i z_k z_i L_{ki} = z_k \sum_i z_i L_{ki}/\sum_l \sum_m z_l z_m L_{lm}. \qquad 13.06.10$$

Comparing (5) with (10) we have

$$F\,dE/dy = \sum_k t_k(-z_k^{-1}\,d\mu_k/dy + z_{Cl^-}^{-1}\,d\mu_{Cl^-}/dy)\sum_i z_i L_{ik}/\sum_i z_i L_{ki}. \qquad 13.06.11$$

Finally by use of Onsager's relation

$$L_{ik} = L_{ki} \qquad 13.06.12$$

(11) reduces to

$$F\,dE/dy = \sum_k t_k(-z_k^{-1}\,d\mu_k/dy + z_{Cl^-}^{-1}\,d\mu_{Cl^-}/dy) \qquad 13.06.13$$

which is the same as formula (8.18.16).

We note that in the derivation given in §8.18 instead of Onsager's relation (12) we used the more restrictive assumption

$$L_{ik} = 0 \qquad (i \neq k). \qquad 13.06.14$$

The author believes that the assumption (14) is in fact true although this has not been proved.

§13.07 *Thermoelectricity*

We shall now discuss the most important and most interesting application of Onsager's relations to a non-isothermal system, namely a system in which electric current is coupled with energy flow. This phenomenon is called *thermoelectricity*. The following treatment is similar to that of Callen.*

We consider a straight uniform metallic wire in a non-uniform temperature through which an electric current can result from a flow of electrons. If the wire lies in the y-direction the rate of internal production of entropy per unit volume is given by

$$V^{-1}\,d_iS/dt = -J_{e^-}\,d(\mu/T)/dy + J_U\,d(T^{-1})/dy \qquad 13.07.1$$

where μ is the electrochemical potential of the electrons, $-FJ_{e^-}$ is the electric current density, and J_U is the energy flux.

From the form of (1) we may regard $d(\mu/T)/dy$ and $d(T^{-1})/dy$ as the driving forces corresponding to the fluxes $-J_{e^-}$ and J_U respectively. We accordingly assume the linear relations

$$-J_{e^-} = L'_{11}\,d(\mu/T)/dy + L'_{12}\,d(T^{-1})/dy \qquad 13.07.2$$

$$J_U = L'_{21}\,d(\mu/T)/dy + L'_{22}\,d(T^{-1})/dy. \qquad 13.07.3$$

We could use formulae (2) and (3) as the basis of our discussion and should of course obtain correct results, but by a simple transformation we obtain formulae in which the coefficients L have a more direct physical significance.

* Callen, Thermodynamics, Wiley 1960 ch. 17.

We define J_q by

$$J_q = J_U - \mu J_{e^-} \tag{13.07.4}$$

and substitute into (2) and (3) obtaining

$$-J_{e^-} = L_{11} T^{-1} d\mu/dy + L_{12} d(T^{-1})/dy \tag{13.07.5}$$

$$J_q = L_{21} T^{-1} d\mu/dy + L_{22} d(T^{-1})/dy \tag{13.07.6}$$

where

$$L_{11} = L'_{11} \tag{13.07.7}$$

$$L_{12} = L'_{12} + L'_{11}\mu \tag{13.07.8}$$

$$L_{21} = L'_{21} + L'_{11}\mu \tag{13.07.9}$$

$$L_{22} = L'_{22} + (L'_{12} + L'_{21})\mu + L'_{11}\mu^2. \tag{13.07.10}$$

We can also verify that the determinant

$$D = L_{11}L_{22} - L_{12}L_{21} = L'_{11}L'_{22} - L'_{12}L'_{21}. \tag{13.07.11}$$

From (8) and (9) we see that Onsager's relation $L'_{21} = L'_{12}$ is equivalent to $L_{21} = L_{12}$.

We next determine the physical interpretation of L_{11} and L_{22}. If we consider the case of an isothermal flow of current we obtain for the *electric conductivity*

$$\sigma = -F^2 J_{e^-}/(d\mu/dy) \quad (dT/dy = 0) \tag{13.07.12}$$

so that by use of (5)

$$T\sigma/F^2 = L_{11}. \tag{13.07.13}$$

Similarly if we consider a flow of energy with zero electric current we obtain for the *thermal conductivity* κ

$$\kappa = -J_U/(dT/dy) = -J_q/(dT/dy) \quad (J_{e^-} = 0) \tag{13.07.14}$$

so that from (5) and (6)

$$T^2\kappa = D/L_{11}. \tag{13.07.15}$$

From (13) and (15) we derive

$$T^3\kappa\sigma/F^2 = D. \tag{13.07.16}$$

§13.08 *Seebeck effect and thermoelectric power*

When there is no electric current we have according to (13.07.5)

$$d\mu/dy = (L_{12}/L_{11}T)dT/dy \tag{13.08.1}$$

and integrating from one end of the wire denoted by the subscript 1 to the other end denoted by the subscript 2 we have

$$\mu_2 - \mu_1 = \int_{T_1}^{T_2} (L_{12}/L_{11} T) dT \qquad 13.08.2$$

or

$$\mu_2 - \mu_1 = -F \int_{T_1}^{T_2} \varepsilon \, dT \qquad 13.08.3$$

where ε, called the *thermoelectric power*, is given by

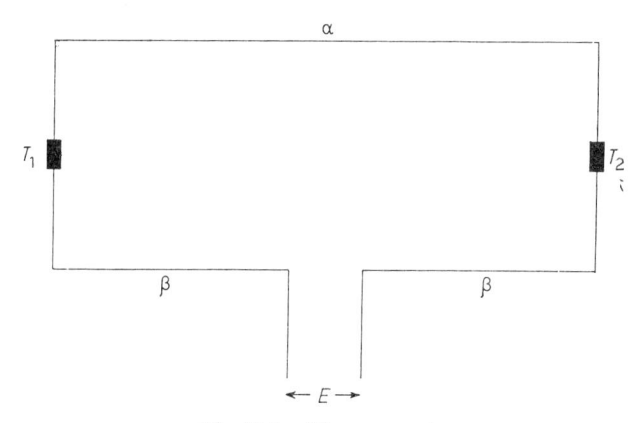

$$FT\varepsilon = -L_{12}/L_{11}. \qquad 13.08.4$$

Fig. 13.1. Thermocouple

We now consider the thermocouple, shown in figure 13.1, consisting of a pair of wires of different metals α and β with their two junctions 1 and 2 at temperatures T_1 and T_2. Since μ is continuous at both metal–metal junctions, it follows that the difference between the values μ_l and μ_r of μ at the left and right terminals is given by

$$\mu_r - \mu_l = F \int_{T_1}^{T_2} (\varepsilon^\beta - \varepsilon^\alpha) dT. \qquad 13.08.5$$

But since the two terminals are at the same temperature T' the electric potential difference between the two terminals or the electromotive force E of the thermocouple is given by

$$E = -(\mu_r - \mu_l)/F = \int_{T_1}^{T_2} (\varepsilon^\alpha - \varepsilon^\beta) dT. \qquad 13.08.6$$

The occurrence of this electromotive force in a thermocouple is called the *Seebeck effect*.

From (4) and (13.07.13) we deduce

$$-T^2\sigma\varepsilon/F = L_{12}. \qquad 13.08.7$$

§13.09 *Peltier effect*

Consider an isothermal junction of two metals α and β through which flows an electric current of density $-FJ_{e^-}$. Then the flux of energy is discontinuous across the junction and the difference in energy flux appears as heat at the junction. Since both μ and J_{e^-} are continuous across the junction it follows from (13.07.4) that

$$J_U^\alpha - J_U^\beta = J_q^\alpha - J_q^\beta \qquad (T^\alpha = T^\beta) \qquad 13.09.1$$

and from (13.07.5) and (13.07.6) that

$$(J_U/J_{e^-})^\beta - (J_U/J_{e^-})^\alpha = (J_q/J_{e^-})^\beta - (J_q/J_{e^-})^\alpha = (L_{21}/L_{11})^\alpha - (L_{21}/L_{11})^\beta.$$
$$13.09.2$$

The *Peltier coefficient* $\pi^{\alpha\beta}$, defined as the heat that must be supplied per unit time to the junction to keep its temperature constant when unit electric current passes from α to β, is given by

$$F\pi^{\alpha\beta} = (L_{21}/L_{11})^\alpha - (L_{21}/L_{11})^\beta. \qquad 13.09.3$$

§13.10 *Kelvin's second relation*

We recall that the *thermoelectric power* ε of a metal is given by (13.08.4)

$$FT\varepsilon = -L_{12}/L_{11} \qquad 13.10.1$$

so that for a pair of metals

$$FT(\varepsilon^\beta - \varepsilon^\alpha) = (L_{12}/L_{11})^\alpha - (L_{12}/L_{11})^\beta. \qquad 13.10.2$$

The *Peltier coefficient* $\pi^{\alpha\beta}$ is given by (13.09.3)

$$F\pi^{\alpha\beta} = (L_{21}/L_{11})^\alpha - (L_{21}/L_{11})^\beta. \qquad 13.10.3$$

When we introduce *Onsager's relation*

$$L_{21} = L_{12} \qquad 13.10.4$$

into (2) and (3) we obtain

$$\pi^{\alpha\beta} = T(\varepsilon^\beta - \varepsilon^\alpha) \qquad 13.10.5$$

which is called *Kelvin's second relation*. This has been verified within an accuracy of one per cent or better for about twenty pairs of metals.*

---

* Miller, Chem. Rev. 1960 **60** 19.

§13.11 Thomson effect

Hitherto we have considered a thermocouple on open circuit only. We now consider a thermocouple balanced by a different cell with an electromotive force opposite to and differing infinitesimally from that of the thermocouple. We then have a flow of current which is reversible apart from the Joule effect which may be neglected because it is proportional to the square of the current and is consequently a second-order small quantity. When the electric current flows in a temperature gradient there is a flow of heat between the wire and its surroundings of the form τdT called the *Thomson heat* superposed on the negligible Joule heat.

We now apply the first law of thermodynamics to this thermocouple when the two metal–metal junctions are at temperatures T and $T+dT$. The heat q absorbed is

$$q = -\pi^{\alpha\beta}(T) + \tau^{\alpha} dT + \pi^{\alpha\beta}(T+dT) - \tau^{\beta} dT \qquad 13.11.1$$

and the work done on the thermocouple is

$$w = (\varepsilon^{\beta} - \varepsilon^{\alpha})dT. \qquad 13.11.2$$

By the first law of thermodynamics we have for a steady state

$$q + w = 0. \qquad 13.11.3$$

Combining (1), (2), and (3) we obtain

$$d\pi^{\alpha\beta}/dT + \tau^{\alpha} - \tau^{\beta} - \varepsilon^{\alpha} + \varepsilon^{\beta} = 0. \qquad 13.11.4$$

We emphasize that this relation is a consequence of the *first law* only. When we combine (4) with (13.10.5) we obtain

$$T d(\varepsilon^{\beta} - \varepsilon^{\alpha})/dT = \tau^{\beta} - \tau^{\alpha}. \qquad 13.11.5$$

Formula (5) is called *Kelvin's first relation*. The author is not aware of its having been verified experimentally.

§13.12 Interdiffusion of two fluids

We shall now consider briefly the interdiffusion along the y-axis, at constant temperature and constant pressure, of two fluids denoted by the subscripts $_1$ and $_2$ respectively. We denote the mass fractions of the two components by $\tilde{x}_1$ and $\tilde{x}_2$ so that

$$\tilde{x}_1 + \tilde{x}_2 = 1. \qquad 13.12.1$$

We denote the specific fluxes or fluxes per unit mass by $\tilde{J}_1$ and $\tilde{J}_2$. We denote the specific chemical potentials or chemical potentials divided by the proper masses by $\tilde{\mu}_1$ and $\tilde{\mu}_2$. We take as origin for the coordinate y the centre of mass of the fluid. This implies that

$$\tilde{J}_1 + \tilde{J}_2 = 0. \qquad 13.12.2$$

The Gibbs–Duhem relation may be written as

$$\tilde{x}_1 d\tilde{\mu}_1 + \tilde{x}_2 d\tilde{\mu}_2 = 0. \qquad 13.12.3$$

From (2) and (3) it follows that

$$-\tilde{x}_2 \tilde{J}_1/(d\mu_1/dy) = -\tilde{x}_1 \tilde{J}_2/(d\tilde{\mu}_2/dy) \qquad 13.12.4$$

and we call this quantity L. We then have, using (1),

$$\tilde{J}_1 = -\tilde{J}_2 = L(d\tilde{\mu}_2/dy - d\tilde{\mu}_1/dy) = -(L/\tilde{x}_2)d\tilde{\mu}_1/dy = (L/\tilde{x}_1)d\tilde{\mu}_2/dy. \qquad 13.12.5$$

From (5) we see that $\tilde{J}_1$, $\tilde{J}_2$ are related to $d\tilde{\mu}_1/dy$, $d\tilde{\mu}_2/dy$ by the single coefficient L. Consequently in this system there is no reciprocal relation.

§13.13 Interdiffusion of two solutes in dilute solution

Having seen that the interdiffusion of two fluids is describable by a single coefficient L, we now turn to a system of three fluids and shall derive the reciprocal relations. We use the same notation as in §13.12 with the subscripts $_1$, $_2$, and $_3$ relating to the three components. Taking as origin the centre of mass of the fluid we have by analogy with (13.12.1), (13.12.2), and (13.12.3)

$$\tilde{x}_1 + \tilde{x}_2 + \tilde{x}_3 = 1 \qquad 13.13.1$$

$$\tilde{J}_1 + \tilde{J}_2 + \tilde{J}_3 = 0 \qquad 13.13.2$$

$$\tilde{x}_1 d\tilde{\mu}_1 + \tilde{x}_2 d\tilde{\mu}_2 + \tilde{x}_3 d\tilde{\mu}_3 = 0. \qquad 13.13.3$$

The rate of entropy production σ per unit volume is given by

$$T\sigma = -\tilde{J}_1 d\tilde{\mu}_1/dy - \tilde{J}_2 d\tilde{\mu}_2/dy - \tilde{J}_3 d\tilde{\mu}_3/dy. \qquad 13.13.4$$

Substituting from (1), (2), and (3) into (4) we obtain

$$-T\sigma = (\tilde{J}_2 + \tilde{J}_3)\{(\tilde{x}_2/\tilde{x}_1)d\tilde{\mu}_2/dy + (\tilde{x}_3/\tilde{x}_1)d\tilde{\mu}_3/dy\} + \tilde{J}_2 d\tilde{\mu}_2/dy + \tilde{J}_3 d\tilde{\mu}_3/dy$$
$$= \tilde{J}_2\{(1 + \tilde{x}_2/\tilde{x}_1)d\tilde{\mu}_2/dy + (\tilde{x}_3/\tilde{x}_1)d\tilde{\mu}_3/dy\}$$
$$+ \tilde{J}_3\{(\tilde{x}_2/\tilde{x}_1)d\tilde{\mu}_2/dy + (1 + \tilde{x}_3/\tilde{x}_1)d\tilde{\mu}_3/dy\}. \qquad 13.13.5$$

If we take as origin for the coordinate y the centre of mass of the component 1, which we now call the solvent, instead of the centre of mass of the mixture,

we have instead of (2)
$$\tilde{J}_1 = 0 \qquad 13.13.6$$
and instead of (4) we have
$$T\sigma = -\tilde{J}_2 \, d\tilde{\mu}_2/dy - \tilde{J}_3 \, d\tilde{\mu}_3/dy. \qquad 13.13.7$$

For the sake of brevity and simplicity we shall henceforth confine ourselves to the case of a dilute solution of 2 and 3 in the solvent 1. We then have
$$\tilde{x}_2 \ll \tilde{x}_1 \qquad \tilde{x}_3 \ll \tilde{x}_1 \qquad 13.13.8$$
and we may accordingly replace (5) by the approximation
$$T\sigma = -\tilde{J}_2 \, d\tilde{\mu}_2/dy - \tilde{J}_3 \, d\tilde{\mu}_3/dy \qquad 13.13.9$$
which is identical with (7). Moreover if we had chosen a slightly different origin for y formula (9) would be unaffected.

We now introduce the linear relations
$$\tilde{J}_2 = -L_{22} \, d\tilde{\mu}_2/dy - L_{23} \, d\tilde{\mu}_3/dy \qquad 13.13.10$$
$$\tilde{J}_3 = -L_{32} \, d\tilde{\mu}_2/dy - L_{33} \, d\tilde{\mu}_3/dy \qquad 13.13.11$$
or the equivalent relations
$$J_2 = -L_{22} \, d\mu_2/dy - L_{23} \, d\mu_3/dy \qquad 13.13.12$$
$$J_3 = -L_{32} \, d\mu_2/dy - L_{33} \, d\mu_3/dy \qquad 13.13.13$$
where J_2 and J_3 denote the fluxes per unit amount of 2 and 3 respectively. Onsager's reciprocal relation is
$$L_{23} = L_{32}. \qquad 13.13.14$$

Whereas the equation (14) relates to a solution of given composition, with values of L_{23} or L_{32} dependent on the composition, the available experimental measurements give only values of diffusion coefficients averaged over a wide range of composition. Consequently the attempt* to verify (14) is unconvincing. The most that can be said is that there is no experimental evidence at variance with (14).

---

* Miller, Chem. Rev. 1960 **60** 19.

CHAPTER 14

SYSTEMS IN MOTION

§14.01 Introduction

Except in chapter 13 we have hitherto tacitly assumed that each system considered was at rest or that its kinetic energy was negligible. In this chapter we shall briefly describe what happens when this restriction is removed. We shall use the formulae of special relativity since these are more revealing and not much more difficult than their prerelativistic approximations.

§14.02 Mechanics and hydrodynamics

We begin by quoting without derivation some important relativistic formulae in the field of mechanics and hydrodynamics. We denote by $\boldsymbol{u}$ the velocity of the system relative to a chosen frame and by c the constant speed of light in a vacuum. We define the auxiliary parameter γ by

$$\gamma = (1 - \boldsymbol{u}^2/c^2)^{\frac{1}{2}} < 1. \qquad 14.02.1$$

When several parts of an isolated system interact, the energy E of the whole system remains unchanged and the linear momentum $\boldsymbol{L}$ of the whole system remains unchanged. We use the subscript $_0$ to denote the value taken when $\boldsymbol{u} = 0$. We have the standard relation

$$E = \gamma^{-1} E_0 = \gamma^{-1} m_0 c^2 \qquad 14.02.2$$

where m_0 denotes the rest-mass of the system. $E_0 = m_0 c^2$ differs from U of previous chapters only in having an absolute value whereas the zero of U is arbitrary. We also have the standard relation

$$\boldsymbol{L} = \gamma^{-1} m_0 \boldsymbol{u}. \qquad 14.02.3$$

When we differentiate (2) we obtain

$$dE = \gamma^{-3} m_0 \boldsymbol{u} \, d\boldsymbol{u} + \gamma^{-1} dE_0 \qquad 14.02.4$$

and when we differentiate (3) we obtain

$$\boldsymbol{u} \, d\boldsymbol{L} = \gamma^{-3} m_0 \boldsymbol{u} \, d\boldsymbol{u} + \gamma^{-1} \boldsymbol{u}^2 \, dm_0 = \gamma^{-3} m_0 \boldsymbol{u} \, d\boldsymbol{u} + \gamma^{-1} (\boldsymbol{u}^2/c^2) dE_0. \qquad 14.02.5$$

Eliminating d*u* from (4) and (5) we obtain

$$dE = \gamma dE_0 + \boldsymbol{u} d\boldsymbol{L}. \qquad 14.02.6$$

We also have the relations

$$dV = \gamma dV_0 \qquad 14.02.7$$

$$P = P_0. \qquad 14.02.8$$

§14.03 Entropy

Since E_0 differs from U only by an arbitrary constant, we have for a closed system at rest

$$\begin{aligned}dE_0 &= (\partial E_0/\partial S_0)dS_0 + (\partial E_0/\partial V_0)dV_0 \\ &= (\partial E_0/\partial S_0)dS_0 - P_0 dV_0.\end{aligned} \qquad 14.03.1$$

Substituting (1) into (14.02.6) we obtain

$$dE = \gamma(\partial E_0/\partial S_0) - \gamma P_0 dV_0 + \boldsymbol{u} d\boldsymbol{L}. \qquad 14.03.2$$

From the statistical mechanical interpretation of entropy it follows that S is independent of $\boldsymbol{u}$. Consequently we have

$$S = S_0 \qquad 14.03.3$$

and (2) may be rewritten

$$dE = \gamma(\partial E_0/\partial S)dS - \gamma P_0 dV_0 + \boldsymbol{u} d\boldsymbol{L}. \qquad 14.03.4$$

§14.04 Thermal equilibrium

We now consider two identical systems α and β moving relative to each other with different but constant values of $\boldsymbol{L}$. Then by repeating the argument of §1.17 we obtain as the condition for thermal equilibrium

$$[\gamma(\partial E_0/\partial S)]^\alpha = [\gamma(\partial E_0/\partial S)]^\beta. \qquad 14.04.1$$

§14.05 Temperature

Formula (14.04.1) is a complete and unambiguous statement of the condition for thermal equilibrium between two identical systems in relative motion. There is no need to mention temperature and indeed the property of temperature will depend on its precise definition. We may define temperature T by

$$T = \gamma(\partial E_0/\partial S) \qquad 14.05.1$$

and the condition for thermal equilibrium then becomes
$$T^\alpha = T^\beta. \qquad 14.05.2$$

§14.06 Fundamental equations

If we substitute (14.05.1) into (14.03.4) we obtain
$$dE = TdS - \gamma P_0 dV_0 + \boldsymbol{u}\, d\boldsymbol{L}. \qquad 14.06.1$$

Using (14.02.7) and (14.02.8) we obtain the fundamental equation
$$dE = TdS - PdV + \boldsymbol{u}\, d\boldsymbol{L}. \qquad 14.06.2$$

If we define the Helmholtz function $\mathcal{F}$ by
$$\mathcal{F} = E - TS \qquad 14.06.3$$

and use this with (2) we obtain the second fundamental equation
$$d\mathcal{F} = -SdT - PdV + \boldsymbol{u}\, d\boldsymbol{L}. \qquad 14.06.4$$

AUTHOR INDEX

Adam, 186, 211, 212
Adcock, 198, 200
Bakker, 45
Bjerrum, 227, 268, 279, 280
Blackman, 112, 115
Blaisse, 126
Born, 11, 59
Bragg, 257
Brewer, 3
Brillouin, 353, 357, 358
Broer, 345
Brønsted, 280, 281, 287, 288, 290
Brown, 198
Callen, 18, 371
Carnot, 10, 17, 44
Casimir, 349
Clarke, 252
Clausius, 11, 17, 18
Clusius, 115
Courant, 22
Debye, 112, 281, 285, 356
De Donder, 37
Defay, 17, 37, 362
De Groot, 362, 365, 367
Denbigh, 362
Donnan, 306
Duhem, 25
Eastman, 251
Ehrenfest, 259, 264
Einstein, 116
Eötvos, 164
Everett, 17, 37
Ewald, 198
Ferguson, 164
Fletcher, 107
Flory, 206

Fowler, 95, 105, 108, 151, 157, 357
Freundlich, 369
Gerke, 314
Giauque, 83
Gibbs, 23, 24, 25, 34, 45, 48, 49, 55, 238, 300, 313
Glew, 252
Gorsky, 257
Griffiths, 315
Guggenheim, 18, 45, 50, 55, 72, 95, 98, 105, 108, 135, 136, 142, 151, 157, 166, 176, 186, 196, 203, 205, 211, 212, 252, 261, 287, 288, 290, 300, 326, 345, 357, 369
Habgood, 126
Harned, 285
Harteck, 115
Heine, 349
Heitler, 198
Helmholtz, 10, 313, 369
Henderson, 1
Herington, 194
Herzberg, 105, 108
Hilbert, 22
Hildebrand, 25, 198
Hills, 319
Holborn, 101
Hückel, 281, 285
Ives, 319
Joule, 10, 86, 92
Katayama, 164
Keesom, 91, 259, 260
Kelley, 245, 315
Kelvin, 9, 17, 18, 92
Kister, 194, 196
Koenig, 58
Kondo, 56

Landau, 107
Lange, 315
Lewis, 3, 38, 97, 176, 181, 282
Lodge, 1
Longuet-Higgins, 142
Macleod, 166
Margules, 185
Massieu, 24
Maxwell, 7, 39, 130, 131, 365
McBain, 167
McGavock, 251
McGlashan, 136, 176, 190, 198, 200, 205
Michels, 126
Miller, 368, 374, 377
Milner, 280
Mochel, 198
Moutier, 45
Nernst, 155
Ogg, 24
Onnes, 22, 91, 348, 355
Onsager, 365
Otto, 101
Owen, 285
Pauling, 149
Perrin, 369
Pitzer, 3, 135, 315
Planck, 1, 12, 24, 124
Porter, 22, 198
Potter, 136
Prausnitz, 166
Prigogine, 17, 37, 362
Ramberg, 340, 344
Randall, 3, 38, 176, 282
Rankine, 7
Raymond, 194

Redlich, 194, 196
Robinson, 289
Rossini, 245, 248
Rowlinson, 98, 137
Scatchard, 194, 196, 198
Schneider, 126
Scott, 25, 198
Shaw, 41
Simon, 155, 157, 356
Slater, 105
Sommerfeld, 1, 340, 344
Sprow, 166
Staveley, 137
Stille, 87
Stokes, G. G., 365
Stokes, R. H., 289
Stoner, 353
Stout, 83
Stratton, 339, 344
Taylor, 326
Thomson, J., 130
Thomson, W., *see* Kelvin
Tolman, 58
Turgeon, 288
Van der Waals, 45, 141, 142
Van Vleck, 352, 353
Vaughen, 115
Verschaffelt, 45
Von Zawidzki, 189
Webb, 314, 315
Widom, 142
Williams, 257
Wingrove, 190
Woltjer, 348, 355
Wood, 198

SUBJECT INDEX

Absolute activities, 73, 95
  of ions, 270, 302
  Standard, 243
Acid–base equilibria, 290
Acidity constant, 291
Activities, *see* Absolute activities *and* Relative activities
Activity coefficients,
  in dilute solutions, 226
  in electrolyte solutions, 273
Adiabatic,
  change in closed system, 26
  change of a gas, 101
  change of radiation, 360
  compressibility, 89
  demagnetization, 355
  process, 7
Adsorption, 167
Affinity, 35, 240
Allotropic changes, 249
Amount of substance, 2
Antisymmetric eigenfunctions, 73
Apparent quantities, 221
Athermal mixtures, 205
Avogadro constant, 70, 94
Azeotropy, 178

Bohr's magneton, 352
Boiling point,
  of dilute solution, 232
  of electrolyte solution, 272
  of liquid mixture, 179
Boltzmann constant, 67, 94
Boltzmann factor, 67
Boltzmann's relation, 63
Boltzmann statistics, 61, 75

Bose–Einstein statistics, 61, 74
Boyle's law, 90, 100
Boyle temperature, 100, 137

Calorie, 87
Capacitor, 333
Capillary rise, 53
Carnot's cycle, 44
Cells, Electrochemical, 307 *et seq.*, *see also* Electrochemical cells
Celsius scale of temperature, 18
Changes,
  Adiabatic, 26
  Isothermal, 27
  Natural, 14, 26
  Reversible, and reversible processes, 13
Charge numbers of ions, 269
Chemical content of a phase, 19
Chemical equilibrium, 35, 240 *et seq.*
  Acid–base, 290
  between gases and solids, 246
  between pure solids or liquids, 248
  Gaseous, 245
  Heterogeneous, involving solutions, 252
  Homogeneous, in solutions, 251
  in electrolyte solutions, 277, 289
  in gravitational field, 332
Chemical potentials, 21, 32
  Conventional, 102
  in electrolyte solutions, 270
  in gravitational field, 328
  Standard, 243
Chemical reactions, 34, 240 *et seq.*, *see also* Chemical equilibrium
  Extent of, 37
Clapeyron's relation, 119

SUBJECT INDEX

Classical and statistical thermodynamics, 59
Classical degrees of freedom, 77
Closed phase, 14
Closed system, 14
Components, 24, 33
  in systems of chemically reacting species, 265
Compressibility,
  Adiabatic, 89
  Isothermal, 38
Compression factor, 95
Concentration, Volume, 239, 252
Condensed phases, 108
  compared with gases, 90
Conjugate acid and base, 290
Contact potential, 304
Content, Chemical, of a phase, 19
Continuity between liquid and gas, 129
Conventional chemical potential, 102
Conventional entropy, 102
Cooperative systems, 255
Corresponding states,
  of crystals, 116
  of fluids, 135 et seq.
  of solids, 140
  of surface tension, 166
Critical mixing, 195
  in simple mixtures, 200
  in symmetrical mixtures, 203
Critical point, 126
Critical properties, 126, 128, 136
Cryoscopic constant, 231
Crystals,
  at very low temperatures, 111
  Debye's formulae for, 112
  Einstein's formulae for, 116
  Zero-temperature entropy of, 144, 148
  See also Solids
Curie point, 260, 351
Curie's law, 354
Cycle,
  Carnot's, 44
  Isothermal, 44
  Reversible, 43
  Thermodynamic efficiency of, 44

Dalton's law of partial pressures, 175
Debye–Hückel limiting law for electrolyte solutions, 281
  Extension of, to less dilute solutions, 284
Debye's formulae for crystals, 112
Degree Kelvin, 18
Degree of order, 257
Degrees of freedom,
  of a chemically reacting system, 265
  of a molecule, 77
  of a phase, 24
  of a system, 34
Demagnetization, Adiabatic, 355
Demagnetizing coefficient, 350
Diamagnetic substances, 351
Dielectric constant, 335, 343
Diffusion, 375
Dilute solutions, 220 et seq., see also Solutions
  of electrolytes, 272 et seq., see also Electrolyte solutions
Displacement, Electric, 334, 342
Distribution, Nernst's, law, 233
  in electrolyte solutions, 275
Donnan membrane equilibrium, 306
Double layer, Electric, 368
Duhem–Margules equation, 185

Ebullioscopic constant, 232
Efficiency, Thermodynamic, of a cycle, 44
Eigenfunctions, 61, 73
Einstein's formulae for crystals, 116
Electric,
  conductivity, 367, 372
  displacement, 334, 342
  double layer, 368
  field strength, 333, 339
  insulators and conductors, 362
  moment, 341
    of molecule, 336
  polarization, 342
  work, 334, 344
Electrochemical cells, 307 et seq.
  without transference, 316
  with transference, 321, 326, 369
Electrochemical potentials, 299
Electrochemical systems, 298 et seq.
Electrode potentials, 309
Electrokinetic effects, 366
Electrolytes,

SUBJECT INDEX 385

Strong, 268
Weak, 293
Electrolyte solutions, 268 *et seq.*
 Chemical equilibrium in, 277, 289
 Ideal dilute and real, 272
 Limiting behaviour of, 280
 Limiting law for, 281
 Specific interactions in, 286
 Surface tension of, 293
Electromotive force, 308, 311
 Standard, 317
 Standard, of half-cell, 320
 Temperature dependence of, 313
Electronic degrees of freedom, 105
Electro-osmotic pressure, 367
Electrostatic systems, 333 *et seq.*
Electrostriction, 335
Energy, 9
 of system in motion, 378
 Surface, 210
Enthalpy, 22
 of chemical reaction, 241
  in electrolyte solution, 277
 of combustion, 242
 of evaporation, 121
 of formation, 242
 of fusion, 84, 120
 of sublimation, 121
 of transition, 249
 Standard, 243
Entropy, 14
 and heat, 16
 Calorimetric, 145, 245
 Conventional, 102
 Dependence of, on P, 87
 Dependence of, on T, 82
 of evaporation, 121
 of fusion, 85, 120
 of mixing in ideal mixture, 188, 219
 of mixing of isotopes, 150
 of hydrogen, 151
 of paramagnetic substance, 355
 of sublimation, 121
 of system in motion, 379
 of transition, 249
 Spectroscopic, 146, 245
 Standard, 243
 Zero-temperature, of crystals, 144, 148

Equations of state, 141
Equilibrium,
 and reversible change, 14
 Chemical, 35, 240 *et seq.*, *see also*
  Chemical equilibrium
 Conditions for, 27
 constant, 247, 251, 277, 291
 Contact, 307
 distribution between phases, 32
 Donnan membrane, 306
 for two phases of a pure substance, 118
  at different pressures, 132
 Frozen, 34
 Hydrostatic, 31
 Liquid–vapour, for mixtures, 177
 Membrane, 34
  Non-osmotic, 304
  Osmotic, 305
 Metastable, 28
 Osmotic, 182
 product, 241
 Stable, 28
 Thermal, 8, 15
  for systems in motion, 379
Evaporation, Enthalpy and entropy of, 121
Excess functions, 191
Expansivity, Thermal, 38
Extensive properties, 18
Extent of reaction, 37, 62

Faraday constant, 298
Fermi–Dirac statistics, 61, 74
Ferromagnetic substances, 351
First law of thermodynamics, 9
Fluctuations, 65
Fluids,
 Corresponding states of, 135
 Equations of state for, 141
 Interdiffusion of, 375
 See also Gases *and* Liquids
Freezing point,
 of dilute solutions, 230
 of electrolyte solutions, 272
 See also Melting point
Frozen equilibrium, 34
Fugacities,
 in dilute solutions, 227
 in gas mixtures, 177

SUBJECT INDEX

in ideal mixtures, 188
in liquid mixtures, 180
of condensed phases, 134, 180
of gases, 97
of saturated solutions, 236
Fusion, Enthalpy and entropy of, 84, 120

Galvanic cells, 307 *et seq.*, *see also* Electrochemical cells
Gas constant, 94
Gases,
    Adiabatic changes of, 101
    at high temperatures, 97
    compared with condensed phases, 90
    Corresponding states of, 135
    Equations of state for, 90, 141
    Fugacity of, 93
    in gravitational field, 329
    Interdiffusion of, 375
    Isothermal behaviour of, 90
    Joule–Thomson throttling of, 92
    Mixtures of, *see* Gas mixtures
    Perfect, 95
        in electric field, 336
    Slightly imperfect, 98
    Virial coefficients of, 90
Gas mixtures,
    Fugacities in, 177
    Perfect, 173
    Slightly imperfect, 175
Gibbs–Duhem relation, 25
    for dilute solutions, 222
    for electrolyte solutions, 277
    for surface phase, 48
Gibbs function, 22
Gibbs geometrical surface, 49
Gibbs–Helmholtz relation, 40, 313
Gibbs' phase rule, 33
Gravitational field, 327 *et seq.*

Heat, 9
    and entropy, 16
    So-called mechanical equivalent of, 86
    theorem, Nernst's, 154, 313
Heat capacities,
    at constant pressure, 85
    at constant volume, 88
    at saturation, 122

of two phases in equilibrium, 121
    Relation between, 89
Helmholtz function, 22
Hess' law, 242
Hydrogen, Entropy of, 151
Hydrolysis, 289
Hydrostatic equilibrium, 31
Hysteresis, Magnetic, 344, 351

Ideal dilute solutions, 233, *see also* Solutions
    of electrolytes, 272
Ideal mixtures, 186, *see also* Mixtures
    Solid, 218
Imperfect gases, 90, *see also* Gases
Imperfect gas mixtures, 175, *see also* Gas mixtures
Interaction parameters (electrolytes), 286
Interfaces, 45, 159, 207, 237, 293
    Curved, 50, 54 *et seq.*
    Gas–solid, 166
    Gibbs–Duhem relation for, 48
    in one-component systems, 159 *et seq.*
    in mixtures, 207 *et seq.*
    in solutions, 237
    in solutions of electrolytes, 293
    Plane, 45
    Thermodynamic functions for, 47
Interfacial tension,
    between two solutions, 238
    Determination of, 53
    of curved interface, 50
    of plane interface, 46
    Temperature dependence of, for mixture, 214
    See also Surface tension
Internal degrees of freedom, 77
Inversion temperature of a gas, 100
Ionic diameter, 285
Ionic solutions, *see* Electrolyte solutions
Ionic strength, 282
Ionization product of water, 292
Irreversible processes, Thermodynamics of, 362 *et seq.*
Isothermal,
    change in closed system, 26
    compressibility, 38
    cycle, 44
Isotopic mixing, 150

SUBJECT INDEX

Jacobians, Use of, in thermodynamics, 41
Joule–Thomson coefficient, 99
Joule–Thomson experiment, 92
Junction potentials, 309

Kelvin's relations for thermocouple, 374
Kirchhoff's relations, 243

Lambda point, 255, 257
  Pressure dependence of, 262
  Ferromagnetic, 351
Legendre transformation, 22
Liquid junction potentials, 309
Liquids,
  Corresponding states of, 135
  Equation of state for, 141
  Fugacities of, 135, 180
  in gravitational field, 329
  Interdiffusion of, 375
  Mixtures of, 177 et seq., see also Mixtures
  Phase equilibria involving, 119 et seq.
  Relative activities in mixtures of, 180
  Surface tension of, 159
  Thermodynamics of, 108, 110

Macroscopic system, 63, 68
Magnetic,
  field intensity, 343
  induction, 339
  moment, 342
  saturation, 348
  systems, 338 et seq.
  vector potential, 340
  work, 344
Magnetization, 343
  Permanent, 351
Magneton, Bohr's, 352
Magnetostriction, 346
Massieu function, 24
  in statistical thermodynamics, 67
Maxwell's equal area rule, 130
Maxwell's relations, 39
  Analogues of, for electrostatic systems, 335
Melting point,
  of pure solid, 120
  Stationary, of solid mixture, 218

  See also Freezing point
Membrane, Semi-permeable, 182, 190, 206, 229, 304
Membrane equilibrium, 34
  Donnan, 306
  Non-osmotic, 304
  Osmotic, 305
Metastability, 28, 125, 130, 194
Miscibility, Partial, 195 et seq.
Mixing,
  Critical, 195 et seq.
  Entropy of, in ideal mixture, 188, 219
  Entropy of, of isotopes, 150
  Excess functions for, 191
  Functions of, 191
  See also Mixtures
Mixtures, 170 et seq., see also Solutions
  Athermal, 205
  Gaseous, 173 et seq.
  Ideal, 186
    Solid, 218
  in gravitational field, 406
  Interfacial layers in, 207
  Liquid, 177 et seq.
  Non-ideal, 190
  of isotopes, 150
  of large and small molecules, 205
  Partial miscibility in, 195, 200
  Simple, 197
  Solid, 217
  Surface tension of, 207 et seq.
  Symmetrical, 196
  Unsymmetrical, 203
Molalities, 3, 220
  Ionic, 268
Mole, 2, 94
Mole fractions, 24, 170
Mole ratios, 220
  Ionic, 268
Motion, Systems in, 378 et seq.
Moutier's theorem, 45

Natural process, 12
Nernst's distribution law, 233, 275
Nernst's heat theorem, 154, 313
Non-ideal mixtures, 190, *see also* Mixtures
Nuclear spins, 151

SUBJECT INDEX

Onsager's reciprocal relations, 362 *et seq.*, 365
Open phase, 14
Open system, 14
Order, Degree of, 257
Ortho hydrogen, 151
Osmotic coefficient of solvent, 226
 in electrolyte solutions, 273, 305
Osmotic equilibrium, 182
Osmotic membrane equilibrium, 305
Osmotic pressure, 183
 Electro-, 367
 of athermal mixtures, 206
 of dilute solutions, 229
 of electrolyte solutions, 273, 305
 of ideal mixtures, 190

Para hydrogen, 151
Paramagnetic substances, 351
Partial areas of surface, 212
Partial miscibility, 195 *et seq.*
Partial pressures, 177
 Dalton's law of, 175
Partial quantities, 20, 171
 at high dilution, 173, 223
 in dilute solutions, 221
Partition functions, 63 *et seq.*
 of units, 76
Peltier effect, 374
Perfect gases, 95, *see also* Gases
Perfect gas mixtures, 173 *see also* Gas mixtures
Permeability, 334
 of a vacuum, 339
Permittivity, 334, 343
 of a vacuum, 333, 339
Phase rule, 33
Phases, 6
 at different pressures, 132
 Chemical content, of 19
 Closed and open, 14
 Condensed, 90, 180, *see also* Liquids and Solids
 Degrees of freedom of, 24
 Electrically charged, 298
 Equilibrium distribution between, 32
  at different pressures, 132
 Equilibrium of three pure, 124

Equilibrium of two pure, 118
Gaseous, *see* Gases
 in gravitational field, 327
 Internal stability of, 30, 33, 194
 Liquid, *see* Liquids
 Solid, *see* Solids
 Surface, 45, *see also* Interfaces
Planck function, 24
 in statistical thermodynamics, 69
Planck's formula for radiation, 359
Polarization, Electric, 342
Pressure,
 Critical, 128, 136
 Osmotic, 183, *see also* Osmotic pressure
 Partial, 177
 Standard, 243
 Triple-point, 124
 Vapour, 121, 177
Principle of corresponding states, *see* Corresponding states
Probability, Thermodynamic, 63, 72
Processes, 7, 12
 Irreversible, Thermodynamics of, 362 *et seq.*
Proper quantities, 3, 20, 171
Properties, Intensive and extensive, 18, 19

Radiation, 357 *et seq.*
Raoult's law, 189
 Deviations from, 191 *et seq.*
Reaction,
 Chemical, 34, 240 *et seq.*, *see also* Chemical equilibrium
 Extent of, 37, 62
Reciprocal relations, Onsager's, 362 *et seq.*
Relative activities, 180
Relative volatility, 192
Relativistic formulae, 378
Reversible,
 change, 13
 cycle, 43
 process, 12, 312
Rotational degrees of freedom, 104 *et seq.*, 151

Second law of thermodynamics, 17
 Statistical basis of, 66
Second order transitions, 252, 351

Second virial coefficient of a gas, 91
  at high temperatures, 97
  Dependence of, on temperature, 99
  Dependence of, on composition, 175
Seebeck effect, 372
Separable degrees of freedom, 77
Simple mixtures, 197, *see also* Mixtures
Slightly imperfect gas mixtures, 175
Solenoid, 341
Solids, 108, 111 *et seq.*, *see also* Crystals
  Chemical reactions between, 248
  Corresponding states of, 116, 140
  Mixtures of, 217
Solubility,
  in dilute solutions, 233
  in electrolyte solutions, 276
Solubility product, 276
Solutions, especially dilute solutions, 220 *et seq.*, *see also* Mixtures
  Chemical reactions in, 251, 277
  Ideal dilute, 223
  in gravitational field, 331
  Interdiffusion of, 376
  of electrolytes, 268 *et seq.*, *see also* Electrolyte solutions
  Real, 226
  Surface tension of, 237, 293
Solvation, 289
Solvolysis, 289
Sorption, Temperature dependence of, 167
Specific interaction of ions, 286
Specific quantities, 3
Spin, Nuclear, 151
Stability and metastability, 28, 125, 130, 194
Stability, Internal, of a phase, 28 *et seq.*, 194
Standard,
  electromotive force, 317, 320
  thermodynamic quantities, 243
State,
  Continuity of, 129
  Critical, 128, 136
  Equations of, for fluid, 141
  Thermodynamic, 6
States, Corresponding, *see* Corresponding states
Statistical thermodynamics, 59, 61 *et seq.*
Stefan–Boltzmann law, 360
Stoichiometric numbers, 36

Streaming current, 367
Streaming potential, 367
Sublimation, Enthalpy and entropy of, 121
Sulphur, Transitions of, 250
Surface, Gibbs geometrical, 49
Surface energy, 210
Surface excess per unit area, 238
Surface phases, *see* Interfaces
Surface tension, 46, *see also* Interfacial tension
  Corresponding states of, 166
  of dilute solutions, 238
  of electrolyte solutions, 293
  of liquids, 159
  of liquid mixtures, 207
Symmetry numbers, 104, 107, 151
Symmetry of eigenfunctions, 61, 73
Systems,
  Closed and open, 14
  Cooperative, 255
  Electrochemical, 307 *et seq.*, *see also* Electrochemical
  Electrostatic, 333 *et seq.*
  in gravitational field, 327 *et seq.*
  in motion, 378 *et seq.*
  Macroscopic, 63, 68
  Magnetic, 338 *et seq.*, *see also* Magnetic
  of a single component, 82 *et seq.*, *see also* Gases *and* Liquids *and* Solids
  of chemically reacting species, 240 *et seq.*, *see also* Chemical equilibrium
  of several components, 170 *et seq.*, 220 *et seq.*, 268 *et seq.*, *see also* Mixtures *and* Solutions *and* Electrolyte solutions

Temperature, 5, 8
  Absolute, 9
  Boyle, 100, 137
  Celsius scale of, 18
  Corresponding, of fluids, 135 *et seq.*
  Corresponding, of solids, 116, 140
  Critical, 126, 136
  Critical mixing, 195
  Curie, 260, 351
  Debye's characteristic, 112
  Einstein's characteristic, 116
  for systems in relative motion, 380
  in statistical thermodynamics, 66

Inversion, 100
Kelvin scale of, 9, 18
Lambda-point, 255
Measurement of thermodynamic, 93
Rotational characteristic, 104
Thermodynamic, 9, 15, 93
Transition, 250
Triple-point, 124
Unattainability of zero, 157, 356
Vibrational characteristic, 105
Thermal,
  conduction, 8
  conductivity, 372
  contact, 8
  equilibrium, 8, 15
    for systems in relative motion, 380
  expansivity, 38
  insulation, 7
  internal stability, 30
  power, 372
Thermochemical tables, 244
Thermochemistry, 241
Thermocouple, 371
Thermodynamic efficiency of cycle, 44
Thermodynamic functions,
  defined, 22
  Dependence of, on T, P, 39
  Dependence of, on T, V, 41
  for magnetic systems, 347
  for surface phases, 47
  for systems in motion, 378 et seq.
  related to partition functions, 63 et seq., 72, 76
  Standard, 96, 243
Thermodynamic probability, 63, 72
Thermodynamic process, 7
Thermodynamics,
  Basis of laws of, 58
  First law of, 9, 58
  of irreversible processes, 362 et seq.
  Scope of, 5
  Second law of, 17, 58
  Statistical, 59, 61 et seq.
  Statistical basis of laws of, 66, 80
  Third law of, 60, 80, 154 et seq.
  Zeroth law of, 8, 58
Thermodynamic state, 6
Thermodynamic temperature, 9, 15, 93, see also Temperature

Thermodynamic tables, 244
Thermoelectricity, 371
Thermoelectric power, 373
Thermometer, 8
  Gas, 93
Thermostat, 8
Third law of thermodynamics, 60, 80, 154 et seq.
  and Nernst's heat theorem, 154
  and unattainability of zero temperature, 157, 356
  Exceptions to, 148
  Statistical basis of, 80
Thomson effect, 375
Throttling, 92
Transitions,
  between pure solids or liquids, 124, 248
  of higher order, 264
  of second order, 252, 351
Translational degrees of freedom, 77, 78, 103 et seq.
Transport numbers of ions, 324
Triple point, 124
  of water, 18
Trouton's rule, 140

Unattainability of zero temperature, 157, 356
Unexcited degrees of freedom, 77
Units, 1, 18
Unnatural process, 12

Van der Waals' equation, 141
Vapour pressure, 121, 177
Vibrational degrees of freedom, 104 et seq.
Virial coefficients,
  of gas mixture, 175
  of pure gas, 90, 97, 137
Volatility, Relative, 192
Volta potential, 304
Volume, Critical, 128, 136
Volume fractions, 205

Work, 9
  Conversion of, to heat, 11
  Electric and magnetic, 344
  of charging capacitor, 334

Zeroth law of thermodynamics, 8, 58
  Statistical basis of, 66